Information Literacy Skills and the Role of Social Media in Disseminating Scholarly Information in the 21st Century

C. Baskaran
Alagappa University, India

S. Dhanavandan
Central University of Tamil Nadu, India

A volume in the Advances in Information Quality and Management (AIQM) Book Series

IGI Global
PUBLISHER of TIMELY KNOWLEDGE

Published in the United States of America by
IGI Global
Information Science Reference (an imprint of IGI Global)
701 E. Chocolate Avenue
Hershey PA, USA 17033
Tel: 717-533-8845
Fax: 717-533-8661
E-mail: cust@igi-global.com
Web site: http://www.igi-global.com

Library of Congress Cataloging-in-Publication Data

Names: Baskaran, C., editor. | Dhanavandan, S., editor.
Title: Information literacy skills and the role of social media in disseminating scholarly information in the 21st century / edited by C. Baskaran and S. Dhanavandan.
Description: Hershey, PA : Information Science Reference, [2023] | Includes bibliographical references and index. | Summary: "Information Literacy Skills and the Role of Social Media in Disseminating Scholarly Information in the 21st Century analyzes the various factors of information literacy skill for disseminating scholarly information in the 21st century and increases the awareness level of social media use by researchers for sharing information. Covering topics such as information literate pedagogy, information literacy instruction, and ICT and learning in the knowledge society, this book is ideal for librarians, teachers, research scholars, students of library and information science, knowledge managers, and information scientists"-- Provided by publisher.
Identifiers: LCCN 2023014945 (print) | LCCN 2023014946 (ebook) | ISBN 9781668488058 (hardback) | ISBN 9781668488096 (paperback) | ISBN 9781668488065 (ebook)
Subjects: LCSH: Information literacy. | Electronic information resource literacy. | Communication in learning and scholarship--Technological innovations. | Libraries and social media. | Social media in higher education. | Online social networks--Library applications. | Online social networks--Educational applications. | Information society.
Classification: LCC ZA3088 I54 2023 (print) | LCC ZA3088 (ebook) | DDC 028.7--dc23/eng/20230501
LC record available at https://lccn.loc.gov/2023014945
LC ebook record available at https://lccn.loc.gov/2023014946

This book is published in the IGI Global book series Advances in Information Quality and Management (AIQM) (ISSN: 2331-7701; eISSN: 2331-771X)

British Cataloguing in Publication Data
A Cataloguing in Publication record for this book is available from the British Library.

For electronic access to this publication, please contact: eresources@igi-global.com.

Advances in Information Quality and Management (AIQM) Book Series

Siddhartha Bhattacharyya
RCC Institute of Information Technology, India

ISSN:2331-7701
EISSN:2331-771X

Mission

Acquiring and managing quality information is essential to an organization's success and profitability. Innovation in information technology provides managers, researchers, and practitioners with the tools and techniques needed to create and adapt new policies, strategies, and solutions for information management.

The **Advances in Information Quality and Management (AIQM) Book Series** provides emerging research principals in knowledge society for the advancement of future technological development. This series aims to increase available research publications and emphasize the global response within the discipline and allow for audiences to benefit from the comprehensive collection of this knowledge.

Coverage

- Mobile Commerce
- Application of IT to Operation
- IT Innovation and Diffusion
- Supply Chain Management
- Electronic Commerce Technologies
- Decision Support and Group Decision Support Systems
- Knowledge Management
- Web Services and Technologies
- Emerging Technologies Management
- Business Process Management and Modeling

IGI Global is currently accepting manuscripts for publication within this series. To submit a proposal for a volume in this series, please contact our Acquisition Editors at Acquisitions@igi-global.com or visit: http://www.igi-global.com/publish/.

The Advances in Information Quality and Management (AIQM) Book Series (ISSN 2331-7701) is published by IGI Global, 701 E. Chocolate Avenue, Hershey, PA 17033-1240, USA, www.igi-global.com. This series is composed of titles available for purchase individually; each title is edited to be contextually exclusive from any other title within the series. For pricing and ordering information please visit http://www.igi-global.com/book-series/advances-information-quality-management/73809. Postmaster: Send all address changes to above address.

Titles in this Series

For a list of additional titles in this series, please visit: www.igi-global.com/book-series

Library and Media Roles in Information Hygiene and Managing Information
Collence Takaingenhamo Chisita (Durban University of Technology, South Africa & University of South Africa, South Africa) Alexander Madanha Rusero (University of Johannesburg, South Africa) Ngoako Solomon Marutha (University of South Africa, South Africa) Josiline Phiri Chigwada (Chinhoyi University of Technology, Zimbabwe) and Oluwole Olumide Durodolu (Department of Information Science, University of South Africa, South Africa)
Information Science Reference • © 2022 • 268pp • H/C (ISBN: 9781799887133) • US $215.00

Using Information Technology Advancements to Adapt to Global Pandemics
Efosa C. Idemudia (Arkansas Tech University, USA) Tiko Iyamu (Cape Peninsula University of Technology, South Africa) Patrick Ndayizigamiye (University of Johannesburg, South Africa) and Irja Naambo Shaanika (Namibia University of Science and Technology, Namibia)
Engineering Science Reference • © 2022 • 276pp • H/C (ISBN: 9781799894186) • US $270.00

Mass Communications and the Influence of Information During Times of Crises
Mohammed Nasser Al-Suqri (Sultan Qaboos University, Oman) Jamal Mattar Alsalmi (University of Nizwa, Oman) and Obaid Said Al-Shaqsi (Media Training Center, Oman)
Information Science Reference • © 2022 • 307pp • H/C (ISBN: 9781799875031) • US $215.00

Social Justice Research Methods for Doctoral Research
Robin Throne (University of the Cumberlands, USA)
Information Science Reference • © 2022 • 397pp • H/C (ISBN: 9781799884798) • US $215.00

Encyclopedia of Information Science and Technology, Fifth Edition
Mehdi Khosrow-Pour D.B.A. (Information Resources Management Association, USA)
Engineering Science Reference • © 2021 • 1966pp • H/C (ISBN: 9781799834793) • US $2,145.00

Handbook of Research on Managing Information Systems in Developing Economies
Richard Boateng (University of Ghana, Ghana)
Business Science Reference • © 2020 • 695pp • H/C (ISBN: 9781799826101) • US $295.00

Legal and Ethical Issues in Information Management
Peter Mazebe II Mothataesi Sebina (University of Botswana, Botswana) Nathan Mwakoshi Mnjama (University of Botswana, Botswana) Kgomotso Hildegard Moahi (University of Botswana, Botswana) Tachilisa Badala Balule (University of Botswana, Botswana) and Balulwami Grand (University of Botswana, Botswana)
Information Science Reference • © 2020 • 320pp • H/C (ISBN: 9781799824220) • US $195.00

IGI Global
PUBLISHER of TIMELY KNOWLEDGE

701 East Chocolate Avenue, Hershey, PA 17033, USA
Tel: 717-533-8845 x100 • Fax: 717-533-8661
E-Mail: cust@igi-global.com • www.igi-global.com

Editorial Advisory Board

Table of Contents

Detailed Table of Contents

The study analyses the academic and research information share through social medias and networks among autonomous arts and science colleges in Chennai, India. The study portrays out of 302 respondents, 149 (49.3%) from female respondents, and 153 (50.7%) responses from male respondents. The p value of the variables Google scholar and Onsite.com, Prospect.com are less than 0.05, at the 5% level of significance. Hence null hypotheses for these variables are accepted. There is a no significant association between the age group of the respondents and their use of Research Gate, Google Scholar, and Prospect. com, as the p value is more than the significant level of 0.05.

This chapter describes the role of social media in libraries in disseminating scholarly information in the 21st century. Social networking sites that connect friends, family, library users, and customers are called social networking. It will also be utilised for social and commercial purposes and plays a vital role in disseminating scholarly information in the 21st century. However, as internet connectivity develops, it is most popular in the professional sphere, notably at academic institutions such as universities, colleges, and schools. In today's culture, using social media has become crucial in academic activity. Social media is generally used for social interaction, sharing of Content, getting up-to-date news & information, and making decisions. It is a powerful and straightforward tool for talking with people, both locally and worldwide, as well as for sharing, generating, and spreading information in the academic environment towards disseminating scholarly information in the 21st century

The study analyses the mapping of the research publications on research ethics, the Scientometrics profile was applied in the field of research during 2000-2022. The data out of 78308 publications, the largest 8751 (10.71%) of publications was distributed by the researchers in 2021. The EGR ranges between 0.068 and 2.044 in the years 2002 and 2019 respectively. The study found ranked authors, out of which twenty-five authors in the research field were concerned. The most prolific author found in the series, Shine, was the top ranked and had published 164 (0.201%) of the publications. The study explored the productivity of the publications contributed among twenty-five researchers. The ranking of the counties in order was England in second with 13353 (16.34%), then Australia 11777 (14.41%), Canada 8424 (10.31%), Peoples Republic of China 5293 (6.48%), Germany 4512 (5.52%), Netherlands 3920 (4.79%), France 2733 (3.34%), Spain 2391 (2.92%) and Sweden 2294 (2.80%).

The role of social media in libraries has been gaining attention in recent years as libraries look for new ways to connect with users and promote their services. This chapter assesses the use of Instagram, a popular photo and video-sharing platform, by libraries as a means of outreach and engagement. Through a literature review and case studies of library Instagram accounts, the chapter explores how libraries use Instagram to showcase their collections, promote events and services, and create a sense of community. It also examines the benefits and challenges of using Instagram as a digital marketing tool and the implications for library practice. The chapter explains that Instagram can effectively reach and engage with users, particularly younger generations.

The study analyses the research impact of scholarly literature on Google scholar; a Scientometric study during 2002 -2022. A total number of 25760 records were retrieved from Web of Science database for the period of study. Out of 25760, the highest 5017(19.47%) of the publications have been witnessed in the year 2021. Following is 4903, then 3835 contributed in the field of Google Scholar during 2022 and 2020 respectively. The top ranking of authors is found to be a Maffulli N with 113 (0.4%) of the publications on Google Scholar. The review articles hold the maximum of publications with 19132 (74.27%). Pharmacology pharmacy was top ranked in subject categories among the listed subjects and holds 2590 (10.05%) of the publications. Tehran University of Medical Sciences is among the most prolific institutions in the listed twenty-five of them. The institute holds 706(2.74%) of the publications in Google Scholar during period of study.

The chapter examines the dissemination of information for teaching and research competence among the faculty members in the Deemed Universities in Coimbatore region by a questionnaire-based survey. The result finds out of 364 respondents, a majority of 26.92% reported from Karunya University. The Boolean searches for access to e-resources by the respondents in the Deemed Universities. The majority 147 (48.8%) of respondents preferred Boolean search technique to very large extent while 74 (20.32%) respondents preferred the large extent. The majority with 132 (36.26%) of the respondents have noticed preferred e-books among the deemed Universities in the Coimbatore region as Excellent towards preferred e-book.

The chapter analyzes the dissemination of scholarly information through online resource use by the respondents in university. The study found that, out of 460 respondents, 234 (50.86%) belong to engineering /architecture—this is the top one category among the categories of disciplines of the study. The majority 435 (94.6%) of the respondents are aware of e-resources. 413 (89.78%), they prefer to make use of e-journals. E-Books are widely used by 426 (92.60%) respondents, followed by the use of e-magazines by 212 (46.08%) respondents. The result shows 185 (40.21%) of the respondents preferred general books. The majority 197 (42.82%) of the respondents preferred to use electronic resources for publishing papers in journals. The majority 142(30.86%) of them have preferred the electronic resources for preparing articles to seminars and conference; the respondents have opined as satisfied.

The study analyses the information literacy skills with access e- contents among faculty members in the deemed universities in Chennai, India. The chapter has shown that a majority of 97.38% of the respondents have been aware of the information literacy skill about access of e-documents. The result shows SRM University and Sathiyabama University have respondents of each 130 (34.21%), 85 (22.36%) respondents are belonging to St. Peters University, 35(9.22%) respondents are from Crescent University. Out of 380 respondents, it is seen that 210 (55.26) respondents are extremely satisfied on OPAC/Web OPAC. 205(53.90) respondents are extremely satisfied with e-databases, 192(50.52) respondents are extremely satisfied on automated circulation services. The majority of respondents are extremely satisfied with the OPAC/Web OPAC (55.26%) provided by university library followed by e-databases (53.90%).

The study analyzes the Impact and usage statistics of social medias platforms in Bangladesh. As the country races to become a mini-Silicon Valley, changes from the traditional way of life to AI-dominated living is inevitable. Investments from global venture capitalists and equity investors are tapping the potential of 166 million Bangladeshis with its big population of youths embracing the future. The study analyzes the majority of the people in Bangladesh preferred all type of social medias platforms; among listed social medias is FaceBook, with percent between 89.69 and 93.79 in August and January in the year 2022. YouTube usage rate follows, at between 2.58 and 5.91, Linked In (0.71 and 1.63), Twitter (0.38 and 1.77), Printerest (0.26 and 0.58), Instagram (0.3 and 1.01), Reddit (0.2 and 0.31), Tumblr (0.01 and 0.04).

The chapter examines information literacy competence on social media use against scholarly communication. A case study among faculty members and scholars in the faculty of management of Alagappa University, India. 340 respondents participated from eight departments, with the majority of them 51(15%) of them from Alagappa Institute of Management. sex and Age wise respondents of study, the higher potion of population being male with 238 (70%), less than half in the respondents 102 (30%) were female. The study witnesses the multiple responses about usage the different social Medias listed in the table. Majority of 324(95.29%) of them preferred face book. The respondents 11.47%, 10%, 0.35%, 0.41%, 0.35%, 0.32% and 0.55% of them have witnessed as ''Not Satisfied'' with Academia, LinkedIn, Research Gate, Mendeley, Google Scholar, Web of Science and Scopus respectively.

The objectives of the present study were to investigate the influences of demographics and social media site preferences. The data were analyzed by using SPSS 23 statistical analysis software package. The results of the study were obtained by applying the normality test, for simple percentage analyses on the obtained data. When the chi-square analysis between the opinions of social science scholars concerning the level of use of social media and the level of use of academic research was examined, a positive and significant analysis was found between them. The sample comprised 623 social science scholars. The numbers of this study were: male 335 (53.8%) and female 288 (46.2%), 13.0%, and 12.9% of respondents preferred WhatsApp at Annamalai University. 12.0% of the respondents prefer YouTube from Alagappa University.

The chapter explores how social media and networks disseminate and share scholarly information among the research scholars in the selected state universities of Tamil Nadu in India. The researcher has adopted a questionnaire method in order to collect data from respondents in the Universities, entertaining only the full-time research scholars. The result of the study demonstrates a majority 87 (14.0%) of them responded from Annamalai University. 47(14.0%) of them were male and 40 (13.9%) of them female. Of a total of 623 respondents, 44 (10.8%) of them responded from Annamalai University in the age group 26-35 years. The majority 594 (95.3%) of the respondents preferred to use "Google Chrome" to search various types of research information; the majority 571(91.7%) of the respondents use of WhatsApp; 499(80.1%) of the respondents used YouTube..

This chapter explores the potential of using open educational resources (OERs) to promote inclusive education settings in Tamil Nādu. The research method adopted for this study is a descriptive survey among with 275 research scholars. The chapter aims to understand the awareness, availability, and utilization of OERs among research scholars in Tamil Nādu, and to identify the opportunities for using OERs to promote inclusive education settings in the region. The survey mode under questionnaire includes questions such as the awareness of OERs, the sources from which research scholars access OERs, and their utilization in teaching and learning. The findings of the survey will help to gain an understanding of the usage of OERs and the potential of using them to promote inclusion in Tamil Nādu. The results of this survey will also be beneficial in understanding the current status of OER usage and the opportunities for utilizing them to improve the quality of education in the region.

The study aims to identify the fundamental principles and utilization of social networking skills (SNS) in information literacy (IL) by research scholars of Alagappa University. It intends to compare the earlier educational level, dissemination of information level after employing SNS in research. It also frameworks the challenges recognized by the scholars in accessing SNS and profound changes in information environments and outlines of generation and necessity of scientific information in new modes of assessment, which predominantly focuses on methods and standards of scientific research work. Data was collected by distributing questionnaires through Google form and print form among research scholars of various departments and received 71 responses. The study aims to identify the fundamental principles and utilization of social networking skills (SNS) in information literacy (IL) by research scholars of Alagappa University. It intends to compare the earlier educational level, dissemination of information level and after employing SNS in research.

This study examines the present usage of ICT and its influence in information literacy skills among postgraduate students of Alagappa University Central Library users. The questionnaire survey was distributed and simple random sampling technique was used in the study. A random questionnaire was distributed to 150 students and the study adopted an exploratory research design. The adoption and use of ICT has enhanced subject knowledge, new educational principles, and technologies. Quantitative research data were entered into an MS Excel file where statistics were uploaded and analyzed using SPSS software; a comparison between postgraduate students' awareness and use of ICT knowledge. Performance difference between gender and age were two groups involved in descriptive statistics, simple percentage, and ANOVA test.

Foreword

I congratulate the editors, Dr. C. Baskaran and Dr. S. Dhanavandan, for bringing out the book *Information Literacy and the Role of Social Media in the Dissemination of Scholarly Information in the 21st Century* by IGI Global Publishers. The current trend is the World Wide Web explosion of information sharing in their hands to the dissemination of scientific knowledge through Social Media/Networks. There are enormous social networks for researchers to spread scholarly information. The editors ensure that the figures are reviewed and followed up for publication. I appreciate their scholarly work at IGI Global Publishers. The main focus area is important for library and information science researchers in the current context, and the authors brought their research ideas from different parts of the world. Social networks such as Facebook and Twitter are new forms of social interaction, dialogue, information exchange and collaboration. Social networks allow users to exchange ideas, share and post updates and comments, or participate in activities and events while sharing their broader interests. From general discussion board to breaking news, from planning a data show, from mild humor to serious research, different communities of users use social networks today for many different reasons. Researchers have long been interested in the dissemination of information about the introduction of innovations. Social media are embedded in people's lives through the construction of a new, innovative information technology-dependent social structure. It has also made online sharing an everyday activity. Widespread use promotes the spread of information, which can produce information transmission effects, that is, the effects and results of information dissemination or the process of information transmission. Widespread adoption results in the phenomenon of vitality or constitutes a social epidemic. These phenomena have become research objectives that influence the formulation and implementation of scientific works and bring benefits to the society. Wide dissemination of information corresponds to an increase in the number of researchers; so publishers can more easily acquire potential academicians or researchers. The research value is also created by eliciting broad information (e.g. social epidemic, vitality), for example helping the young researchers can promote their research ideas and find potential for research area in their focusing field of study at easily and speedily.

V. Sivakumar
Independent Researcher, India

Preface

Social media is any digital technology that allows communication between individuals over online networks and groups. Thanks to social media, students can digitally interact with family, friends, peers, teachers, members of interest groups, and even complete strangers. First, social media gives students, instructors, and parents a smoother, more direct communication method. They can check-in and ask or answer questions. There are more chances for e-learning thanks to social media. Social media platforms can be utilised effectively for collaborative work. For instance, meeting online and going through the task is simpler if given a group research project. These media are used to disseminate news and service alerts, give online users instant service updates, and publish press releases and library news for online users. Promoting libraries' materials and services is made possible by these social media channels. Users can access their friends' profiles on SNSs like LinkedIn, Myspace, Facebook, Twitter, and blogs, widely used globally and extremely well-liked in India.

Using locating tools like catalogues, indexes, and other tools, academic libraries enable students to find the materials they need. When it comes to bringing people and ideas together, libraries play both social and intellectual roles. The goal of the librarians is to offer a setting that meets the needs of all users. In the twenty-first century, social media enables us to stay in touch with distant friends, family members, and others. Social media offers numerous benefits, but if we are not careful, young children could quickly abuse it because they have easy access. We must comprehend and control how young children use social media. This edited volume provides clear perspectives on various technological applications in academic libraries. This volume focuses on the viewpoints of researchers, librarians, information professionals, publishers, and library users in India. This volume provides an excellent overview of innovative librarianship, especially in academic libraries, and it helps readers find, retrieve, read and use the research findings they need. At the same time, it allows authors to enlarge their audience and amplify their impact.

In these circumstances, the editors have incorporated chapters are received from authors from Library and Information Science Professionals at Global arena. They are titled as *Role of Social Media in Libraries to Disseminating Scholarly Information in the 21st Century*. The book has significant implications for the larger scholarly community as researchers use the strategies outlined or listed to discover ones that might otherwise go undiscovered. Continuous increases in online accessibility, the usage of digital environments, and advancements in information technology encourage users of libraries to investigate the inherent value, weight, and quality of services.The author will be delighted if academicians, researchers, library personnel, and other professionals benefit from the book.

The book elucidates different areas are covered in the filed of Library and Information Science, besides the relevance of the books impact and trend to current phenomena was discussed with Social Media has been evolving to dsiisminate Scholarly information in the 21st Centruty .

The chapters decribes the significant information of the suitable studies are reflected as given below.

Digital and ICT Literacy in Distance Education: A Systematic Review of Definitions and Transformations

The Covid-19 pandemic has made these two concepts even more critical, as all educational activities are now carried out using technological tools. This chapter aims to define digital and ICT literacies and highlight the differences between these two concepts. Additionally, it will discuss the importance and definitions of these literacies in the context of distance education in the 21st century (Chen & Chen, 2019; Lin & Zheng, 2016).

Information Literacy Skill for Accessing Electronic Resources Among the Faculty Members in the Institutions in Kancheepuram District, Tamilnadu

The chapter describes the totally 309 respondents the majority of 136(44.017%) of the respondents were highly satisfied with finding the relevant information from Institutions in Kancheepuram District in Tamilnadu. The majority of the respondents 118 (24.05%) of the respondents have noticed Preferred that using e-journals made it easier for them to complete research-related tasks and publish their findings in journals and conference proceedings.

Research Social Networks Share Scholarly Communication Among Scholars in the Autonomous Institutions in Chennai, India

The Chapter analysed the Research Social networks share Scholarly Communication among scholars in the Autonomous institutions in Chennai. The result discuss about gender category, out of 302 respondents, 149 (49.3%) from female respondents, and 153 (50.7%) responses from male respondents. More or less both the gender has participated equally. The highest representation was the department of Library and Information Science (35.1%), 12.3%were Mathematics and about the (10.6percent) held of Management Studies.

Role of Social Media in Libraries to Disseminating Scholarly Information in the 21st Century

The chapter analysed the most popular in the professional sphere, notably at academic institutions such as universities, colleges, and schools. In today's culture, using social media has become crucial in academic activity. Social media is generally used for social interaction, sharing of Content, getting up-to-date news & information, and making decisions. It is a powerful and straightforward tool for talking with people, both locally and worldwide, as well as for sharing, generating, and spreading information in the academic environment towards disseminating scholarly information in the 21st Century

Mapping of the Research Publications on Research Ethics: A Scientometrics Profile

The chapter discussed the EGR ranges between 0.068 and 2.044 in the years 2002 and 2019 respectively. The study is found ranked authors out of twenty five authors in the research filed concerned. The prolific author are found in the series, Shine was the top ranked authors, who had published 164 (0.201%) of the publications. The study was explored on the productivity of the publications were contributed by the researchers among twenty five research areas have been given in the study. The ranking of the counties in the order as second ranked was England with 13353 (16.34%), Australia 11777 (14.41%), Canada 8424 (10.31%), Peoples R china 5293 (6.48%), Germany 4512 (5.52%), Netherlands 3920 (4.79%), France 2733 (3.34%), Spain 2391 (2.92%) and Sweden 2294 (2.80%).

Evolving Role of Instagram as a Digital Marketing Tool in Libraries

The chapter discussed the role of social media in libraries has been gaining attention in recent years as libraries look for new ways to connect with users and promote their services. This paper assesses the use of Instagram, a popular photo and video-sharing platform, by libraries as a means of outreach and engagement. Through a literature review and case studies of library Instagram accounts, the paper explores how libraries use Instagram to showcase their collections, promote events and services, and create a sense of community.

Research Impact of the Scholarly Literature on Google Scholar: A Scientometric Study

The chapter analysed the research impact of the scholarly literature on Google scholar: a Scientometric study during 2002 -2022. Total no. of records 25760 retrieved from Web of Science database for the period of study. Out of 25760, highest 5017(19.47%) of the publications have been witnessed in the year 2021, it followed 4903, 3835 contributed in the field of Google Scholar during 2022 and 2020 respectively. The top ranking of authors is found to be a Maffulli N with 113 (0.4%) of the publications on Google Scholar. The review articles hold the maximum of publications with 19132(74.27%).

A Study on Dissemination of Information for Teaching and Research Competence Among Faculty Members in the Deemed Universities in Coimbatore Region

The chapter analysed the respondents among four deemed Universities in the Coimbatore Region out of 364 respondents majority 26.92% of the respondents reported from Karunya University. The Boolean searches use for accessing e-resources by the respondents in the Deemed Universities. The majority 147 (48.8%) respondents preferred Boolean search technique to Very Large Extent while 74 (20.32%) respondents to the large extent. The majority of 132 (36.26%) of the respondents have noticed preferred e-books among the deemed Universities in the Coimbatore region as Excellent towards preferred e-book.

A Study on Disseminates Scholarly Information Through Online Resources in the Universities: A Questionnaire Based Study

The chapter discussed the Disseminate Scholarly information through online resources in the Universities: A Questionnaire based study in the Selected Universities in Kerala. Out of 460 respondents, 234 (50.86%) of the respondents belong to Engineering /Architecture; this is the top one category among the category of Disciplines of the study. The majority 435 (94.6%) of the respondents aware on e-resources. 413 (89.78%), they prefer to make use of E-Journals. E-Books are widely used by 426 (92.60%) respondents followed by the use of E-Magazines by 212 (46.08%) respondents. The result shows 185 (40.21%) of the respondents preferred General Books.

Information Literacy Skills on Access E- Contents Among Faculty Members in the Deemed Universities in Chennai, India: The Questionnaire Based Survey

The chapter analysed the result shows SRM University and Sathiyabama University have respondents of each 130 (34.21%), 85 (22.36%) respondents are belonging to St. Peters University, 35(9.22%) respondents are from Crescent University. Out of 380 respondents it is seen that 210 (55.26) respondents are extremely satisfied on OPAC/Web OPAC. 205(53.90) respondents are extremely satisfied on E-Databases, 192(50.52) respondents are extremely satisfied on Automated circulation services. The majority of respondents are extremely satisfied on OPAC/Web OPAC (55.26%) provided by University Library followed by E-Databases (53.90%).

A Quantitative Study on Usage Measure the Social Medias Platforms in Bangladesh

The chapter analyzed the majority of the people in Bangladesh preferred all type of Social Medias Platforms, among listed Social Medias, the Face Book with percent between 89.69 and 93.79 in August and January in the year 2022. It follows YouTube usage rate between 2.58 and 5.91, Linked In (0.71 and 1.63), Twitter (0.38 and 1.77), Printerest (0.26 and 0.58), Instagram (0.3 and 1.01), Reddit (0.2 and 0.31), Tumblr (0.01 and 0.04).

Information Literacy Competence on Social Media Use Against Scholarly Communication: A Case Study Among Faculty Members and Scholars in the Faculty of Management of Alagappa University

The chapter discussed the respondents participated out of 340 for eight departments, majority of them 51(15%) of them from Alagappa Institute of Management. sex and Age wise respondents of study, the higher potion of population being male with 238 (70%), less than half in the respondents 102 (30%) were female. The study witnesses the multiple responses about usage the different social Medias listed in the table. Majority of 324(95.29%) of them preferred face book.

Users' Attitude on Sharing Social Medias Among the Research Scholars Belongs to Social Science Background in the Universities: A Questionnaire Based Study

The chapter examined the results of the study were obtained by applying the normality test, for Simple percentage analyses on the obtained data. When the chi-square analysis between the opinions of Social Science Scholars concerning the level of use of Social media and the level of use of academic research was examined, a positive and significant analysis was found between them. The sample comprised 623 social science scholars. Male 335 (53.8%) and 288 (46.2%) of the respondents, 13.0%, and 12.9% of respondents preferred WhatsApp at Annamalai University. 12.0% of the respondents prefer YouTube from Alagappa University.

Preference and Opinion on Social Medias for Disseminating Scholarly Information Among Research Scholars in the Universities Context in Tamil Nadu: A Study

The chapter discussed the Social Medias and Networks disseminate and share Scholarly information among the Research Scholars in the selected State Universities of Tamil Nadu in India. The researcher has adopted Questionnaire method, in order to collect data from respondents in the Universities, on entertain only the full time research scholars. The result of the study demonstrates majority 87 (14.0%) of them responded from Annamalai University.

Exploring Opportunities for Utilizing Open Educational Resources (OERs) Towards Promote Inclusive Education Setting in Tamilnadu

The chapter analysed under questionnaire includes questions such as the awareness of OERs, the sources from which research scholars access OERs, and their utilization in teaching and learning. The findings of the survey will help to gain an understanding of the usage of OERs and the potential of using them to promote inclusion in Tamilnadu. The results of this survey will also be beneficial in understanding the current status of OER usage and the opportunities for utilizing them to improve the quality of education in the region.

Social Networking Skills in Information Literacy Towards Disseminating Scholarly Communication Among Research Scholars at Alagappa University: A Case Study

The chapter analysed the frameworks about the challenges recognized by the scholars in accessing SNS and profound changes in information environments and outlines of generation and necessity of scientific information in new modes of assessment which predominantly focuses on methods and standards of scientific research work. Data collected by distributing questionnaires through Google form and print form among research scholars of various departments and received 71 responses. The study aims to identify the fundamental principles and utilization of social networking skills (SNS) in information literacy (IL) by research scholars of Alagappa University. It intends to compare the earlier educational level, dissemination of information level and after employing SNS in research.

Utilization and Impact of ICT in Information Literacy Expertise Among Post Graduate Students of Alagappa University

The chapter analysed a random questionnaire was distributed to 150 students and the study adopted an exploratory research design. The adoption and use of ICT has enhanced subject knowledge, new educational principles and technologies. Quantitative research data were entered into an MS Excel file where statistics were uploaded and analyzed using SPSS software; a comparison between postgraduate students awareness and use of ICT knowledge. Performance difference between gender and age were two groups involved in Descriptive statistics, Simple percentage and ANOVA test.

The contents are edited suitably in order to understand even by novice professionals. Most of the chapters focus Information Literacy Skill, Scoial Medias Useage, Electronic resources access through Information Communication Technology and Scientometrics. The different variety of user perspectives has been put forth is this part. As the title of book *states Information Literacy Skills and the Role of Social Media in Disseminating Scholarly Information in the 21st Century*, this volume narrates the technological changes that pervade in the library and information Science discipline. This volume is not just an additional literature in the field, provides a nascent trend in the social networking tools, enabling library professionals to enrich their knowledge.

The book's readers will believes it would be a significant addition for which the increasing collection of aslo be the value reference material. The editors are expressed their heartfelt appreciation to the authors everyone who has contributed chapters to this book and they have made precious effort to have a value one.

C. Baskaran
Alagappa University, India

S. Dhanavandan
Central University of Tamil Nadu, India

Acknowledgment

We acknowledge the people who involved directly and directly to bring out the Book, we must grateful to the Faculty Colleagues, Officials, Staff members and Research Scholars of the Alagappa University and Central University of Tamil Nadu for their supports and encouragements for complete the project.

We convey our sincere acknowledge to the all authors and reviewers more specifically, for taken the part in the review process, without their support, this book would not have become a reality.

We would like to thank each one of the authors for their contributions. Our sincere gratitude goes to the chapter's authors who contributed their time and expertise to this book at timely submission of their work to this book.

We would like to acknowledge the valuable contributions of the reviewers regarding the improvement of quality, coherence, and content presentation of chapters. Most of the authors also served as referees; we highly appreciate their dual task.

Finally, we acknowledge our sincere gratitude to IGI Global for the encouragements for provided us this project.

Chapter 1
Digital and ICT Literacy in Distance Education:
A Systematic Review of Definitions and Transformations

Ebru Solmaz
https://orcid.org/0000-0003-4893-450X
Gazi University, Turkey

Seher Özcan
https://orcid.org/0000-0001-8473-3807
Gazi University, Turkey

Burcu Karabulut Coşkun
https://orcid.org/0000-0001-5287-2239
Kastamonu University, Turkey

ABSTRACT

Some studies have shown that the definitions of digital and ICT literacies need clarification, as they are sometimes considered the same thing. Digital literacy refers to the ability to use software or operate a digital device, while ICT literacy involves using digital technology, communication tools, and/or networks to access, manage, integrate, evaluate, create, and transmit information in order to function in an information society. The Covid-19 pandemic has made these two concepts even more critical, as all educational activities are now carried out using technological tools. This chapter aims to define digital and ICT literacies and highlight the differences between these two concepts. Additionally, it will discuss the importance and definitions of these literacies in the context of distance education in the 21st century.

INTRODUCTION

According to the fundamental paradigms that have been guiding learning and teaching processes for centuries, the most critical skills needed by the industrial age are obtaining information, structuring

DOI: 10.4018/978-1-6684-8805-8.ch001

information, synthesizing the structured information, and integrating it into production (Hazar, 2019). The way people access and manage information and communicate with others through technology has made them increasingly addicted to it in day-to-day life. While technology makes many sources available for helping people access information on any subject, if a person cannot manage and utilize these resources efficiently, this resource richness will remain underutilized (ETS, 2002). Similarly, owning a computer is not enough in today's world. Being exposed to technology increasingly does not lead to the skills necessary to use it. Today, the success criterion is how well a person can evaluate, manage, and transmit all information in a technological environment. Additionally, being a member of a digital society cannot offer equal technology opportunities to all individuals in that society.

For this reason, it is not enough to have skills in a single literacy type alone; it is necessary to have multiple literacy and communication skills (Cabero-Almenara et al., 2020). Acquisition of these skills, named skills of Information and Communication Technologies (Tyler, 2005), can be achieved by going through some phases. All of the acquisitions in these phases form the concept of literacy (Eshet, 2004).

These skills, which were initially generalized as "digital literacy," have now been named 21st-century skills and encompass various skills such as information and communication technologies, information literacy, digital literacy, and media literacy (Eshet, 2004). The concept of literacy has undergone several changes from the past to the present, and at the beginning of the 21st century, people's perception of literacy evolved once more. The effect of technology on daily human life is now beyond imagination and could not have been envisioned even 25 or 10 years ago. Hence, the concept of "literate people" has been expanded by policymakers, business leaders, and educators to include the skills and qualifications people need in this increasingly technological world. Additionally, the growing acceptance of lifelong learning has increased the opinions and requests about literacy. Literacy is no longer just a situation that someone has or lacks. It is seen as a continuum of the information, skills, and strategies that individuals acquire from their lives in various situations and through communication with their peers or the broad communities they attend (International ICT Literacy Panel, 2002).

PURPOSE

This study aims to define digital and ICT literacy within the framework of skills that can measure the effectiveness of technology education or technology skills. In this direction, a series of guidelines, standards, evaluation lists, and definitions are included in the study (Fraillon et al., 2014). Although it aims to define a comprehensive competence framework for digital and ICT literacy, a standardized definition will likely only partially be established due to the variable nature of technology and the differences in its significance perceived by many groups (Riel et al., 2012). In this study, the definitions of digital and ICT literacy were synthesized, and a conceptual structure was created for future studies within the framework of the following questions (Goktas et al., 2016).

1. What is the definition of ICT Literacy?
2. What is the definition of Digital Literacy?
3. What is the frame of ICT Literacy and Digital Literacy in Distance Education?

METHODOLOGY

The literature review methodology was used in this research. The authors surveyed multiple databases, including ERIC, ProQuest, EBSCOhost, and Google Scholar, using 21st-century, internet, digital, and ICT literacy keywords. The obtained theoretical information was synthesized to create definitions and frameworks for digital and ICT literacy, the competencies suitable for these literacies, and the conditions required for their development. The study also made suggestions for developing the relationship between these concepts.

Definitions and Frameworks

The nature of knowledge is constantly evolving and changing due to the increasing availability and widespread use of Information and Communication Technologies (ICT). The primary evidence of this shift is the dramatic increase in the amount of digital information available, as opposed to information in printed form. However, if individuals are unable to access reliable and accurate sources, ICT can become a disadvantage rather than an advantage. Thus, the importance of information literacy has become increasingly apparent. Information literacy was initially defined as the ability to obtain, synthesize, and integrate accurate information from reliable sources. However, it has evolved into e-information literacy, which is defined as the ability to obtain, synthesize, and integrate accurate information from electronic and digital sources (Bilawar & Pujar, 2011). Given this, it is crucial to prioritize the examination of information literacy.

INFORMATION LITERACY

Paul Zurkowski coined the term "information literacy" in 1974. Its fundamental definition includes the ability to search, find, obtain, evaluate, and use information (Choukhande Vaishali, 2009). This definition is widely accepted across all fields, including education. According to Bilawar and Pujar (2011), information literacy forms the foundation of various types of literacies, such as basic literacy, library literacy, media literacy, computer literacy, ICT literacy (Information and Communication Technology), visual literacy, tool literacy, resource literacy, social-structural literacy, research literacy, publishing literacy, emerging technology resource literacy, digital literacy, network literacy, and scientific literacy. These literacies can be illustrated as follows:

Literacies play a vital role in individuals' attainment of social, academic, and business goals. Technological literacy has become increasingly important with the pervasive introduction of technology into all aspects of life. With the COVID-19 pandemic, it has become clear that most of these literacies are now a necessity rather than an option. In other words, although information literacy has remained essential since its inception in 1974, its significance needs to be reinterpreted and evaluated within the context of digital skills as digital life becomes increasingly widespread (Leaning, 2017). To this end, this study explicitly examines the concepts of ICT and Digital Literacies.

Furthermore, it is essential to delve deeply into ICT and Digital Literacy to enhance efficiency in educational environments and understand why and how these literacies are applied in school settings. In doing so, it is expected to gain a better understanding of how concepts and skills will be shaped with accurate definitions and contribute to developing advanced digital skills with a far-reaching impact and

Figure 1. Components of information literacy
(Bilawar & Pujar, 2011)

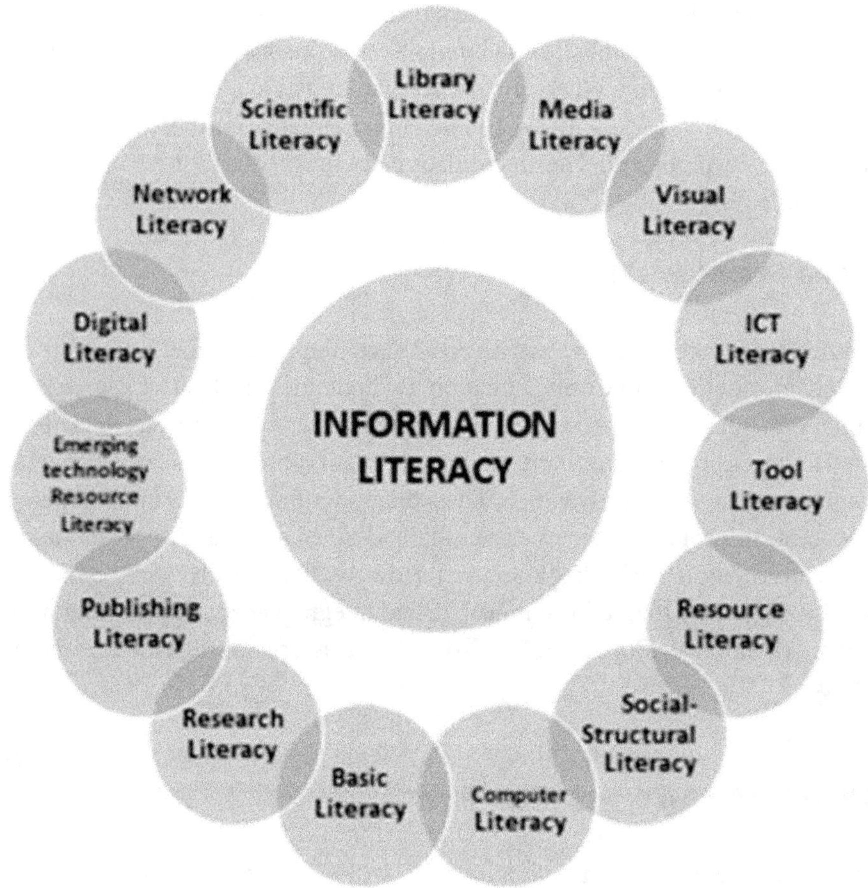

ensuring productivity in the system cycle. In other words, comprehending what digital literacy is and why it is essential now and, in the future, provides a rigid outline of critical elements on the subject (Riis, 2017).

Overall, the current pandemic has made it apparent that digital skills and literacies are essential for survival in the digital age. Understanding the concepts and skills associated with digital literacy and ICT is crucial in developing these skills and ensuring that individuals are equipped to navigate the digital landscape effectively.

ICT LITERACY

ICT literacy is one of the most commonly studied literacy forms. This term is defined as the proficiency necessary for successfully performing ICT and cognitive information-based tasks (Markauskaite, 2007). ICT proficiency involves the adequate use of networks, communication objectives, and digital technologies to solve information problems in an information society. This proficiency includes technology use skills as a tool to transfer, evaluate, organize, and search for information and comprehend ethical and legal subjects embracing access to the information and the use of information (ETS, 2003). According to the International ICT Panel (2002), ICT proficiency involves integrating and applying cognitive and technical skills. In this conference, ICT literacy was defined as "using digital technology, communications tools, and/or networks to access, manage, integrate, evaluate, and create information in order to function in a knowledge society." As can be understood from definitions, ICT literacy is thought of as combining two proficiencies within ICT proficiency. These proficiencies are named cognitive proficiency and technical proficiency by the International Panel (2002) and general cognitive and technical capabilities by Markauskaite (2007). The International ICT Literacy Panel (2002) described cognitive proficiency as essential skills required in daily life at school, home, and work. It was indicated that literacy, numeracy, problem-solving, and spatial/visual literacy explain these skills. Technical proficiency predicates essential components of digital literacy and contains basic information on digital technology, networks, hardware, and software application elements.

In addition to the definitions indicated above, it is seen that ICT literacy divides into various components. The International ICT Literacy Panel (2002) specified these from simple to complex according to their cognitive complexity level: access, manage, integrate, evaluate, and create. These components are seen as the process of information, skill, and strategies acquired from individuals' experiences. ETS (2003) specified these components by adding two more: define and communicate, making them access, manage, integrate, evaluate, create, define, and communicate. These components reflect the wide variety of information and communication technologies. Markauskaite (2007) divides ICT literacy into nine components: plan, access, manage, integrate, evaluate, create, communicate, collaborate, reflect, and judge. Markauskaite (2007) developed this ICT model, including nine components, by combining the ETS model (Markauskaite, 2005; Markauskaite et al., 2006), some standards of information literacy (Bundy, 2004), technology literacy (ISTE, 1998), and frameworks of problem-solving (Eisenberg & Johnson, 2002). The primary areas of this model contain the sufficiencies that are required to execute critical steps of knowledge-based problem-solving processes (for example, to plan, access, manage, integrate, evaluate, create, and communicate information), interpersonal sufficiencies (for example, to communicate and collaborate), and metacognitive sufficiencies (for example, to judge and reflect on performance) (Markauskaite, 2007). The components of these three models and the skills these contain are shown in detail in the table below:

ICT literacy is broadly defined by Claro et al. (2012) within the scope of 21st-century skills as "the capacity to solve problems of information, communication, and knowledge in digital environments." The characteristics of this definition are generally stated below:

1. Speciality in ICT applications, such as functional ICT skills, is required to perform cognitive duties in a digital environment. With technical skills to use ICT applications, students can perform cognitive duties in digital environments (Claro et al., 2012).
2. Skills are not related to technology and do not refer to using specific software. Rather, they refer to the information in an ICT environment and problem-solving skills in communication and knowledge tasks (Claro et al., 2012).
3. Substantial high-level thinking processes are supported and limited by ICT applications (Claro et al., 2012).
4. Skills are about cognitive processes that support students' ongoing learning, and hence, these skills are about learning content in the information society (Claro et al., 2012).

Table 1. Details of the components of the models

Resources	Components	Skills	
International ICT Literacy Panel (2002)	Access	knowing about and knowing how to collect and/or retrieve information.	
	Manage	to apply an existing organizational or classification scheme.	
	Integra	interpreting and representing information. It involves summarizing, comparing and contrasting.	
	Evaluate	making judgments about the quality, relevance, usefulness, or efficiency of information.	
	Create	generating information by adapting, applying, designing, inventing, or authoring information.	
ETS (2003)	Define	Using ICT tools to identify and appropriately represent information need	
	Access	Knowing about and knowing how to collect and/or retrieve information	
	Manage	Organizing information into existing classification schemes	
	Integrate	Interpreting, summarizing, comparing and contrasting information using similar or different forms of representation	
	Evaluate	Reflecting to make judgments about the quality, relevance, usefulness, or efficiency of information	
	Create	Generating new information and knowledge by adapting, applying, designing, inventing, or representing information	
	Communicate	Conveying information and knowledge to various individuals and/or groups	
Markauskaite (2007)		**Technical**	**General cognitive**
	Plan	To use planning and decision-support tools, etc.	To plan a solution to a cognitive task, e.g., identify key concepts, develop potential strategies
	Access	To work within the desktop environment, navigate and search digital resources, maintain a computer, etc.	To select appropriate techniques and tools, obtain information from various media and sources, etc.
	Manage	To perform common operations within software packages, manage data using spreadsheets, design databases, etc.	To apply existing organizational or classification schemes, categorize and store information, keep records of information and its sources, etc.
	Integrate	To solve problems using spreadsheets and modelling software, manipulate databases, etc.	Summarize, compare and contrast different sources and concepts, understand interconnections, etc.
	Evaluate	To evaluate the relevance of digital resources, information and tools, etc.	To define evaluation criteria, judge the quality, usefulness and relevance of information, make choices, etc.
Resources	**Components**	**Skills**	
Markauskaite (c) (2007)		**Technical**	**General cognitive**
	Create	To create graphics, documents, presentations and web pages, etc.	To apply, adapt and generate information, propose new ideas, design artefacts and produce other outputs
	Communicate	To publish and deliver results of research activity using ICT presentation tools and networks, etc.	To convey solutions in a variety of forms and to different audiences, respect copyright, privacy, etc.
	Collaborate (interpersonal capabilities)	To communicate via e-mail and other network tools, collaborate in virtual learning environments, etc.	To collaborate and communicate with various people in a variety of contexts, work in a team, adapt to various learning contexts, etc.
	Reflect and Judge (metacognitive capacities)	To use personal management and reflection tools, etc.	To judge the final product and reflect on the problem-solving process employed

DIGITAL LITERACY

Even though digital literacy is seen as a new concept, it was defined by Paul Glister in 1997 as "the use and comprehension of various information through sources supported by computers" (Glister, 1997). In later studies, this definition was supported by the synthesis of information using technological devices (Eshet, 2004). Since Glister's definition, socio-cultural needs, technological developments, and social changes have caused changes in some definitions. The most important reason is that different reflections of technological developments affect every area.

Today, the definition of digital literacy has changed as a result of these reflections to "the skill of coping with electronic substructure and tools" (Eshet, 2004). It is seen as a skill that both students and teachers should have in technology-aided education environments, and it has become even more critical than basic literacy. Individuals start using digital tools and gaining information about them before actual education and training activities in today's world.

Digital literacy requires various cognitive thinking strategies to use digital knowledge. Individuals read instructions to follow on graphic interfaces and create products like copy/paste using computer-based instructions. It does not limit individuals to create through hypertexts, but it gives instructions for the use of tools or the environment and enables the evaluation of digital knowledge by the individual (Eshet, 2004). The most important aspect of digital literacy is the contribution of integrating tools into the education environment by realizing new opportunities (dis)advantages thanks to the necessity of following the developments closely (Jones & Flannigan, 2006).

Another characteristic of this literacy type is learning necessary acquisitions without being aware of it, as it is in language literacy. This type of literacy does not occur only in the education environment, and it contains a process from the first opening of a newly bought electronic device to learning how to use it by reading user manuals. Individuals mostly need to be made aware of the process. In this way, the individual must realize that s/he starts learning the content as in lessons (Andersen, 2002).

In light of this information, the characteristics of digital literacy can be specified as follows:

1. The necessity of following trends
2. Learning without being aware of acquisitions
3. Need for cognitive thinking strategies

In a study conducted by Riel et al. (2012), literacy skills are defined and coded into a category according to their characteristics. They adapted the defined skills to four essential literacy areas common to the literature. Accordingly, digital literacy skills can be defined in the following categories: 1) Tools and Interface; 2) Information and Data; 3) Sharing and Creating; and 4) Historical and Cultural Context. The categories are examined in *Table 2* below.

Table 2. Classification of digital literacies

Tools & Interface	Information & Data	Sharing & Creation	Historical & Cultural Context
*Computational Basics *Computer Hardware *Computer Software & *Applications *Networks *Design *Augmentation	*Representation *Search *Assembly *Analysis & Judgment *Synthesis *Archiving a Navigation	*Inventive Thinking *Documents (text) *Multimedia *Communication *Online Personal *Productivity *Sharing& Collaboration	*Digital Citizenship *Diversity *Intellectual Property *Privacy and Identity *Programmed Agendas *Technology Impact

(Riel et al., 2012)

As seen in Table 2, digital literacy has been defined as the ability to use digital tools appropriately to fulfill daily life skills (Ozdamar-Keskin et al., 2015). Additionally, it involves the ability to locate, manage, combine, evaluate, analyze/synthesize, and transfer information from digital sources, discover new media types, and communicate with others using digital tools (Martin, 2008). According to Shariman et al., (2012), digital literacy encompasses multiple literacy skills, including information literacy, computer literacy, media literacy, communication literacy, visual literacy, and technology literacy (Ozdamar-Keskin et al., 2015). The sub-disciplines of digital literacy are summarized in Table 3 below.

Table 3 shows that the subdisciplines of Digital Literacy are Information Literacy, Computer Literacy, Media Literacy, Communication Literacy, Visual Literacy, and Technology Literacy. Information Literacy refers to finding the correct information and being able to synthesize it for use (ACRL, 2016). Computer Literacy is an understanding of how to use computers and software applications (Gruber & Kumar, 2019). Media Literacy is the ability to access, analyze, evaluate and communicate information from media (Buckingham, 2019). Communication Literacy is the ability to communicate effectively (Kress & Van Leeuwen, 2006). Visual Literacy is the ability to 'read,' interpret, and understand the information presented in visual materials (Lupton & Phillips, 2015). Technology Literacy refers to the ability to use computers and other technology (ISTE, 2016).

DIGITAL AND ONLINE LEARNING

With the announcement of the COVIDd-19 pandemic in the world, starting from the spring semester of the 2019-2020 academic year, all educational activities in all countries had to be transformed from formal education to distance education (Daniel, 2020). This situation brought an unexpected and sudden obligation for both students and teachers. Educational institutions and teachers working in these institutions have been caught unprepared and searched for how the lessons can be done with distance education and various online tools.

Table 3. Subdisciplines of digital literacy

Sub-Discipline	Definition
Information Literacy	Finding and locating sources, analyzing and synthesizing the material, evaluating the credibility of the source, using and citing ethically and legally, focusing on topics and formulating research questions in an accurate, effective, and efficient manner.
Computer Literacy	An understanding of how to use computers and application software for practical purposes.
Media Literacy	A series of communication competencies, including the ability to access, analyze, evaluate and communicate information in a variety of forms including print and non-print messages.
Communication Literacy	Learners must be able to communicate effectively as individuals and work collaboratively in groups, using publishing technologies (word processor, database, spreadsheet, drawing tools...), the Internet, as well as other electronic and telecommunication tools.
Visual Literacy	The ability to 'read,' interpret, and understand the information presented in pictorial or graphic images; the ability to turn information of all types into pictures, graphics, or forms that help communicate the information; a group of competencies that allows humans to discriminate and interpret the visible action, objects, and/or symbols, natural or constructed, that they encounter in the environment.
Technology Literacy	Computer skills and the ability to use computers and other technology to improve learning, productivity, and performance.

(Ozdamar-Keskin et al., 2015)

Many tools that should be used to sustain distance education were never experienced by most teachers before. The most efficient use of vehicles is directly related to individuals having many technology usage skills. In this context, considering that teachers and institutions have not yet fully acquired the skills mentioned above, it can be predicted that distance education cannot be carried out flawlessly with the rapid transition (Karip, 2020; Çetinkaya Aydın, 2020; Durak et al., 2020). In this context, examining the skills needed to achieve maximum efficiency for all stakeholders in distance education is crucial.

ICT and Digital Literacies in the Context of Distance Education

The daily development of information and communication technologies made it necessary for developing countries to integrate ICT Technologies into their education programs (Bahrini & Qaffas, 2019; Tella & Adu, 2009). Many higher education institutions were adapting ICT teaching-learning technologies to their learning environment to make their students and instructors acquire knowledge and learn cooperatively and see this as a critical element for education reform (Lawrence & Tar, 2018; Peeraer & Van Petegem, 2011). The fall in the prices of these technologies and the making of multimedia tools accessible increased the use of these technologies not only in face-to-face education but also in the distance education process (Ifinedo, 2006).

Distance education became widespread in providing learning needs by removing the place and time restrictions for people with the help of computers and the internet (Zang & Kenny, 2010). Distance education became an invaluable opportunity with web-based technologies for people to have the training and for various companies to train their personnel (Wang et al., 2010). However, there is an increase in its prevalence does not mean that it is effective. Gulati (2008) stated that one of the biggest problems that developing countries face in distance education is the problems that they face when it comes to the efficient use of these technologies in training and learning. Because more than access to technology is needed to reach socioeconomic goals, essential skills necessary to use these technologies efficiently should also be acquired (Farjon et al., 2019; Yusuf, 2006; Buckingham, 2010). Especially in distance education, learners are expected to be more independent and take more responsibility for their learning (Hammond & Trapp, 2001). In other words, learners need to make the information meaningful through their interactions in the learning environment and structure it. Developing digital and ICT literacy skills is vital to use e-learning effectively and become productive individuals (The E-learning Task Force, 2002).

These skills require the critical and safe use of information systems, active participation in social networks and cooperative environments, awareness of security threats, and innovative and creative use of these information systems (Ololube et al., 2007; Buckingham, 2010). With the development of Web 2.0, the transfer of distance education to digital environments, even being named e-learning and web-based learning, has caused an increase in the importance of these skills for distance education day by day. Accordingly, individuals who have these skills at a high level are more self-managed in their education and can continue their educational activities with fewer suspensions compared to those who do not have these skills (Hargittai, 2010).

"Interaction", the most critical element in every type of education, is also significant in distance education (Alqurashi, 2019; Repman & Logan, 1996; Patsula, 2002; Uşun, 2003). In conducted studies, it is evident that no matter how distance education is approached, it is a system based on interaction (Moore & Kearsley, 1996; Collins & Berge, 1996; Richardson & Ting, 1999). Therefore, examining distance education within the context of interaction means the analysis of the whole distance education system.

There are four types of interactions: learner-content, learner-teacher, learner-learner, and learner-interface. In learner-content interaction, the content type acquired by the student through the course material can be in the form of text, voice, video tape, CD-ROM, computer program, or online communication. Interaction occurs when the individual combines old and new knowledge and creates a new message of integrity (Bouhnik & Marcus, 2006). When we approach this interaction type within the scope of ICT and digital literacy, it is seen that the learner should first have the technical knowledge to use these materials efficiently and access the content knowledge without any data loss. Additionally, learners are expected to organize and plan the content according to their learning strategies, compare the information given in different sources, and make connections between them. When considering all these ICT components, they match up to plan, access, manage, integrate, and evaluate components.

While the interaction between the learner and the instructor is based on an environment that encourages learners to understand the content better, the interaction between the learner and the learner is the information exchange among students in the presence or absence of the instructor (Sher, 2009). During this process, the learner interacts with the instructor and other students, communicating with them using different communication tools and working with them cooperatively when required. During this process, communicating and collaborating components of ICT gain importance. Because the learner, who is in the center of learning communities, is influenced by technology and interaction with others, every individual in the community can structure their knowledge (Thoms & Eryilmaz, 2014).

Today, the quality of developing online technologies has been effective in the learner interface, becoming a vital interaction type. Hillman et al. (1994) emphasized that this interaction form occurs between the learner and the technology used for online education. When examined within the scope of online learning, individuals with low digital literacy skill levels do not like using web-based learning environments, while those with a high level of digital literacy skill can easily access information systems and use them based on their needs and priorities. This situation increases their performance (Ferro et al., 2011; Mohammadyari & Singh, 2015). At the same time, digitally literate people are familiar with new tool norms, access options, terminology, and interfaces, which reduces their cognitive burden and allows them to focus more on their learning.

Consequently, ICT and digital literacy have become equally important when approached in the framework of distance education, just as they have become an irreplaceable part of face-to-face education. Moreover, because of distance education, learners are more independent individuals, which increases the importance of these skills. Therefore, individuals who will continue their distance education should be supported to acquire these skills. Otherwise, it will become challenging to keep these individuals within the process, and the expected efficiency cannot be obtained from training programs.

CONCLUSION

This chapter aims to describe the digital and ICT skills that individuals need in distance education, which has become a necessity and the only method for sustaining educational activities during the Covid-19 pandemic. In this context, the types of competencies required to sustain educational activities in a productive, safe, and ethical manner are defined. The importance of this in relation to future classroom roles is also emphasized, as it helps individuals in education systems acquire information from digital resources in safe, secure, and sustainable ways and increase their capacity to benefit from the opportunities offered correctly.

REFERENCES

Alqurashi, E. (2019). Predicting student satisfaction and perceived learning within online learning environments. *Distance Education*, *40*(1), 133–148. doi:10.1080/01587919.2018.1553562

Andersen, N. (2002). New Media and New Media Literacy: The horizon has become the landscape - new media are here. Cable in the Classroom.

Bahrini, R., & Qaffas, A. A. (2019). Impact of information and communication technology on economic growth: Evidence from developing countries. *Economies*, *7*(1), 21. doi:10.3390/economies7010021

Bilawar, P. B., & Pujar, S. M. (2011). *Information literacy models: correlation and conceptual model for higher education*.

Bouhnik, D., & Marcus, T. (2006). Interaction in distance-learning courses. *Journal of the American Society for Information Science and Technology*, *57*(3), 299–305. doi:10.1002/asi.20277

Buckingham, D. (2010). In B. Bachmair (Ed.), *Defining Digital Literacy Medienbildung in neuen Kulturräumen* (pp. 59–71). doi:10.1007/978-3-531-92133-4_4

Bundy, A. (Ed.). (2004). *Australian and New Zealand information literacy framework. Principles, standards and practice* (2nd ed.). Australian and New Zealand Institute for Information Literacy.

Burnett, K., Bonnici, L. J., Miksa, S. D., & Kim, J. (2007). Frequency, Intensity and Topicality in Online Learning: An Exploration of the Interaction Dimensions that Contribute to Student Satisfaction in Online Learning. *Journal of Education for Library and Information Science*, *48*(1), 21–35.

Cabero-Almenara, J., Gutiérrez-Castillo, J. J., Palacios-Rodríguez, A., & Barroso-Osuna, J. (2020). Development of the teacher digital competence validation of DigCompEdu check-in questionnaire in the university context of Andalusia (Spain). *Sustainability (Basel)*, *12*(15), 6094. doi:10.3390u12156094

Çetinkaya Aydın, G. (2020). COVID-19 Salgını Sürecinde Oğretmenler. *Tedmem*. https://tedmem.org/covid-19/covid-19-salgini-surecinde-ogretmenler

Choukhande Vaishali, G. (2009). Information Literacy curriculum in academic libraries. *Information Age*, *3*(3), 5.

Claro, M., Preiss, D. D., San Martín, E., Jara, I., Hinostroza, J. E., Valenzuela, S., Cortes, F., & Nussbaum, M. (2012). Assessment of 21st century ICT skills in Chile: Test design and results from high school level students. *Computers & Education*, *59*(3), 1042–1053. doi:10.1016/j.compedu.2012.04.004

Collins, M. P., & Berge, Z. L. (1996). *Mailing lists as a venue for adult learning. Continuing and Distance Education Research Conference*, Pennsylvania State University, University Park, PA.

Çubukçu, A & Bayzan, B. (2013). Türkiye'de Dijital Vatandaşlık Algısı ve Bu Algıyı İnternetin Bilinçli, Güvenli ve Etkin Kullanımı ile Artırma Yöntemleri. *Middle Eastern & African Journal of Educational Research*, 5-148.

Daniel, S. J. (2020). Education and the COVID-19 pandemic. *Prospects*, *49*(1-2), 91–96. doi:10.100711125-020-09464-3 PMID:32313309

Durak, G., Çankaya, S., & İzmirli, S. (2020). COVID-19 pandemi döneminde Türkiye'deki üniversitelerin uzaktan eğitim sistemlerinin incelenmesi. *Necatibey Eğitim Fakültesi Elektronik Fen ve Matematik Eğitimi Dergisi*, *14*(1), 787–809. doi:10.17522/balikesirnef.743080

E-learning Task Force. (2002). *Get on with IT*. Department for Education and Skills.

Eisenberg, M. B., & Johnson, D. (2002). *Learning And Teaching İnformation Technology – Computer Skills İn Context*. ERIC Digest, ED465377.

Eshet, Y. (2004). Digital Literacy: A Conceptual Framework for Survival Skills in the Digital era. [Norfolk, VA: Association for the Advancement of Computing in Education] [AACE]. *Journal of Educational Multimedia and Hypermedia*, *13*(1), 93–106.

ETS. (2003). *Succeeding in the 21st century. What higher education must do to address the gap in information and communication technology proficiencies. Assessing literacy for today and tomorrow*. Educational Testing Service.

Farjon, D., Smits, A., & Voogt, J. (2019). Technology integration of pre-service teachers explained by attitudes and beliefs, competency, access, and experience. *Computers & Education*, *130*, 81–93. doi:10.1016/j.compedu.2018.11.010

Ferro, E., Helbig, N. C., & Gil-Garcia, J. R. (2011). The role of IT literacy in defining digital divide policy needs. *Government Information Quarterly, 28*(1), 3e10.

Glister, P. (1997). *Digital Literacy*. John Wiley&Sons Inc.

Gulati, S. (2008). Technology-enhanced learning in developing nations: A review. *International Review of Research in Open and Distance Learning*, *9*(1). doi:10.19173/irrodl.v9i1.477

Hammond, N. V., & Trapp, A. L. (2001). *How can the web support the learning of Psychology. Learning and Teaching on the World Wide Web*. Academic Press.

Hargittai, E. (2010). Digital na(t)ives? Variation in internet skills and uses among members of the "net generation". *Sociological Inquiry, 80*(1), 92e113.

Hazar, E. (2019). A Comparison between European digital competence framework and the Turkish ICT curriculum. *Online Submission*, *7*(4), 954–962. doi:10.13189/ujer.2019.070406

International ICT Literacy Panel. (2002). *Digital transformation: A framework for ICT literacy (A report of the International ICT Literacy Panel)*. Princeton, NJ: Educational Testing Service. http://www.ibertic.org/evaluacion/sites/default/files/biblioteca/32_digitaltransformation.pdf

ISTE. (1998). *The national educational technology standards for students*. International Society for Technology in Education.

Jones, B., & Flannigan, B. L. (2006). *Connecting the digital dots: Literacy of the 21st century*. Educause Quarterly. http://www.nmc.org/pdf/Connecting%20the%20Digital%20Dots.pdf

Karip, E. (2020). COVID-19: Okulların Kapatılması ve Sonrası. *Tedem*. https://tedmem.org/vurus/covid-19-okullarin-kapatilmasi-ve-sonrasi,

Keegan, D. (1996). *Foundations of distance education.* Psychology Press.

Lawrence, J. E., & Tar, U. A. (2018). Factors that influence teachers' adoption and integration of ICT in teaching/learning process. *Educational Media International, 55*(1), 79–105. doi:10.1080/09523987.2018.1439712

Markauskaite, L. (2005). *From a static to dynamic concept: A model of ICT literacy and an instrument for self-assessment.* In P. Goodyear, D. G. Sampson, D. J.-T. Yang, Kinshunk, T. Okamoto, R. Hartley, & N.-S. Chen (Eds.), ICALT 2005. The *5th IEEE international conference on advanced learning technologies Proceedings* (pp. 464–466). IEEE Computer Society. 10.1109/ICALT.2005.161

Markauskaite, L. (2007). Exploring the structure of trainee teachers' ICT literacy: The main components of, and relationships between, general cognitive and technical capabilities. *Educational Technology Research and Development, 55*(6), 547–572. doi:10.100711423-007-9043-8

Markauskaite, L., Reimann, P., Reid, D., & Goodwin, N. (2006). *Exploring the fit of an information technology course to the ICT literacy of trainee teachers.* In C. M. Crawford, R. Carlsen, K. McFerrin, J. Price, R. Weber, & D. A. Willis (Eds.), *SITE 2006 – society for information technology & teacher education. 17th International conference. Proceedings* (pp. 2960–2967). Chesapeake, VA: AACE.

Mohammadyari, S., & Singh, H. (2015). Understanding the effect of e-learning on individual performance: The role of digital literacy. *Computers & Education, 82*, 11–25. doi:10.1016/j.compedu.2014.10.025

Moore, M. (1989). Three types of interaction. *American Journal of Distance Education, 3*(2), 1–6. doi:10.1080/08923648909526659

Moore, M. G., & Kearsley, G. (1996). *Distance education: A systems view.* Wadsworth Publishing.

Ololube, N. P., Ubogu, A. E., & Egbezor, D. E. (2007). ICT and distance education programs in a sub-Saharan African country: A theoretical perspective. *Journal of Information Technology Impact, 7*(3), 181–194.

Ozdamar-Keskin, N., Ozata, F. Z., Banar, K., & Royle, K. (2015). Examining digital literacy competences and learning habits of open and distance learners. *Contemporary Educational Technology, 6*(1), 74–90. doi:10.30935/cedtech/6140

Peeraer, J., & Van Petegem, P. (2011). ICT in Teacher Education in an Emerging Country: Vietnam's Baseline Situation at The Start Of "The year of ICT". *Computers & Education, 56*(4), 974–982. doi:10.1016/j.compedu.2010.11.015

Repman, J., & Logan, S. (1996). Interactions At a Distance: Possible Barriers and Collaborative Solutions. *TechTrends, 41*(6), 35–38. doi:10.1007/BF02818861

Richardson, J., & Ting, E. E. (1999) *Making the most of interaction: what instructors do that most affect students' perceptions of their learning.* ALN. https://www.aln.org/conference/proceedings/1999/ppt/99_richardson.ppt

RielJ.ChristianS.HinsonB. (2012). Charting digital literacy: A framework for information technology and digital skills education in the community college. *Available at* SSRN 2781161. doi:10.2139/ssrn.2781161

Patsula, P. J. (2002). *Practical guidelines for selecting media: An international perspective*. The Usableword Monitor.

Paulsen, M. F. (1995). *The online report on pedalogical techniques for computer-mediated communication.* NKI. http://www.hs.nki.no/~morten/ cmcped.ht doi:10.2139/ssrn.2781161

Sher, A. (2009). Assessing the relationship of student-instructor and student-student interaction to student learning and satisfaction in Web-based Online Learning Environment. *Journal of Interactive Online Learning*, *8*(2).

Tella, A., & Adu, E. O. (2009). Information communication technology (ICT) and curriculum development: The challenges for education for sustainable development. *Indian Journal of Science and Technology*, *2*(3), 55–59. doi:10.17485/ijst/2009/v2i3.12

Thoms, B., & Eryilmaz, E. (2014). How media choice affects learner interactions in distance learning classes. *Computers & Education*, *75*, 112–126. doi:10.1016/j.compedu.2014.02.002

Tyler, L. (2005). ICT literacy: *Equipping students to succeed in an information-rich, technology-based society.* USA: Educational Testing Service. Retrieved from https://www.ets.org/Media/Tests/ICT_Literacy/pdf/ICT_Equipping_Students_to_Succeed.pdf

Uşun, S. (2003). A Review of Communication Elements and Learner Support Services in Turkish Distance Education System. *The Turkish Online Journal of Educational Technology*, *2*(3).

Wang, M., Ran, W., Liao, J., & Yang, S. J. (2010). A Performance-Oriented Approach to E-Learning in the Workplace. *Journal of Educational Technology & Society*, *13*(4), 167–179.

Yusuf, M. O. (2006). Problems and Prospects of Open and Distance Education in Nigeria. *Turkish Online Journal of Distance Education*, *7*(1), 22–29.

Zhang, Z., & Kenny, R. (2010). Learning in an online distance education course: Experiences of three international students. *International Review of Research in Open and Distance Learning*, *11*(1), 17–36. doi:10.19173/irrodl.v11i1.775

Chapter 2
Information Literacy Skill for Accessing Electronic Resources Among the Faculty Members in the Institutions in Kancheepuram District, Tamil Nadu

S. Radhakrishnan
Anna Centenary Library, Chennai, India

ABSTRACT

The study analyses the information literacy skill for accessing electronic resources among the faculty members in the institutions in Kancheepuram district, Tamil Nādu. The result of the study showed about 84 (27.18%) of the respondents had their information literacy skill through orientation, 41 (13.26%) had theirs through tutorial, 49 (15.85%) and 38 (12.29%) had theirs seminars and workshop while 21 (06.79%) had one-on-one discussion. The totally 309 respondents the majority of 136(44.017%) of the respondents were highly satisfied with finding the relevant information from Institutions in Kancheepuram District in Tamil Nādu. The majority of the respondents 118 (24.05%) have noticed that using e-journals made it easier for them to complete research-related tasks and publish their findings in journals and conference proceedings.

INTRODUCTION

The sources of information that are examined through modern ICT devices are referred to as electronic resources. They are established, reformed, and more frequently stored in cyberspace in the most tangible and compressed form and they can be accessed simultaneously by a large number of viewers from infinite sockets. The saying "electronic assets", has comprehensively been characterized as, data got to

DOI: 10.4018/978-1-6684-8805-8.ch002

by a PC, might be useful as bibliographic advisers for potential sources however, at this point, they now and again proceed as referred to references in their specific right. Additionally, electronic resources now include documents in alphanumeric formats that are made available to library patrons via a computer-based information retrieval system. Electronic assets and administrations notice to the assortment of electronic and computerized sources of data open to instructors and understudies inside a theoretical setting. Libraries and information services have undergone significant transformation in recent years in order to successfully deliver electronic resources to the academic community. In recent years, online e-journals, an essential module of e-resources, have gained popularity among library users. The conversion of probable document delivery services from print to electronic occurred very quickly.

Information Literacy Skill: Definition

ALA (1989) cited by Ogundipe (2009) noted that information literacy (IL) is the ability to recognize when information is needed, then to locate it and evaluate the appropriate information and use it effectively. Despite the fact that Oxford Advanced Leaner's Dictionary defines information literacy as the ability to read and write; researches have shown that to be information literate, goes beyond mere reading and writing. For instance, the American Library Association (ALA) Presidential Committee opines that an information literate person must be able to recognize when information is needed and have the ability to locate, evaluate and use effectively the needed information (ALA, 1989).

Considering complexity associated with access to certain information, couple with emerging concepts in ICT and the need to effectively utilize of information resources in any library, a user require certain degree of skills and knowledge to access and use information. With rapid growth in information development, the ability of students to be information literate has become critically important (Ratanya, 2012). Information literacy programmes are usually organized by university library. In most libraries, information literacy programmes cover orientation, tutorial, seminar, workshop, class lecture on use of library, one-on-one discussion, online courses, computer aided instruction, user education (library instruction), library tour and briefing by librarians.

REVIEW OF LITERATURE

Paul G. Zurowski first defined information literacy in 1974, according to Kerns (2002). He said a data educated individual is "any individual who had figured out how to utilize an extensive variety of data sources to take care of issues at work and in their day to day existence. De Jager and Nassimbeni (2002) referred to by Somi and De Jager (2005) figured out that a data proficient individual must groups a few qualities. These incorporate to distinguish data, access data successfully, assess data, integrate chosen data into one's information base, use data successfully to achieve explicit purposes, comprehend the monetary. Adeniran (2005) conducts the use of electronic resources by students at the University of Redeemer in Nigeria. This study shows that the use of electronic resources has a significant impact on the academic performance of Redeemer University undergraduates. However, it is needed to learn more skills to use electronic resources. Information Literacy skills provide the impetus to understand students' skills, identify areas of strengths and weaknesses, help formulate appropriate pedagogical changes and intervention programmes, as well as provide accountability of such an initiative (Oakleaf and Kaske, 2009; Foo, Majid, Mokhtar, Zhang, Chang, Luyt and Theng, 2014). The three major assessment approaches include knowledge and skills tests, performance assessments and informal assessments. The first approach of knowledge and skills tests is most widely employed and reported. Such tests are less resource-intensive and provide a useful basis to allow comparisons across time intervals at the individual and institutional level (Oakleaf and Kaske, 2009).

Many such international IL tests and assessment have been developed and used extensively by many. Examples include Standardized Assessment of Information Literacy Skills (SAILS) (Kent State University Libraries, 2016a), Tool for Real-time Assessment of Information Literacy (TRAILS) (Kent State University Libraries, 2016b), Research Readiness Self-Assessment (RRSA) (Central Michigan University, 2016) and Information Literacy Test (James Madison University, 2016) which are used to assess discrete IL skills. These tests have been developed across different grades of students, including college and university students, basically US based, but are widely adapted and used by other schools and institutions across the world. Madhusudan (2010) conducted a study entitled Use of Electronic Resources by Researchers at the University of Kurukshetra. This study points out that electronic resources have become an integral part of the information needs of researchers at Kurukshetra University. This study also shows that sufficient availability of computer systems and internet speeds make resource usage more effective and efficient. M. Doraswamy (2010) conducted a survey on "Familiarity and Usage of Electronic Information Resources Available to Students in the Faculty of Engineering Library". This study concludes that most of the students are familiar with electronic information resources.

Students primarily use search engines compared to other electronic sources. Shehbaz Husain Naqvi (2014) worked on the use of E-Resources by students of the Graduate School of Engineering at Jamia Millia Islamia University. This study shows that the use of e-Resources is very widespread among graduate students of Jamia Millia Islamia in the Faculty of Engineering. Dr. Jeongsam Beda Devi and Dr. Nil Ratan Roy (2012) conducted a survey of college students at Assam University on Internet use. The study found that the younger generation is embracing the Internet as a means of accessing information related to science and research. The survey also found that the Internet plays an important role in today's life. Baskaran (2019) analyzed that use of social network and Medias by the Research scholars in Selected Universities of Tamilnadu. To highest 87 (31.9%) of respondents Annamalai University, (16.8%) and (15%) of them Male and Female respectively out of 273 total respondents responded to the study. Baskaran (2020) analysed a majority of the publications, 2,207 (18.48%), were published from communication area, on account of the specific nature of the research concentrating on Social networks. Computer communications information systems area scores second with 1,172 (9.81%) of the publications.

METHODOLOGY

The survey was conducted based on Questionnaire method adopted for collection of data for this study. A structured questionnaires questionnaire was designed based on related research studies and in consultation with the research guide. The aim of the study is to discuss about Information Literacy skill among towards access electronic resources among the faculty members in the Institutions in Kancheepuram District, Tamil Nadu. The purpose of study using e-resources, e-resources/services of library, use of electronic resources compared to the print resources and benefits/usefulness of electronic resources. The data was collected in Ten Institutions include one University covered under the study. The questionnaire briefly explained the research topic and guaranteed the respondents that the information provided would be confidential and used for research purpose only. The questionnaires were personally distributed to respondents in the central libraries and departmental libraries of the Universities. The questionnaires 309 (92%) of them distributed and were received back. Out of 335 questionnaires 26 were could not be used for the final analysis as they were incomplete.

RESULTS AND DISCUSSION

Table 1 analyses the Designation wise respondents from the institutions in the Kancheepuram District, Tamilnadu. Out of 309 respondents, the study has been witnessed that Assistant Professors was highest with 38.83% of them, 37.21% of them Associate Professors and 23.94% of them professors have reported from in the Institutions in Kancheepuram District, Tamil Nadu.

Table 2 has been discussed that 84 (27.18%) of the respondents had their information literacy skill through orientation, 41 (13.26%) had theirs through tutorial, 49 (15.85%) and 38 (12.29%) had theirs seminars and workshop while 21 (06.79%) had One-on-one discussion. Further, the study discussed that 20 (06.47%) had training through online course, 23(7.44%) had Computer-aided instruction, 21(0.79%) and 12 (3.88%) had User Education and Explained by librarian (Fig. 2).

Table 1. Designation wise respondents in the institutions in Kancheepuram District, Tamil Nadu

S.No	Designation	No. of Respondents N=461	Percentage
1	Assistant Professor	120	38.83
2	Associate Professor	115	37.21
3	Professor	74	23.94
		309	

Figure 1. Designation wise respondents

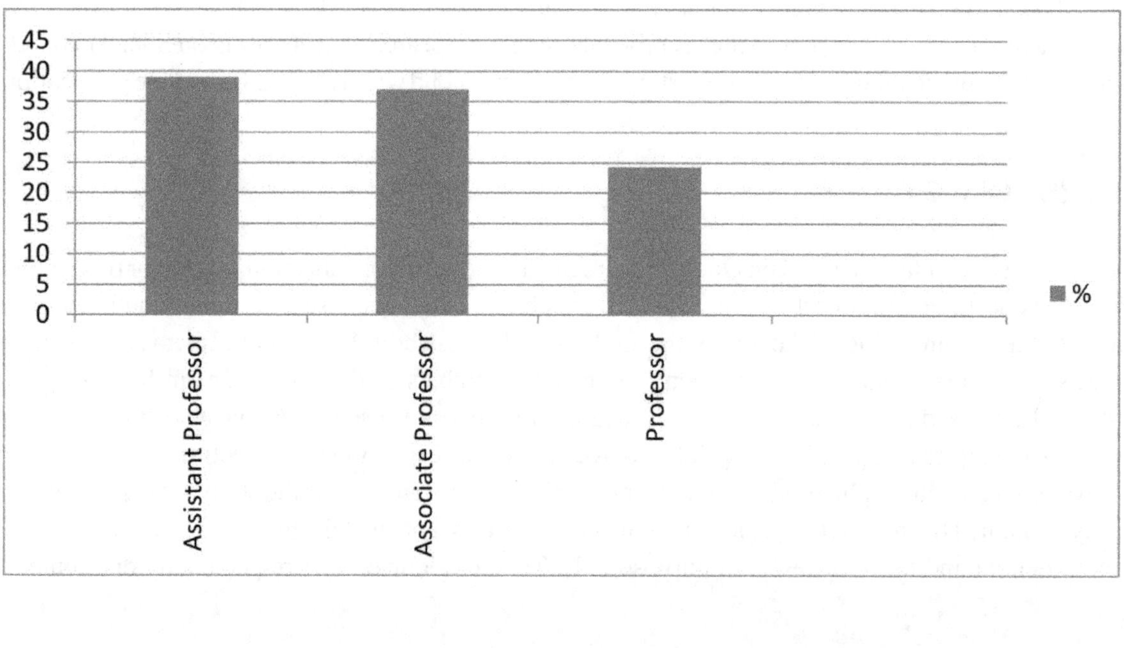

Table 2. Information literacy training acquired by the respondents

S.No	Information Literacy Training	No. of Respondents N=309	Percentage
1	Orientation	84	27.18
2	Tutorial	41	13.26
3	Seminar	49	15.85
4	Workshop	38	12.29
5	One-on-one discussion	21	06.79
6	Online courses	20	06.47
7	Computer-aided instruction	23	07.44
8	User Education	21	06.79
9	Explained by librarian	12	03.88

Figure 2. Information literacy training acquired by the respondents

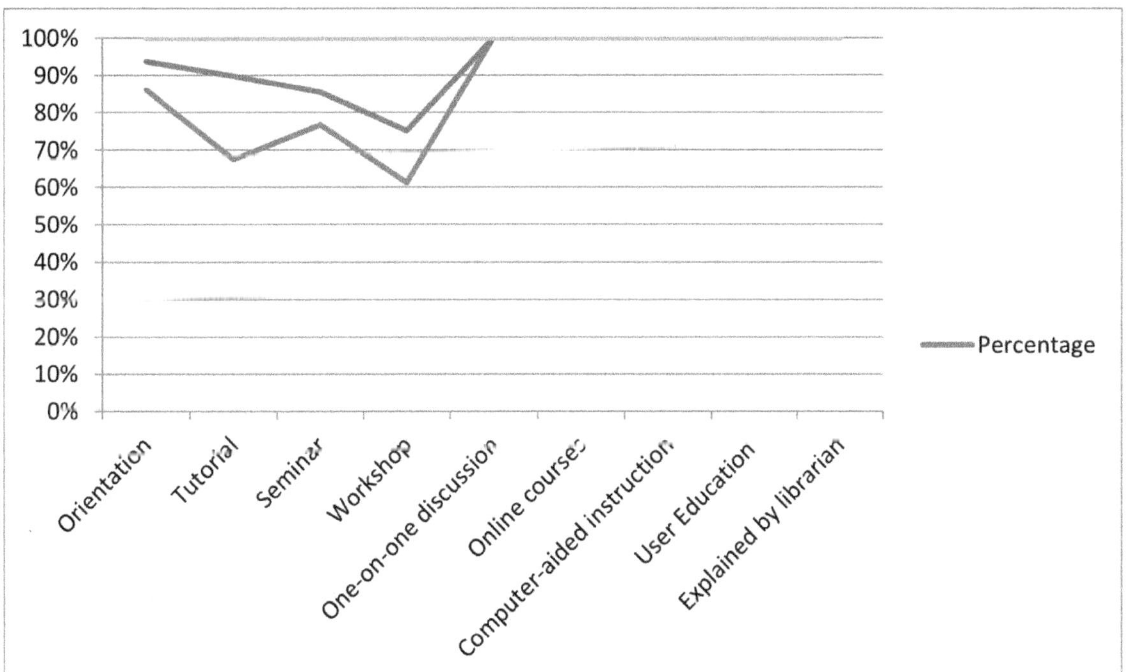

Table 3. Purpose of visiting the library

Purposes of visit	Mean	Rank
Borrow books	3.88	1
Read Journals/periodicals	3.51	2
Consult reference books	2.97	3
Access electronic resources (e-journals/e-books etc.)	2.39	4
Other purposes	2.18	5

Figure 3. Purpose of visiting the University Library

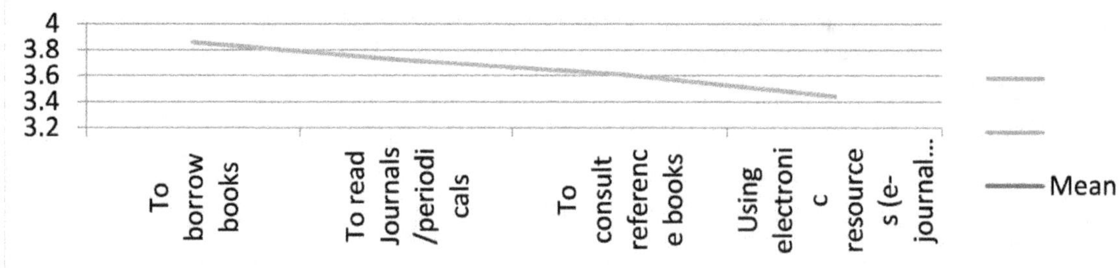

Purpose of Visiting the University Library

Table 3 is observed that mean score of each of the purpose is computed on the basis of this rating and ranks are assigned to them based on these mean. In the result analysis, mean value for 'Borrow books' was 3.88 and assigned the rank one. Followed by ' Read Journals/periodicals' (Mean 3.51) with rank two, 'Consult reference books' (Mean 2.97) rank three, 'Using electronic resources' (Mean 2.39) rank four and for 'Other purposes' we got the mean value 2.18 and ranked five. The result is exhibited in the above figure 3.

Satisfaction Level on Visiting the Websites

Table 4 was analysed that level of satisfaction on visiting the websites, the respondents out of 309 respondents the majority of 136(44.017%) of the respondents were highly satisfied with finding the relevant

Table 4. Level of satisfaction on searching information through online

S.No.	Description	Opinion	Respondents
1	Finding Relevant Information	No Comment	56(18.12)
		Satisfied	117(37.86)
		Highly Satisfied	136(44.01)
2	Career development & Entertainment	No Comment	0
		Not Satisfied	84 (27.18)
		Satisfied	97(31.39)
		Highly Satisfied	128(41.42)
3	Research work	Less Satisfied	61(19.74)
		Satisfied	101(32.68)
		Highly Satisfied	147(47.57)
4	Seminars & Assignments	No Comment	0
		Not Satisfied	0
		Less Satisfied	89 (28.80)
		Satisfied	112(36.24)
		Highly Satisfied	108(34.95)

information from Institutions in Kancheepuram District in Tamilnadu. It is followed by 117(37.86%) of the respondents who stated that were satisfied and 56(18.12%) of the respondents said no comment on ''Finding Relevant Information'' on visit the websites in the Institutions in Kancheepuram District in Tamilnadu respectively. Moreover, the majority of the 128(41.42%) of the respondents were highly satisfied with career development/entertainment through websites in Institutions in Kancheepuram District in Tamilnadu. It is followed by 97(31.29%) of the respondents with satisfied. It is also found that 84(27.18%) of the respondents were not satisfied with career development entertainment through websites.

Further, the study has been witnessed that the respondents visit websites for research work. The Majority 147 (47.51%) of them preferred as Highly satisfied, it followed by 101(32.68%) of the respondents were satisfied and 61 (19.74%) of responded as Less Satisfied on research work upon visiting the websites. The study has been analysed that majority of 112(36.24%) of the respondents visited the websites apart from for preparing seminars and assignment above the task in the selected Institutions in Kancheepuram District in Tamilnadu. It is followed, 112(36.24%) of the respondents noted Satisfied and 89 (28.80%) of them given their response as Less Satisfied.

Prefer the Electronic Resources to Seek the Research Information

The study has been analyzed that 309 respondents, respondents from Institutions in Kancheepuram District in Tamilnadu favoured using electronic resources to find material for their studies. The majority of the respondents 118 (24.05%) of the respondents have noticed Preferred that using e-journals made it easier for them to complete research-related tasks and publish their findings in journals and conference

Table 5. Electronic resources to seek the research information

S.No.	Electronic Resources	Opinion	Respondents
1	E-Journals	No Comment	39 (5.85%)
		Not Prefer	63 (11.29%)
		Prefer	118 (24.05%)
		Highly Prefer	97(21.40%)
2	E-Books	No Comment	86(11.34%)
		Not Prefer	57 (8.49%)
		Prefer	117(46.14%)
		Highly Prefer	109(21.42%)
3	E-Databases	No Comment	57 (8.29%)
		Not Prefer	69 (9.18%)
		Less Prefer	99(33.90%)
		Prefer	117(19.26%)
		Highly Prefer	121(14.85%)
4	E-Theses & Dissertations	No Comment	63 (9.33%)
		Not Prefer	117(25.97%)
		Prefer	96(17.91%)
		Highly Prefer	86(13.77%)

proceedings. It followed by 97 (21.40%), 63 (11.29%) and 39 (5.85%) of the respondents towards opted Highly preferred, Not preferred and No Comment of using e-Journals among the Respondents from the Institutions in Kancheepuram District.

Table 5 analyzes the majority of the 117 (46.14%) of the respondents from Institutions in Kancheepuram District in Tamilnadu preferred the e-books as Preferred from various reputable publishers, It is followed by 109 (21.42%) of the respondents who have given e-books a strong preference. Regarding the use of e-books, 86 (11.34%) of the respondents indicated they had no opinion. The majority of 117 (19.26%) of the respondents from Institutions in Kancheepuram District in Tamilnadu stated that they suggested as Preferred Electronic Database. It followed by 99(33.90%), 69(9.18%), and 57 (8.29%) of the respondents have reported Less preferred, Not Preferred and No Comment of using E-database respectively among the respondents in Institutions in Kancheepuram District in Tamilnadu.

The majority of the respondents from Institutions in Kancheepuram District in Tamilnadu, the majority 117 (25.77%) of the respondents have reported as Not Preferred. It followed by 96 (17.91%), 86 (13.79%), 63(9.33%) of the respondents have given opinion as preferred, Highly preferred and No comment respectively on access the E-Theses and Dissertations among the respondents from Institutions in Kancheepuram District in Tamilnadu,

Table 6. Level of satisfaction of search engines

S.No.	Search Engines	Opinion	Respondents
1	Google	Less Satisfied	24 (7.76%)
		Satisfied	43 (13.91%)
		Highly Satisfied	98(31.71%)
2	Yahoo	No Comment	35(11.32%)
		Less Satisfied	74 (23.94%)
		Satisfied	78 (25.24%)
		Highly Satisfied	69(22.33%)
3	Alta Vista	No Comment	89(28.80%)
		Not Satisfied	67 (21.68%)
		Less Satisfied	113(36.56%)
		Satisfied	117(37.86%)
		Highly Satisfied	83 (26.86%)
4	Hot mail	No Comment	117(37.86%)
		Not Satisfied	67 (21.68%)
		Less Satisfied	98(31.71%)
		Satisfied	117(37.86%)
		Highly Satisfied	61 (17.74%)
5	Mozilla Firefox	No Comment	106 (34.30%)
		Not Satisfied	97 (31.39%)
		Less Satisfied	93 (30.09%)
		Satisfied	69 (22.33%)
		Highly Satisfied	54 (6.64%)

Level of Satisfaction of Search Engines

Table 6 analysed the respondents have used the Institutions in Kancheepuram District in Tamilnadu students were satisfied with how search engines were used to access electronic resources. The majority of the 102(33.71%) of the respondents said they were Less satisfied with Google search engines. It followed by 98(7.76%) of the respondents, 43(13.9%) of the respondents have opted their opinion as Highly satisfied and satisfied respectively towards used Google Search Engines. The majority of the 69 (25.24%) of the respondents have suggested to be satisfied with Yahoo's search engines. 74 (23.94%), 69(22.33%), 35(11.32%) of the respondents replied about use of Yahoo search engine preferred as Less satisfied, Highly Satisfied, No comments respectively with use among the respondents from Institutions in Kancheepuram District in Tamilnadu.

The majority of the respondents, 117 (37.86%) said they were satisfied with Alta vista search engines; however, 113 (36.56%) said they were less satisfied, 83 (26.86%) said they were Highly satisfied, and 89 (28.80%) and 67 (21.68%) of the respondents have replied on use of Alta Vista search engines as No comment and Not satisfied respectively among the respondents in the Institutions in Kancheepuram District in Tamilnadu. The majority 119 (38.80%) of respondents they had no opinion, 117(37.86%) of them with satisfied, 98 (31.71%) with Less satisfied, 61(17.74%) of them with Highly satisfied, 67 (21.68%) of them with Not satisfied on use Hotmail search engines among the respondents in the Institutions in Kancheepuram District in Tamilnadu, respectively.

Regard with Mozilla firebox search engines, the majority 106 (34.30%) of the respondents from Institutions in Kancheepuram District in Tamilnadu respectively said No Comment. It followed, 97 (31.39%), 93 (30.09%), 69 (22.33%) and 54 (6.64%) of the respondents have realized performance of use the search engines with Not Satisfied, Less satisfied, Satisfied and Highly Satisfied on Mozilla firebox search engines among the respondents in the Institutions in Kancheepuram District in Tamilnadu.

CONCLUSION

The study has been analysed about impact usefulness on Information Literacy Skill on Use of Electronic Resources in the Institutions of Kancheepuram District in Tamilnadu. Out of 309 respondents, the study has been witnessed that Assistant Professors was highest with 38.83%.84 (27.18%) of the respondents had their information literacy skill through orientation. The mean values for 'Borrow books' was 3.88 and assigned the rank one. Followed by ' Read Journals/periodicals' (Mean 3.51) with rank two, 'Consult reference books' (Mean 2.97) rank three. The majority of the 128(41.42%) of the respondents were highly satisfied with career development/entertainment through websites in Institutions in Kancheepuram District in Tamilnadu. The study has been witnessed that the respondents visit websites for research work. The Majority 147 (47.51%) of them preferred as highly satisfied, it followed by 101(32.68%) of the respondents were satisfied and 61 (19.74%) of responded as Less Satisfied on research work upon visiting the websites. The majority of the 102(33.71%) of the respondents said they were Less satisfied with Google search engines. It followed by 98(7.76%) of the respondents, 43(13.9%) of the respondents have opted their opinion as highly satisfied and satisfied respectively towards used Google Search Engines.

REFERENCES

Adeniran, P. (2013). Usage of Electronic Resources by Under Graduates at the Redeemer's University, Nigeria. *International Journal of Library and Information Science*, *5*(10), 319–324.

American Library Association. (1989). *Presidential committee on information literacy: final report.* ALA. https://www.ala.org/ala/acrl/acrlpubs/whis/presidential.cfm

Baskaran, C. (2019). Scholarly information share through Social Networks (SNs) and Medias among social science scholars in selected State Universities in Tamilnadu. *International Journal of Library and Information Studies*, *9*(3), 83–92.

Baskaran, C. (2020). Research Patterns on the Social Networks and Media: A Scientometric Portrait. Handbook of Research on Emerging Trends and Technologies in Library and Information Science, pp.189-207. IGI Global. doi:10.4018/978-1-5225-9825-1.ch014

Baskaran, C. (2021). An Exponential trend and Mapping of Research in Web 2.0. *Library Philosophy and Practice (e-journal),* 5008. https://digitalcommons.unl.edu/libphilprac/5008

Central Michigan University. (2016). Research readiness self-assessment (RRSA). CMU. http://rrsa.cmich.edu/cgi-bin/rrsalib.cgi/

Devi, C. B., & Roy, N. R. (2012). Internet Use among University Students: A Case Study of Assam University Silchar. *Pratidhwani- A journal of Humanities and Social Science.* http://www.thecho.in/files/abstract,-nil-ratan.pdf

Doraswamy, M. (2005). *Familiarity and Use of The Available Electronic Information Resources by The Students in Siddhartha Engineering College Library, Vijay Wada: A Survey.* The National Convention On Library and Information Networking (NACLIN), Bangalore.

Fasae, J., & Aladeniyi, A. (2012). Internet Use by Students of Faculty of Science In Two Nigerian Universities. *Library Philosophy and Practice*. https://unllib.unl.edu/LPP

Foo, S., Majid, S., Mokhtar, I. A., Zhang, X., Chang, Y. K., Luyt, B., & Theng, Y. L. (2014). Information literacy skills of secondary school students in Singapore. *Aslib Journal of Information Management*, *66*(1), 54–76. doi:10.1108/AJIM-08-2012-0066

James Madison University. (2016). *Information literacy test, JMU Center for Assessment and Research Studies and JMU Libraries*. James Madison University. www.madisonassessment.com/assessment-testing/information-literacy-test/

Kerns, S. C. (2002). Information literacy instruction: Theory and practice. *Journal of the Medical Library Association: JMLA*, *90*(3), 351. https://www.ncbi.nlm.nih.gov/pmc/articles/PMC116414/

Madhusudhan, M. (2010). Use of Electronic Resources by Research Scholars of Kurukshetra University. *The Electronic Library*, *28*(4), 492–506. doi:10.1108/02640471011033684

Naqvi, S. H. (2014). *The Use of E-Resources by Postgraduate Engineering Students at Jamia Millia Islamia University*. PARIPEX Indian Journal of Research., doi:10.15373/22501991/July2014/33

Oakleaf, M., & Kaske, N. (2009). Guiding questions for assessing information literacy in higher education. *Portal (Baltimore, Md.)*, *9*(2), 273–28.

Somi, N. G., & De Jager, K. (2005). The role of academic libraries in the enhancement of information literacy: A study of Fort Hare Library. *South African Journal of Library and Information Science*, 259–266.

Chapter 3
Academic and Research Information Sharing Through Social Media and Networks Among Autonomous Arts and Science Colleges in Chennai, India

K. Palanivel
Anna Centenary Library, Chennai, India

ABSTRACT

The study analyses the academic and research information share through social medias and networks among autonomous arts and science colleges in Chennai, India. The study portrays out of 302 respondents, 149 (49.3%) from female respondents, and 153 (50.7%) responses from male respondents. The p value of the variables Google scholar and Onsite.com, Prospect.com are less than 0.05, at the 5% level of significance. Hence null hypotheses for these variables are accepted. There is a no significant association between the age group of the respondents and their use of Research Gate, Google Scholar, and Prospect.com, as the p value is more than the significant level of 0.05.

INTRODUCTION

Use of social media and mobile devices provide both advantages as well as challenges, substantially its benefits seen in terms of penetrating course contents, videotape clip, transfer of the educational notes etc. The overall scholars feel that social media and mobile bias are the cheap and accessible tools of carrying applicable information. Studies in western countries have brazened that online social media use for co-operative literacy has a significant donation to scholars' academic performance and satisfaction (Zhu, 2012). The purpose of this research design was to explore how literacy and tutoring conditioning

DOI: 10.4018/978-1-6684-8805-8.ch003

in advanced education institutions were affected by the integration and operation of mobile bias in participating the resource accoutrements, commerce with associates and scholars ' academic performance. The broad thing of this exploration was to contemporise the in- depth perspectives of scholars' perception of mobile bias and social media in literacy and Training conditioning. still, this exploration paper paid attention to only scholars ' get, and their understanding of mobile bias and social media brought changes and its faculty in academic performance.

The Basic technology exploration question of this exploration was, what are the opinions of scholars on social media and mobile bias when it's integrating into advanced education for penetrating, interacting with peers. The general, academic, political, and marketable goods of social media on Autonomous institutions in Iraq are all motifs covered by the quantitative analysis of this study composition's social communication (Aljuboori etal., 2020). The study discusses the use of e-resource for their educational purpose in the library and the quality of collection with respect to books, Journals and e-resources. Yahoo is most popular hunt machine among the stoner for browsing the net. Book lending service is the most prefer by the staff (Manca & Ranieri, 2017). The exploration reported that scholars and Scholars in Italian Autonomous Institutions borrow Social Media spots for scholarly communication for networks in scholars and Education and provocations. SWOT analysis is to organizational knowledge structure of strength, weakness, openings and pitfalls of Scholarly Communication process (Srinivasa Rao, 2018).

REVIEW OF LITERATURE

Lin(2022) discusses the exploration report aims to pierce cooperative planning and implicit challenges of social media functions and challenges of civic planning, including population bias, sequestration enterprises, information credibility, suppression, and opinion polarization. Social media and mobile base allow the researchers and to produce, edit and partake the course contents in textual, videotape or audio forms. Selwyn (2012) analyses the technological inventions give birth to a new kind of literacy societies, learning grounded on the principles of collaborative disquisition and commerce Social media the marvels began in 2005 after the Web2.0 actuality into the reality, defined more easily as " a group of Internet- grounded operations that make on the ideological and technological foundation of web2.0 and allow creation and exchange of stoner- generated contents(Kaplan & Haenlein, 2010). Mobile bias and social media give openings to the scholars for penetrating coffers, accoutrements, course contents, commerce with tutor and associates(Cavus & Ibrahim,(2008, 2009),(Richardson & Lenarcic, 2008). The Social media platform in academic institutions allows scholars to interact with their instructors, pierce their course contents, customisation and to make them scholars communities (Greenhow, 2011a, 2011b). 90 academy going scholars presently use the internet constantly, with further than 75 teenagers using online networking spots fore-learning(DeBell & Chapman, 2006), (Lenhart, Arafeh, & Smith, 2008),(Lenhar etal., 2005). The result of the focus group interview of the scholars in 3 different Autonomous Institutions in the United States verified that use of social media created openings to the learners for cooperative literacy, creating and engaging the scholars in colorful redundant class conditioning(Gikas & Grant, 2013). The Technological invention and increased use of the internet fore-learning by the scholars in advanced education institutions has brought revolutionary changes in communication pattern. A report on 3000 Research scholars in the United States revealed that 90 using Facebook while 37 using Twitter to partake the resource accoutrements as cited in (Elkaseh etal., 2016). A study stressed that the operation of social networking spots in educational institutions has a practical outgrowth on scholars' learning issues (Jackson, 2011).

The empirical disquisition over 252 undergraduate students of business and operation showed that time spent on twitter and involvement in managing social lives and participating information, course- related influences their performance (Evans, 2014). Baskaran (2019) anatomized that use of social network and Medias by the Research Students and Scholars in named Autonomous Institutions of Tamilnadu. Majority 87(31.9) of them responded from Annamalai Autonomous institutions, (16.8) and(15) of them manly and womanish independently out of 273 total repliers responded to the study. Baskaran (2020) anatomized that maturity of the publications, 2,207(18.48), were published from communication area, on account of the specific nature of the exploration concentrating on Social networks. Computer dispatches information systems area scores second with 1,172(9.81) of the publications. Baskaran (2021) explored the maturity 134(1.96) of the publications was contributed by the experimenters from the Autonomous institutions of California systems. Zhang Y was the top author was contributed 16(0.23) of the publications in the field of Web2.0.

OBJECTIVES OF THE STUDY

1. To analyze the use of Social Medias for the needs of the Students and Scholars to share scholarly communication in the Autonomous Arts and Science Colleges in Chennai, India, India;
2. To identify the scholarly communication share by the Students and Scholars through different types Social Medias platforms
3. To analyze the Social Medias Networks use by the respondents of the Age group apply with One Way ANOVA
4. To explore the Social Networks use by the respondents of the Age group apply with Post Hoc–Tukey HSD Test
5. To determine the level of satisfaction among the Students and Scholars for sharing scholarly communication in the Autonomous Arts and Science Colleges in Chennai, India, India.

METHODOLOGY

An online survey was conducted from the respondents Selected by purposive samples techniques from the Selected Autonomous Arts and Science Colleges in Chennai. The research is collecting data from an online questionnaire mode. This survey was conducted through Google Forms; the link online survey was shared through E-Mail and Whattsapp. The Data collection was taken for the study in the selected Autonomous Arts and Science Colleges in Chennai. The online survey was prepared in Google forms through the shared link in Whattsapp. The Participants were students and Research Students and Scholars from Autonomous Institutions in Chennai. The total population size was 302 students and research Students and Scholars of academic positions. The collected data were exported to SPSS 23 version for analysis. Data were analyzed and descriptive statistics were used to research. A survey was analyzed using Microsoft Excel and SPSS 23 version. The results were exported to SPSS 23 version and dependent variable, the analyzed data frequency of Simple %age Analysis and one-way Analysis of Variance

(One Way ANOVA) Test, Post Hoc Tests, and Chi-Square test was used calculated for the previously described of the respondents.

RESULTS AND DISCUSSION

The survey identifies are absolute use of social networks and media for sharing research information among the respondents in the Autonomous Institutions in Chennai. The reason for sharing the social Networks opportunities and activities for Scholarly communication in both of technology accept to the online educational sources. The study Research Social Networks (Academia.edu, Research gate, Google scholar, Researcher ID, LinkedIn, Worldwide Science, Microsoft Academic Search, Myspace.com, Vital networks, Onsite.com, Socialengine.com, Prospect.com, Neetz.com, Xing.com) in, etc., the survey well consider online sources.

It was analysed the study based on use of Social Media for sharing academic and research information among the respondents in the selected Autonomous Colleges of Arts and Science in Chennai. Table 1 analysed the respondents given their responses from four autonomous Colleges of Arts and Science in Chennai.

The majority of the respondents (39.73%) of them have participated from Loyola College, it follows, Madras Christian College (22.51%), Gurunanak College (20.86%) and Govt. Arts College for Men (16.88%).

Table 2 shows the independent variables used by the researcher for the present work. 302 students and Scholars at autonomous institutions participated in the study.

Gender: The gender category, out of 302 respondents, 149 (49.3%) from female respondents, and 153 (50.7%) responses from male respondents. More or less both the gender has participated equally.

Age Group: It is noticed that most 38.4% of the respondents were in the age group of 20-25 years and 36.8% of the are in the respondents' age group of 26-30 years, Whereas 14.6% of the age group of 31-35 years and 10.3% of the respondents from Above 35 years.

Departments: shows the characteristics of the sample. About majority of the sample was the department with the highest representation was the department of Social Work 66 (21.85%), were Mathematics and about the (11.58%), it followed by Management Studies. Whereas 8.9% of the respondents were participated from Management studies (10.59%) of the respondents, Economics with (14.56%), Chemistry (9.27%), Commerce (5.09%), History (3.31%), Geography (3.31%), Physics (1.98%), Computer Science and Corporate Secretaryship (1.65%) of them responded to the study.

Table 1. Institutions wise respondents of the study

S.No	Name of the Autonomous Institutions in Chennai	No. of Respondents	%
1	Loyola College	120	39.73
2	Madras Christian College	68	22.51
3	Gurunanak College	63	20.86
4	Govt. Arts College for Men	51	16.88
	Total	302	

Table 2. Personal factors of the respondents

Independent variables	Frequency	N	%	Valid %	Cumulative %
Gender	Male	153	50.7	50.7	50.7
	Female	149	49.3	49.3	100.0
	Total	302	100.0	100.0	
Age Group	20-25	116	38.4	38.4	38.4
	26-30	111	36.8	36.8	75.2
	31-35	44	14.6	14.6	89.7
	Above 35	31	10.3	10.3	100.0
	Total	302	100.0	100.0	
Departments	Management Studies	32	10.59	10.59	10.59
	Social Work	66	21.85	21.85	32.44
	Economics	44	14.56	14.56	47.0
	History	10	03.31	03.31	50.31
	Geography	10	03.31	03.31	53.62
	Chemistry	28	09.27	09.27	62.89
	Mathematics	35	11.58	11.58	74.47
	Zoology	31	10.26	10.26	84.73
	Physics	6	01.98	01.98	86.71
	Tamil	14	04.63	04.63	91.34
	Commerce	16	05.59	05.59	96.93
	Computer Science	5	01.65	01.65	98.58
	Corporate secretary ship	5	01.65	01.65	100.0
	Total	302	100.0	100.0	
Residing areas	Rural	177	58.6	58.6	58.6
	Urban	53	17.5	17.5	76.2
	Semi-Urban	72	23.8	23.8	100.0
	Total	302	100.0	100.0	

Residing areas: Slightly most of the respondents are hailed from a rural background (177, 58.6%) while (72, 23.8%) of the respondents are from areas of Semi-Urban and (53, 17.5%) of the respondents belong to the Urban areas.

Table 3 displays the % of Cases for the respondents. The research question was multiple choices, the multiple option % of Cases higher than eight. This indicates high use on average. Those the multiple options were Social Networks tools (Facebook, Twitter, and so on), with (14.5%) and (44.6%) of Cases are (6.1%) are (18.8%) of Cases respectively and mobile applications of Whattsapp and Telegram (30.5%) and % of Cases of 93.4%, and (8.2%), (25.1%) of Cases and sharing among the Social Medias are YouTube and Instagram (22.8%) and % of Cases of (70.%) and (11.3%) and % of Cases of 34.5. Whereas (2.8%), (8.7%) % of Cases other social Networks tools use respectively **(Figure 1).**

Table 3. Types of use the social medias

Types of Social Networks/Medias	Responses		% of Cases
	N	%age	
Face book	128	14.5	44.6
Twitter	54	6.1	18.8
Youtube	201	22.8	70.0
Tumblr/Messenger	33	3.8	11.5
Whattsapp	268	30.5	93.4
Instagram	99	11.3	34.5
Telegram	72	8.2	25.1
Others	25	2.8	8.7
Total	880	100.0	306.6%

Figure 1. Types of use the social medias

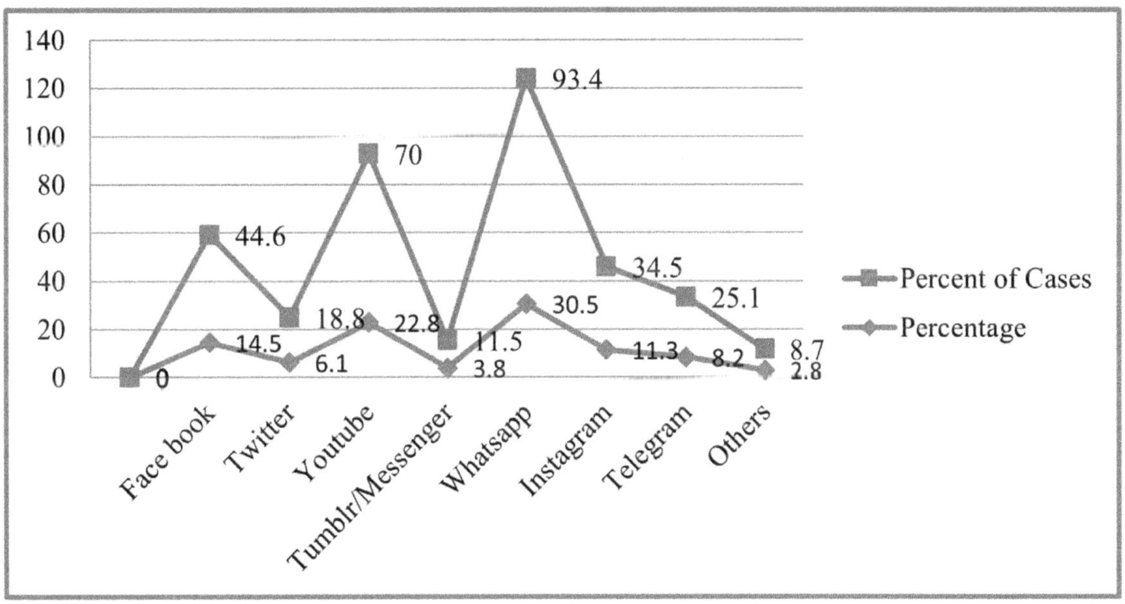

Table 4 Shows the p value of variables namely Academia.edu, Research gate, Researcher ID, LinkedIn, Worldwide Science, Microsoft Academic Search, Myspace.com, Vital networks, Socialengine.com, Neetz.com, Xing.com are less than 0.05, at the 5% level of significance.

Hence null hypotheses are rejected for these variables. It concludes that there is a significance difference between these Research Social Networks variables and Age of the respondents.

The p value of the variables Google scholar and Onsite.com, Prospect.com are less than 0.05, at the 5% level of significance. Hence null hypotheses for these variables are accepted.

It concludes that there is no significance difference between these Research Social Networks variables and Age of the respondents.

Table 4. Social medias networks use by the respondents and age group: One Way ANOVA test

		Sum of Squares	df	Mean Square	F	Sig.
Academia.edu	Between Groups	18.673	3	6.224	4.777	.003
	Within Groups	388.311	298	1.303		
	Total	406.983	301			
Research gate	Between Groups	6.228	3	2.076	2.231	.085
	Within Groups	277.295	298	.931		
	Total	283.523	301			
Google scholar	Between Groups	8.160	3	2.720	2.428	.066
	Within Groups	333.774	298	1.120		
	Total	341.934	301			
Researcher ID	Between Groups	27.483	3	9.161	6.440	.001
	Within Groups	423.898	298	1.422		
	Total	451.381	301			
LinkedIn	Between Groups	10.880	3	3.627	6.095	.001
	Within Groups	177.309	298	.595		
	Total	188.189	301			
Worldwide Science	Between Groups	57.497	3	19.166	13.651	.001
	Within Groups	418.384	298	1.404		
	Total	475.881	301			
Microsoft Academic Search	Between Groups	25.778	3	8.593	5.831	.001
	Within Groups	439.140	298	1.474		
	Total	464.917	301			
Myspace.com	Between Groups	32.333	3	10.778	7.679	.001
	Within Groups	418.266	298	1.404		
	Total	450.599	301			
Vital networks	Between Groups	23.296	3	7.765	6.186	.001
	Within Groups	374.108	298	1.255		
	Total	397.404	301			
Onsite.com	Between Groups	11.375	3	3.792	3.069	.028
	Within Groups	368.174	298	1.235		
	Total	379.550	301			
Socialengine.com	Between Groups	38.670	3	12.890	8.555	.001
	Within Groups	449.002	298	1.507		
	Total	487.672	301			
Prospect.com	Between Groups	6.157	3	2.052	1.395	.244
	Within Groups	438.429	298	1.471		
	Total	444.586	301			
Neetz.com	Between Groups	12.199	3	4.066	3.580	.014
	Within Groups	338.490	298	1.136		
	Total	350.689	301			
Xing.com	Between Groups	45.813	3	15.271	11.691	.001
	Within Groups	389.237	298	1.306		
	Total	435.050	301			

Table 5. Social medias use by the respondents vs. age group by Post Hoc – Tukey HSD test

Dependent Variable	Age of respondents	Age of respondents	Mean Difference (I-J)	Std. Error	Sig.
Academia.edu	20-25	26-30	-.268	.153	.297
		31-35	-.028	.200	.999
		Above 35	.589	.229	.051
	26-30	20-25	.268	.153	.297
		31-35	.240	.200	.626
		Above 35	.857*	.229	.001
	31-35	20-25	.028	.200	.999
		26-30	-.240	.200	.626
		Above 35	.617	.263	.090
	Above 35	20-25	-.589	.229	.051
		26-30	-.857*	.229	.001
		31-35	-.617	.263	.090
Research gate	20-25	26-30	-.036	.129	.993
		31-35	-.020	.169	.999
		Above 35	-.482	.193	.063
	26-30	20-25	.036	.129	.993
		31-35	.016	.169	1.000
		Above 35	-.446	.193	.098
	31-35	20-25	.020	.169	.999
		26-30	-.016	.169	1.000
		Above 35	-.462	.222	.162
	Above 35	20-25	.482	.193	.063
		26-30	.446	.193	.098
		31-35	.462	.222	.162
Google scholar	20-25	26-30	-.357	.141	.058
		31-35	-.201	.185	.700
		Above 35	.004	.212	1.000
	26-30	20-25	.357	.141	.058
		31-35	.156	.185	.833
		Above 35	.362	.212	.323
	31-35	20-25	.201	.185	.700
		26-30	-.156	.185	.833
		Above 35	.205	.244	.834
	Above 35	20-25	-.004	.212	1.000
		26-30	-.362	.212	.323
		31-35	-.205	.244	.834

continued on the following page

Table 5. Continued

Dependent Variable	Age of respondents	Age of respondents	Mean Difference (I-J)	Std. Error	Sig.
Researcher ID	20-25	26-30	-.071	.159	.970
		31-35	-.200	.209	.775
		Above 35	.893*	.239	.001
	26-30	20-25	.071	.159	.970
		31-35	-.128	.209	.928
		Above 35	.964*	.239	.001
	31-35	20-25	.200	.209	.775
		26-30	.128	.209	.928
		Above 35	1.092*	.275	.001
	Above 35	20-25	-.893*	.239	.001
		26-30	-.964*	.239	.001
		31-35	-1.092*	.275	.001
LinkedIn	20-25	26-30	-.366*	.103	.003
		31-35	-.019	.135	.999
		Above 35	-.442*	.155	.023
	26-30	20-25	.366*	.103	.003
		31-35	.347	.135	.052
		Above 35	-.076	.155	.961
	31-35	20-25	.019	.135	.999
		26-30	-.347	.135	.052
		Above 35	-.423	.178	.083
	Above 35	20-25	.442*	.155	.023
		26-30	.076	.155	.961
		31-35	.423	.178	.083
World- wide Science	20-25	26-30	.446*	.158	.026
		31-35	-.494	.208	.083
		Above 35	1.067*	.238	.001
	26-30	20-25	-.446*	.158	.026
		31-35	-.941*	.208	.001
		Above 35	.621*	.238	.046
	31-35	20-25	.494	.208	.083
		26-30	.941*	.208	.001
		Above 35	1.561*	.273	.001
	Above 35	20-25	-1.067*	.238	.001
		26-30	-.621*	.238	.046
		31-35	-1.561*	.273	.001

continued on the following page

Table 5. Continued

Dependent Variable	Age of respondents	Age of respondents	Mean Difference (I-J)	Std. Error	Sig.
Microsoft Academic Search	20-25	26-30	.518*	.162	.008
		31-35	-.141	.213	.910
		Above 35	.594	.243	.072
	26-30	20-25	-.518*	.162	.008
		31-35	-.659*	.213	.011
		Above 35	.076	.243	.989
	31-35	20-25	.141	.213	.910
		26-30	.659*	.213	.011
		Above 35	.735*	.279	.044
	Above 35	20-25	-.594	.243	.072
		26-30	-.076	.243	.989
		31-35	-.735*	.279	.044
Myspace.com	20-25	26-30	.205	.158	.565
		31-35	-.615*	.207	.017
		Above 35	.580	.237	.071
	26-30	20-25	-.205	.158	.565
		31-35	-.821*	.207	.001
		Above 35	.375	.237	.392
	31-35	20-25	.615*	.207	.017
		26-30	.821*	.207	.001
		Above 35	1.196*	.273	.001
	Above 35	20-25	-.580	.237	.071
		26-30	-.375	.237	.392
		31-35	-1.196*	.273	.001
Vital networks	20-25	26-30	.116	.150	.866
		31-35	-.453	.196	.098
		Above 35	.634*	.225	.026
	26-30	20-25	-.116	.150	.866
		31-35	-.569*	.196	.021
		Above 35	.518	.225	.099
	31-35	20-25	.453	.196	.098
		26-30	.569*	.196	.021
		Above 35	1.087*	.258	.001
	Above 35	20-25	-.634*	.225	.026
		26-30	-.518	.225	.099
		31-35	-1.087*	.258	.001

continued on the following page

Table 5. Continued

Dependent Variable	Age of respondents	Age of respondents	Mean Difference (I-J)	Std. Error	Sig.
Onsite.com	20-25	26-30	.071	.149	.963
		31-35	.024	.195	.999
		Above 35	.656*	.223	.018
	26-30	20-25	-.071	.149	.963
		31-35	-.047	.195	.995
		Above 35	.585*	.223	.045
	31-35	20-25	-.024	.195	.999
		26-30	.047	.195	.995
		Above 35	.632	.256	.067
	Above 35	20-25	-.656*	.223	.018
		26-30	-.585*	.223	.045
		31-35	-.632	.256	.067
Socialengine.com	20-25	26-30	-.170	.164	.729
		31-35	-.509	.215	.086
		Above 35	.871*	.246	.003
	26-30	20-25	.170	.164	.729
		31-35	-.339	.215	.394
		Above 35	1.040*	.246	.001
	31-35	20-25	.509	.215	.086
		26-30	.339	.215	.394
		Above 35	1.379*	.283	.001
	Above 35	20-25	-.871*	.246	.003
		26-30	-1.040*	.246	.001
		31-35	-1.379*	.283	.001
Prospect.com	20-25	26-30	-.009	.162	1.000
		31-35	-.184	.212	.821
		Above 35	.379	.243	.403
	26-30	20-25	.009	.162	1.000
		31-35	-.175	.212	.842
		Above 35	.388	.243	.382
	31-35	20-25	.184	.212	.821
		26-30	.175	.212	.842
		Above 35	.564	.279	.183
	Above 35	20-25	-.379	.243	.403
		26-30	-.388	.243	.382
		31-35	-.564	.279	.183

continued on the following page

Table 5. Continued

Dependent Variable	Age of respondents	Age of respondents	Mean Difference (I-J)	Std. Error	Sig.
Neetz.com	20-25	26-30	.071	.142	.959
		31-35	-.057	.187	.990
		Above 35	.656*	.214	.012
	26-30	20-25	-.071	.142	.959
		31-35	-.128	.187	.901
		Above 35	.585*	.214	.033
	31-35	20-25	.057	.187	.990
		26-30	.128	.187	.901
		Above 35	.713*	.245	.020
	Above 35	20-25	-.656*	.214	.012
		26-30	-.585*	.214	.033
		31-35	-.713*	.245	.020
Xing.com	20-25	26-30	-.732*	.153	.001
		31-35	-.417	.200	.162
		Above 35	.366	.229	.381
	26-30	20-25	.732*	.153	.001
		31-35	.316	.200	.393
		Above 35	1.098*	.229	.001
	31-35	20-25	.417	.200	.162
		26-30	-.316	.200	.393
		Above 35	.783*	.263	.017
	Above 35	20-25	-.366	.229	.381
		26-30	-1.098*	.229	.001
		31-35	-.783*	.263	.017
*. The mean difference is significant at the 0.05 level.					

The Post-Hoc-Tukey HSD test is used to test the significant difference between the groups based on the mean difference. Table 5 shows, 20-25 year respondents have a more mean difference compared with up to 26-30 years, 31-35 year respondents and Above 35-year respondents. a chi-square test was conducted. The result of the test and reveals that

1. There is a significant association between the age group of the respondents and their use of Academia.edu, LinkedIn, Researcher ID, Worldwide Science, Microsoft Academic Search, Myspace.com, Vital networks, Onsite.com, Socialengine.com, Neetz.com, and Xing.com, as the p value is less than the significant level of 0.05. Thus, the null hypothesis is rejected.
2. There is a no significant association between the age group of the respondents and their use of Research Gate, Google Scholar, and Prospect.com, as the p value is more than the significant level of 0.05. Thus, the null hypothesis is accepted.

Table 6. Determine the Level of satisfaction among the Students and Scholars for sharing scholarly communication

SC	LOS		Name of Autonomous institutions				Total	Chi-Square Tests	
			LC	MCC	GNC	GAC			
To communicate with academics or researchers at the Autonomous Institutions	VSA	Count	32	20	18	12	82	X2	3.125
		%	26.7%	29.4%	28.6%	23.5%	27.2%		
	SA	Count	38	22	23	14	97		
		%	31.7%	32.4%	36.5%	27.5%	32.1%		
	A	Count	46	24	20	22	**112**	Df	9
		%	38.3%	35.3%	31.7%	43.1%	37.1%		
	LA	Count	4	2	2	3	11	Sig.	.959
		%	3.3%	2.9%	3.2%	5.9%	3.6%		
	NA	Count	0	0	0	0	0		
		%	0%	0%	0%	0%	0%		
Total		Count	120	68	63	51	302		
		%	100.0%	100.0%	100.0%	100.0%	100.0%		
To exchange documents or information about issues or topics in an area of research.	VSA	Count	36	23	20	13	92	X2	3.318
		%	30.0%	33.8%	31.7%	25.5%	30.5%		
	SA	Count	32	18	18	12	80		
		%	26.7%	26.5%	28.6%	23.5%	26.5%		
	A	Count	48	25	22	25	**120**	Df	9
		%	40.0%	36.8%	34.9%	49.0%	39.7%		
	LA	Count	4	2	3	1	10	Sig.	.950
		%	3.3%	2.9%	4.8%	2.0%	3.3%		
	NA	Count	0	0	0	0	0		
		%	0%	0%	0%	0%	0%		
Total		Count	120	68	63	51	302		
		%	100.0%	100.0%	100.0%	100.0%	100.0%		
To provide or obtain updates on research information	VSA	Count	30	16	13	16	75	X2	3.460
		%	25.0%	23.5%	20.6%	31.4%	24.8%		
	SA	Count	28	16	16	10	70		
		%	23.3%	23.5%	25.4%	19.6%	23.2%		
	A	Count	46	27	25	19	**117**	Df	12
		%	38.3%	39.7%	39.7%	37.3%	38.7%		
	LA	Count	14	8	9	5	36	Sig.	.991
		%	11.7%	11.8%	14.3%	9.8%	11.9%		
	NA	Count	2	1	0	1	4		
		%	1.7%	1.5%	0.0%	2.0%	1.3%		
Total		Count	120	68	63	51	302		
		%	100.0%	100.0%	100.0%	100.0%	100.0%		

continued on the following page

Table 6. Continued

SC	LOS		Name of Autonomous institutions				Total	Chi-Square Tests	
			LC	MCC	GNC	GAC			
To keep current in an area of research.	VSA	Count	42	25	21	18	**106**	X2	.479
		%	35.0%	36.8%	33.3%	35.3%	35.1%		
	SA	Count	42	23	22	18	105		
		%	35.0%	33.8%	34.9%	35.3%	34.8%		
	A	Count	28	16	15	11	70	Df	9
		%	23.3%	23.5%	23.8%	21.6%	23.2%		
	LA	Count	8	4	5	4	21	Sig.	1.000
		%	6.7%	5.9%	7.9%	7.8%	7.0%		
	NA	Count	0	0	0	0	0		
		%	0%	0%	0%	0%	0%		
Total		Count	120	68	63	51	302		
		%	100.0%	100.0%	100.0%	100.0%	100.0%		
To learn about conference announcements	VSA	Count	26	15	14	11	66	X2	3.820
		%	21.7%	22.1%	22.2%	21.6%	21.9%		
	SA	Count	28	17	11	13	69		
		%	23.3%	25.0%	17.5%	25.5%	22.8%		
	A	Count	58	32	34	21	**145**	Df	12
		%	48.3%	47.1%	54.0%	41.2%	48.0%		
	LA	Count	4	2	2	3	11	Sig.	.986
		%	3.3%	2.9%	3.2%	5.9%	3.6%		
	NA	Count	4	2	2	3	11		
		%	3.3%	2.9%	3.2%	5.9%	3.6%		
Total		Count	120	68	63	51	302		
		%	100.0%	100.0%	100.0%	100.0%	100.0%		
To communicate with publishers	VSA	Count	22	13	13	7	55	X2	1.943
		%	18.3%	19.1%	20.6%	13.7%	18.2%		
	SA	Count	26	15	14	11	66		
		%	21.7%	22.1%	22.2%	21.6%	21.9%		
	A	Count	52	29	24	24	**129**	Df	12
		%	43.3%	42.6%	38.1%	47.1%	42.7%		
	LA	Count	14	7	9	6	36	Sig.	.999
		%	11.7%	10.3%	14.3%	11.8%	11.9%		
	NA	Count	6	4	3	3	16		
		%	5.0%	5.9%	4.8%	5.9%	5.3%		
Total		Count	120	68	63	51	302		
		%	100.0%	100.0%	100.0%	100.0%	100.0%		

LC: Loyola College, MCC: Madras Christian College, GNC: Guru Nanak College, GAC: Government Arts College
Note 2: VSA: Very Strongly Agree, **SA:** Strongly Agree, **A:** Agree, **LA:** Less Agree, **NA:** Not Agree.

The level of satisfaction of Students and the Scholars in the Selected Autonomous Institutions in Chennai, in terms of communicating with academics or researchers at the Autonomous Institutions. To provide or obtain updates on research information and keep current in an area of research, to learn about conference announcements and To communicate with publishers.

Overall Analysis: A majority of 35.1% (106) of the respondents is Very Strongly Agree with To keep current in an area of research. Whereas 37.1% (112) and 39.7% (120), 38.7% (117), 48.0% (145) and 42.7% (129) of the respondents are Agree level of To communicate with academics or researchers at the Autonomous Institutions and To exchange documents or information about issues or topics in an area of research, To provide or obtain updates on research information, To learn about conference announcements, To communicate with publishers.

Chi-Square Analysis: Table 5 examines whether there is a relationship between Selected Autonomous Institutions of Research Students and Scholars and whether they are Very Strongly Agree, Strongly Agree, Agree, Less Agree, and Not Agree with To communicate with academics or researchers at the Autonomous Institutions, To exchange documents or information about issues or topics in an area of research, To provide or obtain updates on research information, To keep current in an area of research, To learn about conference announcements and To communicate with publishers. a chi-square test was performed.

1. There is a no significant association between Selected Autonomous Institutions of the respondents and their level of satisfaction with to communicate with academics or researchers at the Autonomous Institutions (x2= 3.125, df=9 and p= .959), To exchange documents or information about issues or topics in an area of research (x2= 3.318, df=9 and p= .950), To provide or obtain updates on research information (x2=3.460, df=12 and p= .991), To learn about conference announcements (x2=3.820, df=12 and p= .986), To communicate with publishers (x2=1.943, df=12 and p= .999) as the p value is less than the significant level of 0,05. Thus, the null hypothesis is accepted.
2. There is a significant association between Selected Autonomous Institutions of the respondents and their level of satisfaction on the purpose of Social Medias to update To keep current in an area of research (x2=.479, df=9, p= 1.000) as the p value is more than the significant level of 0.05. Thus, the null hypothesis rejected.

CONCLUSION

The present study investigates the uses of social media among the scholars and students to attain scholarly communication quantitative study in the independent institutions in Chennai. Scholarly communication has analysed the factors similar as particular information of gender, age group, departments, hearthstone areas, and types of use social media, use of exploration social networks, and satisfaction with scholarly communication. Beyond scholarly publishing, the use of social media enhances exploration work and four selected Autonomous Institutions in Chennai. The use of social media in exploration on scholars and students to they have gained experiences about literacy for search scholarly and academic information in the Autonomous Institutions in the environment of analysis explains scholarly information in social media.

The study examines the relationship between the exploration scholar's uses of social media and how they're applied in scholarly communication and practical exploration workshop. According to the study, the use of social media by the experimenters explains the types of research on social networks, the choices of multi-communication to use of Social Medias tools. Some significant studies have corroborated the results of the current study on the use of social networking tools for scholarly communication is still in its early stages. Further, the was discussed above that academics are getting familiar with social networking tools and relinquishment is adding, also analogous to the current study Gu and Widen-Wulff (2011).

REFERENCES

Al-Aufi, A. (2007). *Networked Research and Scholarly Communication in a Developing Arabic Country: An Investigation of Sultan Qaboos Autonomous institutions.* [Unpublished Doctoral Dissertation, Curtin Autonomous institutions of Technology, Australia].

Aljuboori, A. F., Fashakh, A. M., & Bayat, O. (2020). The impacts of social media on Autonomous institutions students in Iraq. *Egyptian Informatics Journal, 21*(3), 139–144. doi:10.1016/j.eij.2019.12.003

Baskaran, C. (2019). Scholarly information share through Social Networks (SNs) and Medias among social science Students and Scholars in selected State Autonomous Institutions in Tamilnadu. *International Journal of Library and Information Studies, 9*(3), 83–92.

Baskaran, C. (2020). Research Patterns on the Social Networks and Media: A Scientometric Portrait. Handbook of Research on Emerging Trends and Technologies in Library and Information Science, pp.189-207. IGI Global. doi:10.4018/978-1-5225-9825-1.ch014

Baskaran, C. (2021). An Exponential trend and Mapping of Research in Web 2.0. *Library Philosophy and Practice (e-journal),* 5008. https://digitalcommons.unl.edu/libphilprac/5008

Cavus, N., & Ibrahim, D. (2008). *A mobile tool for learning English words*. Libes Proxy. http://libezproxy.open.ac.uk/login?url=http://search.ebscohost.com/login.aspx?direct=true&db=eric&AN=ED504283&site=ehost-live&scope=site

Chen, B., & Bryer, T. (2012). Investigating instructional strategies for using social media in context of social media. *The Electronic Library, 29*(6), 762–776. doi:10.1108/02640471111187999

Elkaseh, A. M., Wong, K. W., & Fung, C. C. (2016). Perceived ease of use and perceived usefulness of social media for e-learning in Libyan higher education: A structural equation modelling analysis. *International Journal of Information and Education Technology (IJIET), 6*(3), 192–199. doi:10.7763/IJIET.2016.V6.683

Evans, C. (2014). Twitter for teaching: Can social media be used to enhance the process of learning? *British Journal of Educational Technology, 45*(5), 902–915. doi:10.1111/bjet.12099

Gikas, J., & Grant, M. M. (2013). Mobile computing devices in higher education: Student perspectives on learning with cell phones, smartphones & social media. *The Internet and Higher Education, 19*, 18–26. doi:10.1016/j.iheduc.2013.06.002

Gu, F., & Widén-Wulff, G. (2011). Scholarly communication and possible changes in the Jackson, C. (2011). Your students love social media… and so can you. *Teaching Tolerance, 39*, 38–41.

Kaplan, A. M., & Haenlein, M. (2010). Users of the world, unite! The challenges and opportunities of social media. *Business Horizons*, *53*(1), 59–68. doi:10.1016/j.bushor.2009.09.003

Lenhart, A., Arafeh, S., & Smith, A. (2008). *Writing, technology and teens*. Pew Internet & American Life Project.

Lenhart, A., Madden, M., & Hitlin, P. (2005). *Teens and technology*. Pew Charitable Trusts.

Lin, Y. (2022). Social Media for collaborative planning: A typology of support functions an challenges. *Cities (London, England)*, *125*, 103641. doi:10.1016/j.cities.2022.103641

Manca, S., & Ranieri, M. (2017). Networked Students and Scholarship and motivations for social media use in scholarly communication. *International Review of Research in Open and Distance Learning*, *18*(2). Advance online publication. doi:10.19173/irrodl.v18i2.2859

Miller, R. C., & Tsai, C. J. (2020). Scholarly Publishing in the wake of COVID-19. *International Journal of Radiation Oncology, Biology, Physics*, *108*(2), 491–495. doi:10.1016/j.ijrobp.2020.06.048 PMID:32890542

Ogdanov, E., Limpens, F., Li, N., El Helou, S., Salzmann, C., & Gillet, D. (2012). A social media Platform in higher education. In *Proceedings of the 2012 IEEE Global Engineering Education Conference (EDUCON)*. IEEE.

Procter, R., Williams, R., Stewart, J., Poschen, M., Snee, H., Voss, A., & Asgari-Targhi, M. (2010). Adoption and use of web 2.0 in scholarly communications. *Philosophical Transactions of the Royal Society A: Mathematical, Physical and Engineering Sciences*, *368*(1926), 4039–4056. https://doi.org/doi:10.1098/rsta.2010.0155

Srinivasa Rao Y. (2018). Scholarly Communication Cycle: SWOT Analysis. *AIJR Abstracts*, 9.

Zhu, C. (2012). Student satisfaction, performance and knowledge construction in online collaborative learning. *Journal of Educational Technology & Society*, *15*(1), 127–136.

Chapter 4
Role of Social Media in Libraries in Disseminating Scholarly Information in the 21st Century

S. Dhanavandan
https://orcid.org/0000-0002-1333-972X
Central University of Tamil Nadu,Thiruvarur, India

ABSTRACT

This chapter describes the role of social media in libraries in disseminating scholarly information in the 21st century. Social networking sites that connect friends, family, library users, and customers are called social networking. It will also be utilised for social and commercial purposes and plays a vital role in disseminating scholarly information in the 21st century. However, as internet connectivity develops, it is most popular in the professional sphere, notably at academic institutions such as universities, colleges, and schools. In today's culture, using social media has become crucial in academic activity. Social media is generally used for social interaction, sharing of Content, getting up-to-date news & information, and making decisions. It is a powerful and straightforward tool for talking with people, both locally and worldwide, as well as for sharing, generating, and spreading information in the academic environment towards disseminating scholarly information in the 21st century

INTRODUCTION

The recent growth and development of social media and the technology that goes along with it has made life simpler for librarians and other library professionals. Social media is the most efficient communication in today's society, where everyone can contact one another with a click. To use in their libraries, library professionals are also becoming familiar with social media and the tools that go along with it. Social media is being implemented by library staff to build a virtual platform for user interaction, and it is also being used to reach out to specific audiences and clients. Librarians use social media to connect with library customers and to sell their sources and services. Several reasons influence the adoption of social media in libraries because library professionals believe that social media is the greatest way

DOI: 10.4018/978-1-6684-8805-8.ch004

to bring library users closer together. Aside from these facts, library professionals face many obstacles and issues when using social media in their various libraries. These obstacles and concerns must be overcome to use social media more effectively and consistently in libraries. Library personnel use social media to communicate with potential consumers. Academic libraries are increasingly adopting social media platforms to connect users, felicitate services, promote resources, reach out to new audiences, and increase exposure. This exploratory study will look into the acceptance rate of social media as an innovative medium for service delivery and the influencing factors and the effectiveness of using social media tools in Indian business school libraries to benefit users. According to the findings, libraries should use evidence-based, innovative strategies to effectively adopt social media to benefit patrons in a collaborative online environment. The library should not be static but rather dynamic to deal with transformation. The internet's function in shifting marketing contexts becomes critical. In this regard, the collaborative online learning platform has developed as an innovative and rapid means for reaching out. The key advancements in information and communication technology have significantly impacted academic libraries. Academic libraries must adopt social media tools to communicate information, advertise services, provide innovative Content, update new knowledge, and increase library visibility.

The web and internet-based services that connect friends, family, library users, and customers are known as social networking. It will also be used for social and business objectives. Because of the widespread availability of internet access, it is most popular in the professional sphere, particularly in universities and colleges. Social networking and social media are becoming increasingly popular in libraries nowadays. Facebook, Web forums, blogging, webinars, Twitter, wikis, social blogs, magazines, social networking, photography, YouTube, weblogs, webcasts, and social tagging are all popular. The process by which individuals join online virtual communities and networks to trade, create, and exchange information and ideas is referred to as social media. Social media, a computer-based collection of materials, is a platform in the twenty-first Century where people can create social relationships with others who share their values, activities, personal issues, etc. Libraries utilize social media to publicize events and images, educate people about their facilities and services, display their collections, and collaborate with other libraries.

SOCIAL MEDIA AND ACADEMIC LIBRARIES

Social networking refers to web and internet-based applications that connect friends, family, library users, and customers. It will also be used for social and business objectives. However, it is most popular in the professional sphere, particularly at academic institutions such as universities, colleges, and schools, as internet connectivity spreads. Social media use has become a necessary everyday activity in today's culture. Social media is primarily used for social contact, accessing the most recent/updated news and information, and making decisions. It is a powerful and simple tool for communicating with others locally and globally and for sharing, creating, and disseminating information. Social media provides an effective platform for users to access and share information with people who live far away. Social media enables library professionals to make things easier for themselves and their readers, increasing their capacity to develop positive interactions between library employees and customers. Social media can foster far deeper relationships between libraries and their customers, regardless of where users are located or how they learn about and use library services and resources. Baskaran (2020) analyzed that majority of the publications, 2,207 (18.48%), were published from communication area, on account of the specific nature of the research concentrating on Social networks. Computer communications information systems area scores second with 1,172 (9.81%) of the publications.

CONCEPT OF SOCIAL MEDIA

Social media refers to the mechanisms individuals connect in virtual communities and networks to produce, share, and/or exchange information and ideas. The major Facebook, Twitter, Instagram, LinkedIn, and YouTube profiles are managed by the Office of Communications and Marketing. Social media is used by billions of people across the world to exchange information and build relationships. On a more personal level, social media allows you to contact with friends and family, learn new things, pursue interests, and be entertained. One of the most major advantages of social media is connectivity. It can connect many people at any time and from anywhere. Through social media and its connection, information may be transmitted internationally, making it easy for individuals to communicate with one another. It leads to worldwide partnerships. (Kaplan, Andreas M. and Michael Haenlein, 2010), Define social media as "a collection of web-based applications that build on the ideological and technological roots of Web 2.0 and enable the creation and exchange of user-generated content."

In virtual communities and networks, social media provides a platform for people to create, share, engage and share information and ideas.

Characteristics of social media:

- Users exchange information in a blink of an eye in direct communication with the material and its producers.
- Web users can run any programme directly from their browsers and have complete control over their data on social networking platforms.
- Users on the web can do more than merely obtain information.
- It is straightforward and convenient for online users to utilize. Therefore, there is no understanding required to use this material by online users.
- Because there is no control over social media under traditional media systems, it is not centralized.

USES OF SOCIAL MEDIA IN LIBRARIES

- Social media may be useful for spreading information, and libraries can utilize it to promote their programmes, collections, and other offerings.
- Social media offers a constant communication channel between the library and its users.
- To gather comments on the library and its offerings to evaluate the institution, encourage discussion, and allow staff members to address patron complaints.
- To provide users with up-to-date information.
- To distribute notifications for services and information.
- To Increase usage by promoting new and current Content in the library's collections.
- Social media is a great tool for promoting library materials and services and drawing in new users.

- Students can post Content to the library's website and choose library materials with the help of social media.
- Libraries can utilize social media to promote their events, resources, and services. Social media may be an efficient tool for information dissemination. •
- To inform those who might not be aware of the services offered by the library.
- The library can start a book discussion group using social media.
- The library can offer reference services through social media, exchange information with other libraries and its users at any time, and engage with them.
- It works wonders for attracting new users' interest.

BENEFITS OF USING SOCIAL MEDIA IN LIBRARIES

- It supports libraries in building a platform for user collaboration and becoming closer to their patrons. Social media is a fantastic tool for attracting new users and promoting library services and resources. Social media thus generates potential library users.
- Anyone may register extremely easily. It enables users to update their profile through text messaging and apps downloaded for specific smartphones and tablets on their mobile devices. The user can create as many accounts as he wants on various social media.
- Social media allows users to find particular answers to inquiries. It is beneficial to generate thoughts and proposals. Thus itenhances reference service.
- Library authorities can promote programmes and activities by rating, reviewing, and sharing them with friends and neighbours. Amazon has used this innovative method, which libraries may also use.
- Messages can be sent to other people or users via social media to access them quickly.
- This is an excellent attempt to promote the concept of librarian-generated reading lists, and in certain circumstances, user-generated reading lists are more useful than librarian-generated reading lists.
- Social media assists students in selecting library materials and making it simple for them to upload Content to the library's website.
- It is not prohibitively pricey. Users can gain access to social media by paying phone service provider fees.
- Users are eager to use social media in libraries and have expressed an interest in the benefits of social media in library resources and services..

PURPOSE OF USING SOCIAL MEDIA IN LIBRARIES

- To attract potential library users through announcements, reference services, networking with other libraries, promoting general library services, providing timely updates to users and their queries, and developing communities.
- To provide information on late arrivals and library news, to promote user debates about new arrivals, and to create discussion groups.
- Users will be offered links to recommended Internet resources, book reviews, new arrivals, etc.

- To communicate with librarians regarding professional growth.
- To establish a reputation in the LIS domain.
- Tomodernizethe Library&Informationcentre.
- To provide a platform for patrons and staff to discuss the library's services and issues.
- Toprovideentertainmentoramusementforusers.

CHALLENGES OF USING SOCIAL MEDIA IN LIBRARIES

- Obstacles are an inevitable part of development, and the following are some of the ones that must be overcome if social media is to be employed in library development.
- Users are concerned about privacy and identity theft when using social media.
- Lack of time management in using social media in libraries, poor Wi-Fi or internet connectivity, insufficient technology, and librarians' lack of passion.
- Library employees are not properly trained to use social media.
- Because there are so many social networking sites, library administrators cannot select the best one for their patrons.
- A misunderstanding of how to use it.
- Library personnel and users must be interested in social media and networking to establish a successful social media profile.
- The appearance and operation of social media platforms, whether a blog or a website, should be straightforward. It should be simple for consumers to get the most out of such programmes.

Figure 1. Libraries' biggest social media challenges

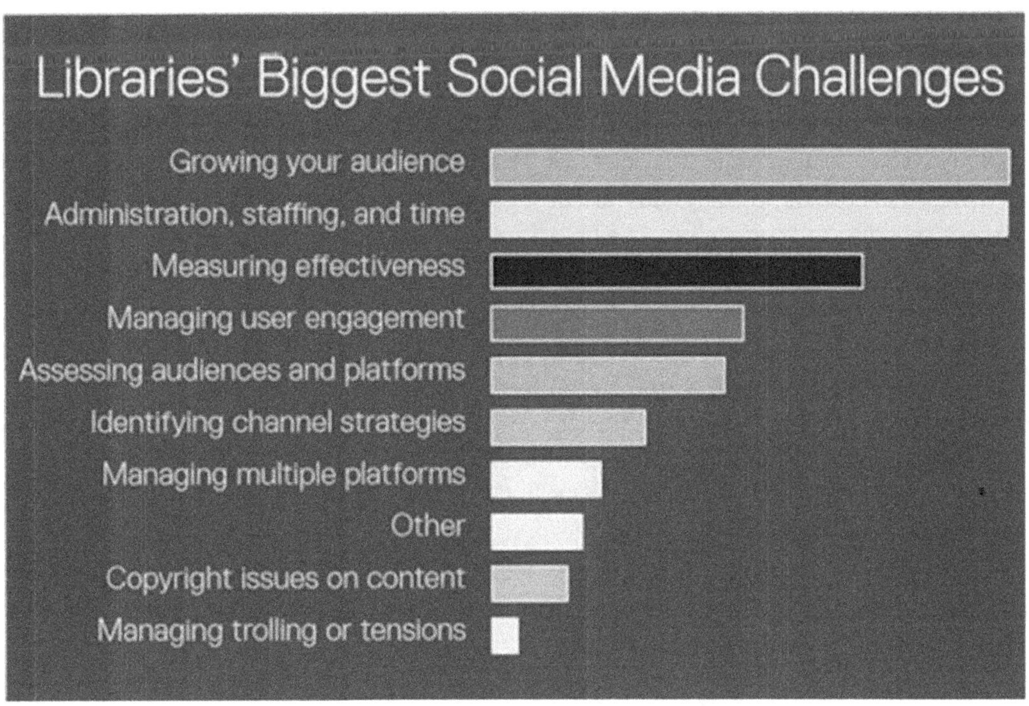

BENEFITS OF SOCIAL NETWORKING SITES USERS' PERSPECTIVE

- The first and main advantage of social networking is that it is free and reaches consumers quickly.
- Social networking disseminates information beyond geographical boundaries.
- Because things are changing rapidly with new technology daily, libraries should also update their in-house activities by utilizing social networking services.
- Accessibility for Library services is quite precise.
- Users/readers can access the library's services remotely/off campus.
- Every day, new information is delivered to the public.
- To promote library services.
- To inform users about the many types of library resources that are available.
- To keep the library's collections up to date.
- To publicize library guides and exhibition guides.
- Complaints, suggestions, inquiries, and feedback are welcome.
- Can retrieve information and actively interact with both the Content and its authors.
- It can promote collaborative work by replacing existing documents or PDF files.
- Can improve service delivery and contribute to professional development.

DISADVANTAGES IN THE USE OF SOCIAL NETWORKING SITES

Lack of privacy and identity theft, information confidentiality, ignorance of correct SNS use, and librarians' lack of enthusiasm in learning and coping with new social media trends. Some downsides of social media in libraries include a lack of time and knowledge among library workers in spending time to SNS.

APPLICATION OF SOCIAL MEDIA IN LIBRARIES

Social networking is a free and open way to communicate with others and market library services. The library only needs to create a page or account on the social networking site and periodically update its activity. Nowadays, social networking sites reflect our society. It encourages and provides ways to improve customer service issues and complaints. The library encourages people to share certain types of Content. They differ in terms of the Content they share (text, photographs, videos, PowerPoint presentations, etc.). Some have grown in popularity due to their ability to organize and share material..For example,

- Text Sharing—BookCrossing.
- PhotoSharing—Flickr,Picasa,PhotoBucker.
- Video Sharing—YouTube, DailyMotion, Metcalfe.
- PresentationSharing—SlideShare,Prezi, Scribd

Social networking sites give the library a new look. Social networking sites encourage library professionals to provide information and communicate with library users. The most commonly used social networking sites in the library are given below:

Facebook is a social networking platform founded in the United States in February 2004. It is the most widely used social networking site. The advantage of using Facebook is that the librarian may connect and converse with its users immediately. Users can also provide feedback and suggestions to the librarian for the library's development. Today, Facebook is the most popular social networking platform in the world. In 2004, Mark Zuckerberg launched Facebook. It allows people to build online communities and groups to trade or convey useful information to others, from the unknown to the well-known.

Libraries mostly use Facebook to promote their services, inform patrons of important information, share images, offer chat support, and maintain a social media presence. The future success of these libraries' Facebook presence is something they are optimistic about. The data gathered at this point in time provide usage benchmarks but are inconclusive in determining whether or not Facebook is a useful and effective application for health sciences libraries, according to the researchers, who believe this is due to the small number of actual libraries using Facebook. This question may be reviewed to arrive at a more precise conclusion if Facebook becomes a more popular programme.

Myspace: In the year 2003, Myspace was launched in California. It is the most widely utilized social networking platform on the planet. Like Facebook, the librarian uses Myspace to exchange information, images, slideshows, and other media. Many libraries use the Myspace website. MySpace, a popular social networking site comparable to Facebook, allows users to create profiles to serve them better. MySpace is the biggest of these websites. People of every age, religion, and nationality can be found on MySpace. But the majority of MySpace users are under 25. (Farkas 2007). "However, even in January 2006, when the number of users on MySpace was hovering at 47 million, Google still received more daily hits than MySpace" (Evans 2006). MySpace has increased steadily since 2006. Libraries have started to see the possibilities MySpace has given them to connect with users, particularly the younger generation. There has been opposition to libraries setting up their own MySpace profile.

YouTube: YouTube was established in 2005. It is a social networking platform widely used for storing and sharing videos. The library's activities and events can be shared on YouTube by the librarian for users. Libraries all across the world commonly use this significant social networking site. This helps libraries advertise informational resources and provide information about their programmes, conferences, and workshops.

One of YouTube's biggest benefits is the freedom to submit videos of any kind in any format. Register for a free account on the Upload page to publish your own film, according to a Slate article. After choosing your file and pressing the "Upload Video" button, you're done! You don't need to know your video's format because YouTube's servers will convert it to a normal one. When you send your aunt a URL, it will run smoothly without flooding the screen with pop-ups or errors.

Twitter is an extremely dynamic social networking platform that enables users to tweet or send messages. It is accessible to everyone. Through tweets, which are brief messages, users of the social networking site Twitter can interact with one another. A library could distribute whatever type of news that its users desire. Brief messages can inform people of upcoming events like readings, talks, book sales, and newly released materials and alterations to the building's hours. A recommendation for locating or accessing information online or inside the building could be shared in a message once a day or once a week. Tweets may provide links to informative news articles on libraries or literacy. For additional in-depth information, the pieces may, when appropriate, connect to a library's own website and blog.

LinkedIn is the most efficient method of contacting academic researchers and other library personnel. Like other websites, LinkedIn likewise provides its users with information. The service is known as professional social networking. This niche social networking platform is a great way to connect library users with people who can help them find information. Librarians can utilize this technology to offer skills like selective information dissemination (SDI).

LinkedIn may be a valuable tool for librarians in terms of career advancement and helping library patrons. Customers of libraries can now download the LinkedIn Learning mobile app. After downloading it from your app store, you can log in through your library by selecting Sign in with your library card. Your library's ID, followed by your library card number and pin, must be entered. Ask questions, provide links, retweet noteworthy tweets from others, and respond to direct messages from users are all ways that libraries on Twitter can get their followers to engage with them.

Blog: A blog is a website where we can post daily on any subject. It resembles an online journal. The library instantly communicates with the greatest number of users by developing a blog. Professionals at the library may quickly share all kinds of information with their users' thanks to blogs. Using blogs, librarians can regularly update readers on library operations, share new information on a particular subject or problem, and invite users to contribute to the Content by authoring articles and getting quick user feedback.

The use of blogs by libraries for marketing purposes is not surprising. A user-generated blog is a website where posts are made in journals and shown in reverse chronological order. They can serve as an online diary or offer commentary and information on a specific topic. A blog helps to establish a connection between a business and its readers. Additionally, it can develop a community, increase engagement and traffic to your website, and impart knowledge on pertinent subjects. It can build a strong basis for related social media platforms if done well.

Instagram; It is an excellent social networking site. The core of Instagram is the sharing of photos and videos. The library can use this website to share details about its programmes and holdings. You can browse a list of your Instagram video, photo, carousel, stories, and IGTV posts under the Content library tab in Creator Studio's Instagram section. Use the search bar to find information fast or the filters to sort it by post status (all, published, archived), date, or both.

A library will build the brand recognition necessary to promote activities and events to reach more community members by continuously producing Content that can be accessed across internet channels, such as Instagram. Instagram has a wide range of advantages, including

Identify users who share your opinions, Make your Content known., Get paid., Send videos and pictures, Your company's brand, You can succeed with free tools, It's simple to engage and Amazing filters.

Figure 2. Growth of using popular social media sites worldwide

WhatsAap: WhatsApp is the most widespread and well-liked social networking platform for the younger generation. The library encourages people to participate in library activities by developing a Whatsapp group. It acts as a bridge between the library's patrons and its resources. The main focus of this study is the accessibility of various social media platforms, including social networking sites like Facebook, Twitter, Myspace, LinkedIn, YouTube, Ning, Library Thing, Wikis, and Blogs.

Figure 3. Global social media users (2017 to 2027)

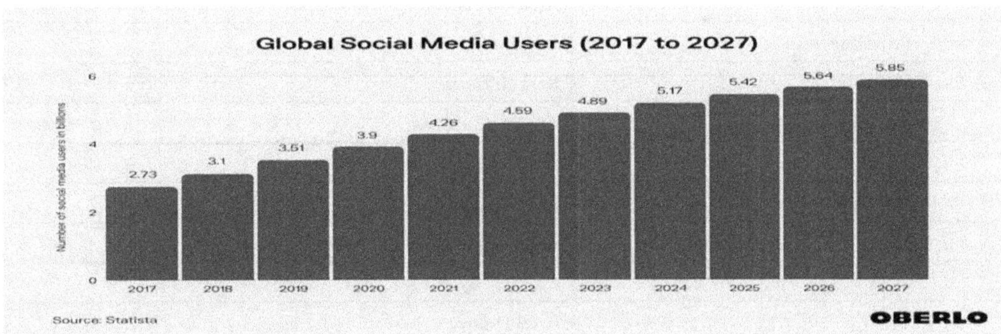

Timeline of Social Media (1973-2021)

Figure 4. Social media timeline

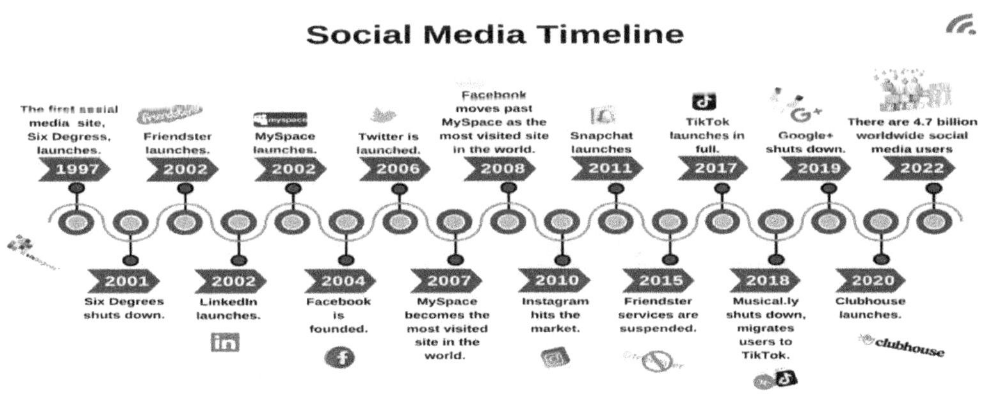

LIBRARIES USING SOCIAL MEDIA TOOLS

To educate and promote what is available, libraries can use social media channels to share insights on certain resources and collections.. The library can use various social media platforms for a variety of purposes. Some of them include knowledge organization, content development, and information sharing (Facebook, Twitter, blogs, etc). (Net vibe, Delicious etc.). Libraries are now adopting social media to promote new initiatives, such as book exhibits, recently added materials to their collections, library events, and services to the public as well as to help other libraries.

Table 1. Social medias platforms (1973-2017)

Year	Platform	Developer/Founder
1973	Talkomatic	Dave Wooly, Douglas Brown
1997	SixDegrees.com	Andrew Weinreich
1997	AOL instant messenger	Barry Appelman, Eric Bosco, Jerry Harris
1999	Yahoo Messenger	Jerry Yang, David Filo
1999	MSN Messenger	Microsoft
1999	LiveJournal	Brad Fitzpatrick
2002	Friendster	Jonathan Abrams
2003	LinkedIn	Reid Hoffman
2003	Myspace	Thomas Anderson
2003	Skype	Microsoft
2004	Facebook	Mark Zuckerberg
2005	Reddit	Aaron Swartz
2006	Twitter	Jack Dorsey
2006	VK	Pavel Durov
2010	Pinterest	Ben Silbermann
2010	Instagram	Kevin Systrom
2011	Snapchat	Evan Spiegel
2011	Google plus	Bradley Horowitz
2011	Twitch	Justin Kan
2012	Tinder	Sean Rad
2013	Google Hangouts	Larry Page, Sergey Brin
2014	musical.ly	Alex Zhu, Luyu Yang
2015	Discord	Jason Citron, Stan Vishnevskiy
2017	TikTok	Zhang Yiming

HOW MANY LIBRARIES USE SOCIAL MEDIA?

"Librarians are extending themselves and enabling library patrons in this complicated, dynamic, and demanding environment." 76% of academic libraries reported utilizing social media in the previous year, with Facebook, blogs, and Twitter leading the list. Academic libraries increasingly use social media to communicate with varied community groups and go beyond the traditional library boundaries (Harrison, Burress, Velasquez and Schreiner 2017) The main advantage of libraries using social media is regarded as being related to its low cost.

The most immediate benefit of having a social media account is that customers may contact with you directly. Individuals can reach the library to inquire about available books, services, and office hours. It is also an opportunity for libraries to receive direct input. Findings also revealed that librarians' use of social media encourages two-way communication, facilitates communication with library users, provides a venue for feedback, and boosts library users / usage. Some difficulties that librarians face were also

Figure 5. How are libraries applying social media?

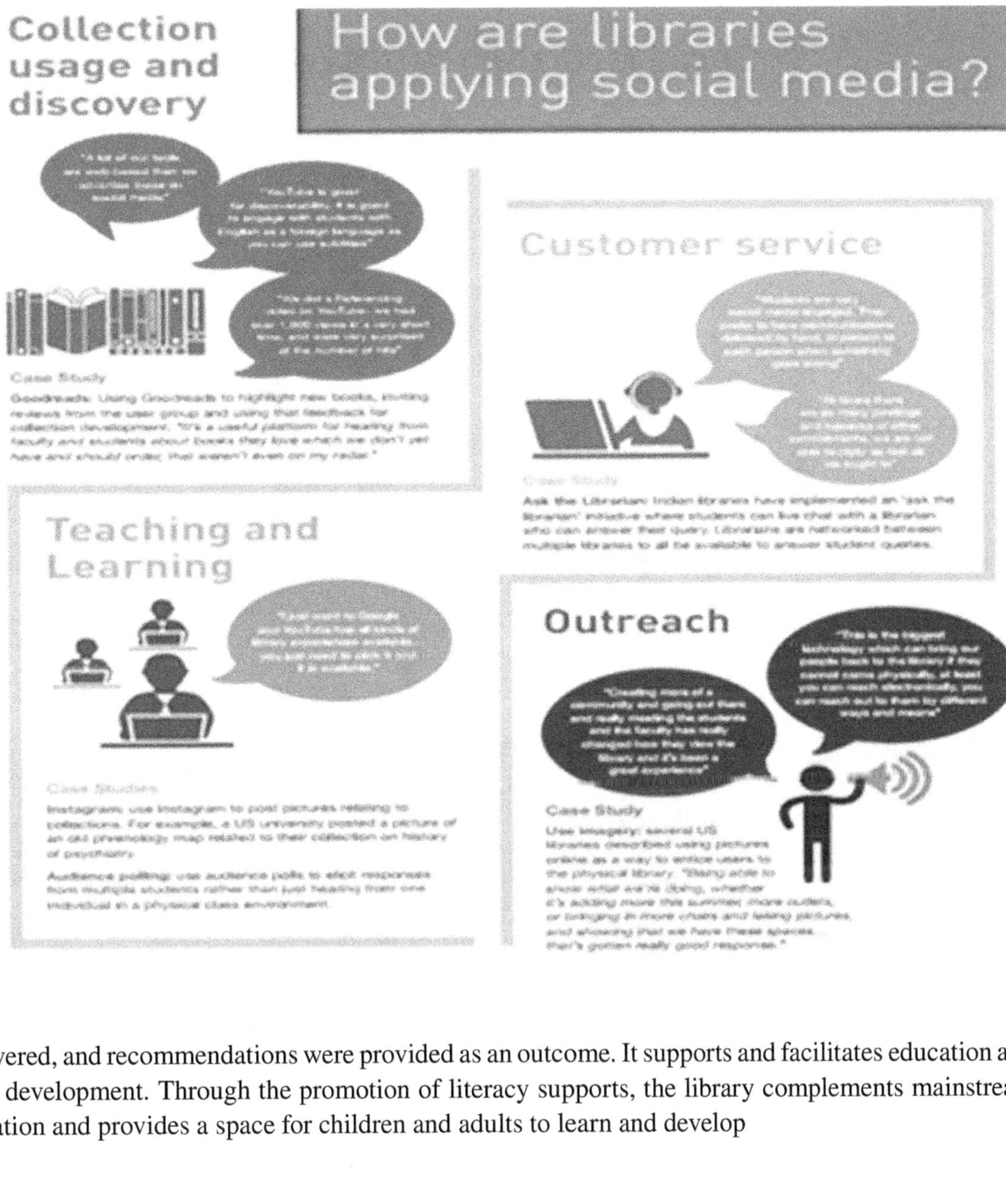

uncovered, and recommendations were provided as an outcome. It supports and facilitates education and skills development. Through the promotion of literacy supports, the library complements mainstream education and provides a space for children and adults to learn and develop

CONCLUSION

The relevance of social media platforms is rising by the day. It is one of the most essential and significant tools for propagating library services. The librarians may start a public outreach session to instruct librarians and users about social networking services and applications. Many different social media tools are available worldwide, but the librarian has to put them to use and find the best option available. Every library should develop a social media marketing strategy and offer social media services. Social media can also be utilized to communicate the library's mission smartly.

Figure 6. The social media users 2012-2022

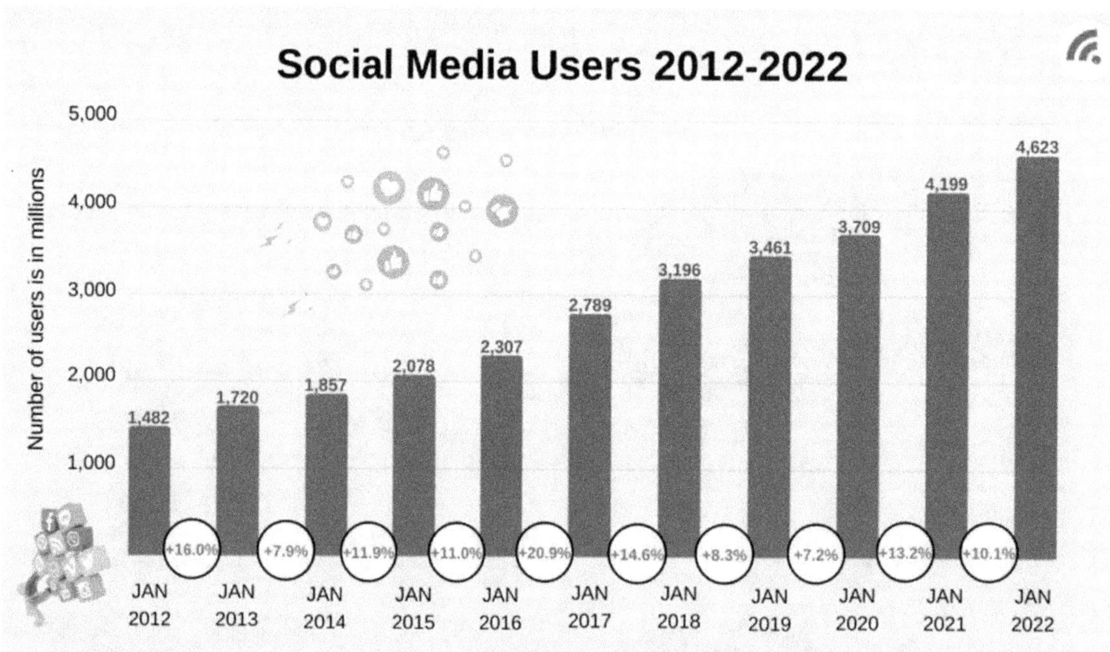

Figure 7. What libraries use social media for

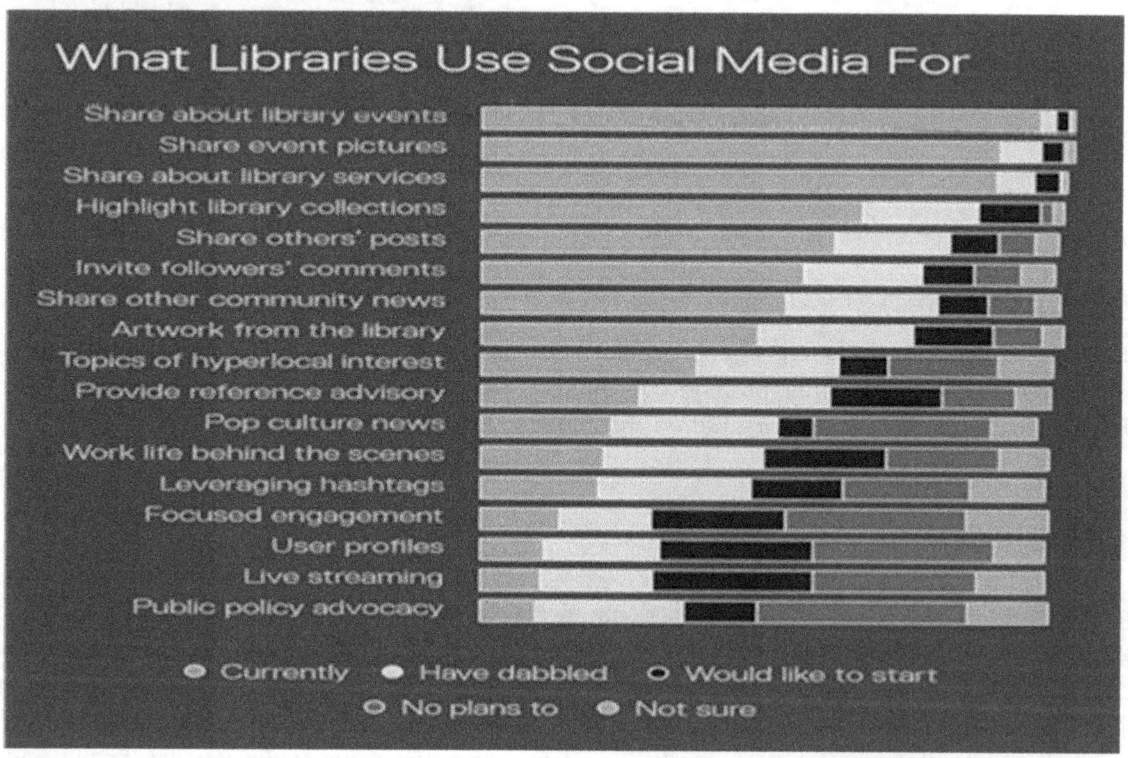

The librarians should organize a public awareness forum to create awareness and educate librarians and users on social networking services and applications. Embracing the current trend is essential to remain relevant and adapt to the changing environment. It is recommended that libraries provide their patrons with tools for accessing social media pages on library websites, which most libraries do nowadays. The interactions are accelerated when using social media to inform the community about the new ideas and activities happening in and around the library. Every library should create its social media marketing plan and social media services. Social media can also be strategically used to communicate the library's mission.

REFERENCES

Akullo, W. N., Akello, C., & Mulumba, O. (2016). *The Role of Social Media In Promoting Library Services in Secondary School in Uganda.* 517-525. SCECSAL. https://www.scecsal.org/publications/papers2016/051_akullo_2016.pdf

s., A., M., A., & B., J. (2019). Use of Social Media in Marketing of Library Information Services in Polytechic University of the Philippines. *IJARIIE, 5*(6), 534-545. http://ijariie.com/AdminUploadPdf/Use_of_Social_Media_in_Marketing_of_Library_Information_Services_in_Polytechnic_University_of_the_Philippines_ijariie11028.pdf

Anwar, M & Tang, Z. (2020). Social Media and Libraries: The Perfect Combination of the Next Generation. *International Journal of Library and Information Studies, 10*(3).

Baskaran, C. (2020). Research Patterns on the Social Networks and Media: A Scientometric Portrait. Handbook of Research on Emerging Trends and Technologies in Library and Information Science, pp.189-207. IGI Global. doi:10.4018/978-1-5225-9825-1.ch014

Canty, N. (2012). *Social Media in Libraries: It's like, Complicated.* Research Gate. https://www.researchgate.net/publication/249371034

Chakrabarti, A. (2016, April-June). Social Media & Libraries: A Symbolic Relationship for 21st Century Librarianship. *International Journal of Digital Library Services, 6*(2).

Dhanavandan, S., Esmail, S. M., & Nagarajan, M. (2012). Access and awareness of ICT resources and services in medical college libraries in Puducherry. *Library Philosophy and Practice (e-journal), 750.*

Dhanavandan, S., Saravanan, G., & Esmail, S. M. (2011). Internet-based resources and services among librarians in Tamil Nadu, India. *Library philosophy and Practice.* Digital Commons. https://digitalcommons.unl.edu/libphilprac/471/

Dhanavandan, S. S. (2020). Open Access Repositories in the World: An Overview. *Library Philosophy and Practice, 3805.*

Divya, V., & Subaveerapandiyan, A. (2019). *Use of Social Media in Libraries and Higher Education.* Research Gate. https://www.researchgate.net/publication/349552956

Faisal, S. (2015). *Use of Social Media in Libraries. Capacity Building of SCERT, DIET and School Librarians to Modernise the Library.* Regional Institute of Education (NCERT). https://www.academia.edu/11980034/Use_of_Social_Media_in_Libraries

Kaplan, A. M., & Haenlein, M. (2010). Users of the World, Unite! The Challenges and Opportunities of Social Media. *Business Horizons*, *53*(1), 59–68. doi:10.1016/j.bushor.2009.09.003

Kasimani, C., & Kasilingam, K. (2018). The Impact of Social Media in the Library Service. *International Journal of Information Studies*, *11*(3), 80–86.

Parabhoi, L., & Pathy, S. (2017). India, Social Media and its Application of Library Services in India. *KIIT Journal of Library and Information Management, 4*(1).

Prabhakar, S.V.R., & Rani, M.S.V. (2017). Influence of Social Networking Sites on Library and Information centers. [IJLIS]. *International Journal of Library and Information Science*, *6*(1), 83–87.

Sahu, R. (2018). *Social Media for Effective Library Services*. JMIR Publications. doi:10.2196/jmir.2912

Shashikiran, M., & Maadu, M. S. (2021). *Library Services Through Social Media.* Research Gate. https://www.researchgate.net/publication/353417222_library_services_through_social_media

Shinde, M. D. B. (2016). Social Media Applications And Its Impact on Libraries. *Prime International Research Journal*, *3*(2), 9–16.

Zhiwei, T., & Muhammad, A. (2021). Why libraries are giving too much importance to social media? A review note. *Library Philosophy and Practice (e-Journal*). https://digitalcommons.unl.edu/libphilprac/5175

Chapter 5
Mapping of the Research Publications on Research Ethics: A Scientometrics Profile

S. Saravanan
Government Arts College, Ariyalur, India

ABSTRACT

The study analyses the mapping of the research publications on research ethics, the Scientometrics profile was applied in the field of research during 2000-2022. The data out of 78308 publications, the largest 8751 (10.71%) of publications was distributed by the researchers in 2021. The EGR ranges between 0.068 and 2.044 in the years 2002 and 2019 respectively. The study found ranked authors, out of which twenty-five authors in the research field were concerned. The most prolific author found in the series, Shine, was the top ranked and had published 164 (0.201%) of the publications. The study explored the productivity of the publications contributed among twenty-five researchers. The ranking of the counties in order was England in second with 13353 (16.34%), then Australia 11777 (14.41%), Canada 8424 (10.31%), Peoples Republic of China 5293 (6.48%), Germany 4512 (5.52%), Netherlands 3920 (4.79%), France 2733 (3.34%), Spain 2391 (2.92%) and Sweden 2294 (2.80%).

INTRODUCTION

Scientometrics is a discipline which examinations logical distributions to investigate the pattern and development of science. The scientometric strategies are utilized to investigate different quantitative or subjective parts of a distribution. A logical field concentrates on the development of science through a few quantitative proportions of logical data, as the quantity of logical articles distributed in a given timeframe, their reference influence, and so on. Scientometrics is related to the fields of science and technology history, philosophy of science, and sociology of scientific knowledge. In 1969, Vassily V. Nalimov and Z. M. Mulchenko authored what might be compared to the term 'scientometrics' ('naukometriya') (Nalimov and Mulchenko, 1969b). This term is mostly used to study all aspects of the science and technology literature, as the name suggests. The term had earned wide respect by the establish-

DOI: 10.4018/978-1-6684-8805-8.ch005

ment in 1978 of the diary Scientometrics by Tibor Braun in Hungary. Scientometrics encompasses all quantitative aspects of the science of science, as well as communication in science and science policy, as stated in its subtitle (Wilson, 2001). Nalimov became the only Consulting Editor shortly after the company was established. The following are some additional early papers by Nalimov that contributed to the development of the nascent field of Scientometrics: Nalimov (1970), Nalimov et al. (1971), and Nalimov & Mulchenko (1969a).

Research Ethics

Research Ethics is characterized here to be the morals of the preparation, direct, and announcing of Research. However, not all researchers use human or animal subjects, and the ethical ramifications of research do not merely revolve around safeguarding the subjects of the study. The space of Research morals is planned to not incorporate anything not exactly the encouraging of examination that safeguards the interests of general society, the subjects of Research, and the actual analysts. Additionally, it is helpful to differentiate between two distinct levels of discussion—or ethical questions—: questions of the first or "ground-level" order and questions of the second order.

Questions of first order morality concern our actions. These kinds of questions can be very broad or very narrow. One might inquire as to whether or not the custom of "senior authorship" should be upheld and protected, or, more generally, what principles should be used to settle the question of "senior authorship." Such inquiries and the considerable recommendations in regards to how to answer them have a place with the space of what moral scholars call 'standardizing morals.' The nature and purpose of morality as a whole is the subject of second-order moral questions. What exactly is the standing of a claim that says data falsification is wrong? What precisely does the word 'wrong' mean in the lead of logical examination? And when we make assertions regarding right and wrong, scientific integrity, and research misconduct, what are we doing? These second-request questions are very not quite the same as the ground-level inquiries concerning how to direct one's private or expert life transcended. They are not about the content of morality—what actions are required, permitted, or prohibited—but rather about its nature. According to Kagan (1998), moral philosophers refer to this as "meta ethics."

Research Ethics and Assessment

Similar to DuBois & Antes (2018), who define research ethics as "doing good science in a good manner," the term "research ethics" is generally used in a broad and inclusive sense. Good manners include, among other things, honest reporting of findings, management of conflicts of interest, protection of human participants and animal subjects, appropriate data storage, and proper citation of sources. Research ethics studies discuss a wide range of ethical concepts and principles that are applied to scientific values and "rigor," such as honesty, objectivity, integrity, carefulness, openness, trust, accountability, respect for colleagues and intellectual property, confidentiality, fairness, efficiency, the protection of human subjects, the care of animals, and so on. However, operational elements (data) that are more clearly defined are needed for evaluation and monitoring. For instance, the logical respectability regarding individual way of behaving, covering logical wrongdoing, like distorting research information, overlooking or dodging significant parts of human subject prerequisites, not appropriately revealing irreconcilable situations, changing the plan, strategy or consequences of a concentrate in light of strain from a source of financial support, improperly relegating creation credit, etc. (Martinson, Anderson, and De Vries, 2005).

REVIEW OF LITERATURE

In 2006, Singh, Mittal, and Ahmad conducted a Bibliometric analysis of digital library-related literature. The significant finding is that 61% of articles are single authored; Except in one instance where there are three articles, author productivity does not match Lotka's Law. In 2003, there were the most articles published, with English being the most popular language; The majority of articles appeared in D-lib Magazine; The distribution of articles almost adheres to Bradford's Law; and the United States came in first place for the most journals. Tiew (2000) saw that as 53% of articles contained diary self-references, and a propensity is seen for writers partnered to the foundation distributing the diary to refer to the diary. Patra, Bhattacharya, and Verma (2006) used data from the Library and Information Science Abstract (LISA) to examine the growth pattern, core journals, and author distribution in the bibliometric field. They discovered that the growth of literature does not follow a clear pattern.

Bakri and Willet (2008) found statistically significant changes in the types of articles, the number of references per article, and the length of the articles when comparing their findings from 2001 to 2006 with those of the previous study. The concentrate likewise uncovered that the two-wrote articles were more noteworthy in number. From 2003 to 2012, Karpagam (2014) used scientometric analysis to evaluate the growth, global publications share and citation impact, share of international collaborative papers, and contributions of major collaborative partner nations in nanobiotechnology research. Baskaran (2019) looked at how researchers at a few Tamil Nadu universities used social media and other media. To most noteworthy 87 (31.9%) of respondents Annamalai College, (16.8%) and (15%) of them Male and Female individually out of 273 all out respondents answered the review. Due to the specific nature of the research focusing on social networks, Baskaran (2020) analyzed that the majority of the publications, 2,207 (18.48%), were published in the communication field. The area of computer communications and information systems comes in second, with 1,172 publications (9.81%). Baskaran (2021) investigated the greater part 134 (1.96%) of the distributions was contributed by the analysts from the College of California frameworks. Zhang Y was the most prolific author, contributing 16 (0.23%) of the publications in the Web 2.0 field.

Brennan (2017) examines various contextual analyses propose that and how CRIS ought to incorporate data on conventions on moral votes. This includes the trading of information among CRIS and the data sets of morals advisory groups from one viewpoint and the stages for the organization of information the executives anticipates the other hand. Klausen, (2016) examined the Assessments of morals councils are especially important in projects in the fields of medication and medical services. Vancauwenbergh (2017; 2018) [requires corresponding competencies on the part of the CRIS operators] to address the question of how the opinions of ethics committees ought to be represented in the systems in a uniform manner, both regionally and nationally, and in a comparable manner, independent of discipline.

METHODOLOGY

The distribution of research output analyzes on the basis of research publications contributed by Scientists in the field of Research Ethics. The required data was collected from Web of Science (WoS) database for the period 2000-2022. A total of 78308 records were retrieved from the database using the search strings "Research Ethics", analysed by using MS-Excel software package on the study and data has been presented in tabular as well as in graphical form. This study incorporates the Relative Growth Rate (RGR) of publications and citation patterns, author productivity and collaboration, most productive journals, Top ranked authors, Institutions and Network Visualization portrait presented through VoS viewer.

RESULTS AND DISCUSSION

Year Wise Publications Distributed in Research Ethics

The year-wise publications were identified based on Web of Science database in the research area of Research Ethics during 2000-2022. Table 1 is found based on analyzed the data out of 78308 publications, the larg-

Table 1. Year wise publications distributed in research ethics

Year	No. of Publications	%
2000	587	0.719
2001	646	0.791
2002	700	0.857
2003	826	1.011
2004	941	1.152
2005	953	1.167
2006	1108	1.356
2007	1166	1.427
2008	1943	2.379
2009	2397	2.934
2010	2632	3.222
2011	2772	3.393
2012	3084	3.775
2013	3445	4.217
2014	3594	4.4
2015	4292	5.254
2016	4603	5.635
2017	5266	6.446
2018	5444	6.664
2019	7045	8.624
2020	7847	9.606
2021	8751	10.713
2022	8266	10.119
Total	78308	

est 8751 (10.71%) of publications was distributed by the researchers in 2021. The study was elaborated on ranked the publications as year wise series 8266 (10.71%), 7847 (9.60%), 7045 (8.62%) of the publications distributed during 2022 and 2020 and followed in the sequence of years 2022, 2020 and 2019 respectively. The result of the study was analyzed less than thousand publications contributed by the researchers during 2000-2005. Above thousand and less than five thousand of publications found during 2006-2016. The study being steadily rose in the publication trend during thorough out study period (Figure 1).

Exponential Growth Rate (EGR) in Research Ethics

Table 2 analyses an Exponential Growth Rate of publications in the field of Research Ethics during 2000-2022. The study is observed the EGR ranges between 0.068 and 2.044 in the years 2002 and 2019 respectively. Further, it was discussed that the negative active EGR was found -0.619 in the year 2022. The study witnessed that the proportion between publications and EGR, as the result when publications rises, the EGR will be increased as default (Figure 2).

Table 3 analyzed the ranking of authors are listed accordingly publications distributed in the field of Research Ethics. The study is found ranked authors out of twenty-five authors in the research filed concerned. The prolific author are found in the series, Shine was the top ranked authors, who had published 164 (0.201%) of the publications, Zhang Y (0.193%), Savulesu (0.18%), LI J (0.164%), Wang Y (0.163%), Anonymous (0.149%), Sugarman (0.147%), Weijer (0.144%), Wang (0.14%) and Racine (0.133%) in the ranking of list are Second, Third, fourth, Fifth,Sixth, Seventh, Eighth, Ninth and Tenth respectively. The study has been discussed that the publications ranges between 164 and 70 was contributed by Shine R and Zhang L respectively.

Figure 1. Year wise publications distributed in research ethics

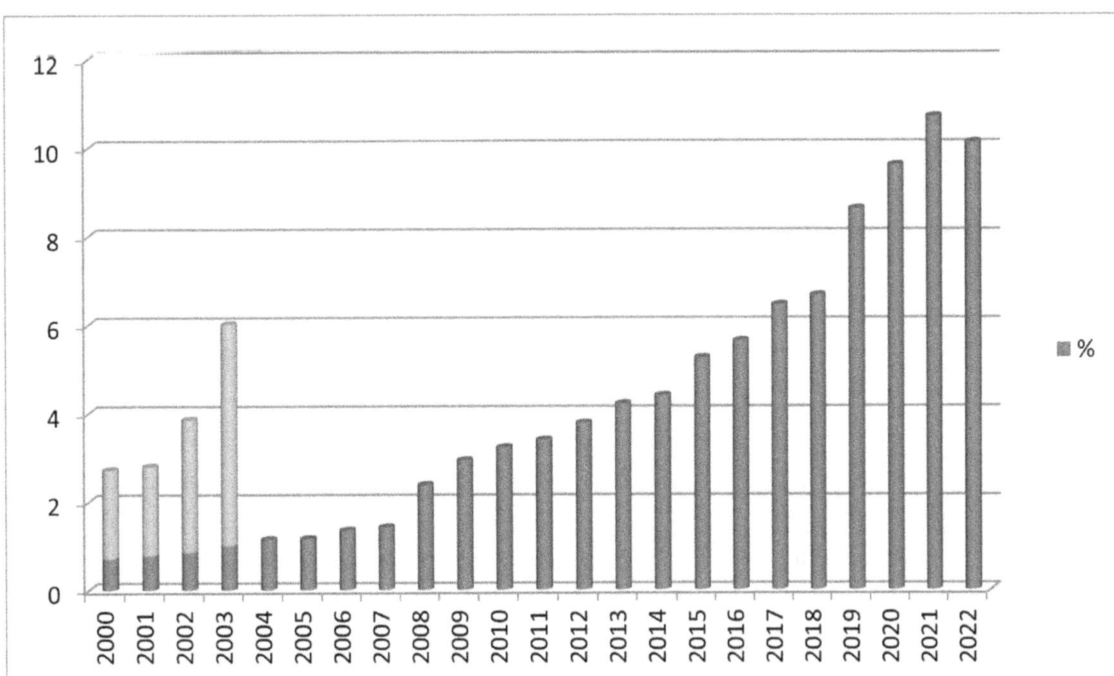

Table 2. Exponential Growth Rate (EGR) in research ethics

YEAR	PUBLICATIONS	EGR(EXPONENTIAL GROWTH RATE)
2000	587	0
2001	646	0.075
2002	700	0.068
2003	826	0.146
2004	941	0.015
2005	953	0.015
2006	1108	0.197
2007	1166	0.076
2008	1943	0.992
2009	2397	0.579
2010	2632	0.300
2011	2772	0.178
2012	3084	0.398
2013	3445	0.461
2014	3594	0.190
2015	4292	0.891
2016	4603	0.397
2017	5266	0.846
2018	5444	0.227
2019	7045	2.044
2020	7847	1.024
2021	8751	1.154
2022	8266	-0.619
Total	78308	

Figure 2. Exponential growth rate (EGR) in research ethics

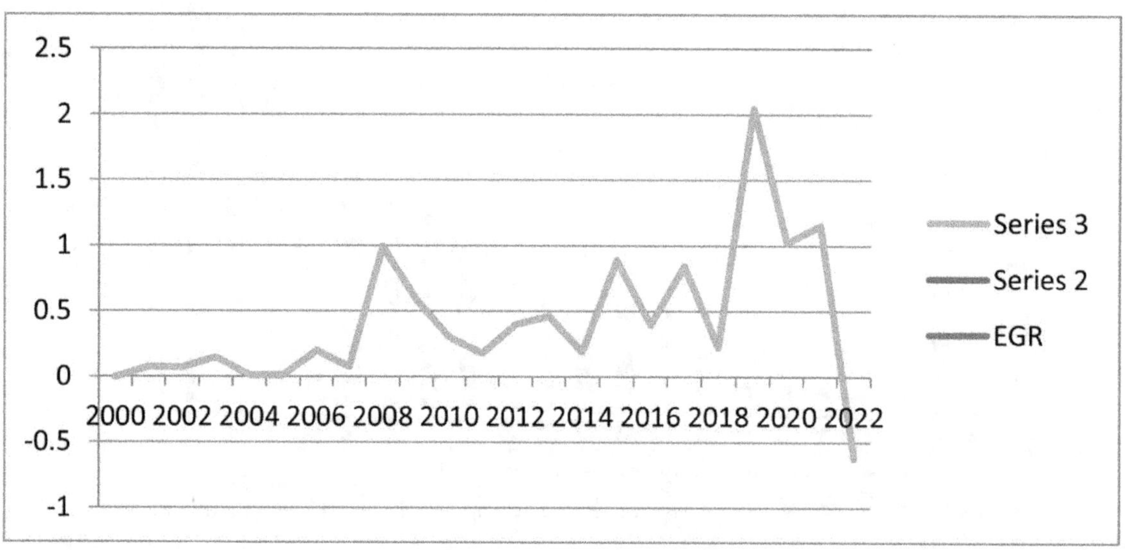

Table 3. Ranking of authors according to publications in the field of research ethics

S.NO	AUTHORS	RECORD COUNT	%
1	SHINE R	164	0.201
2	ZHANG Y	158	0.193
3	SAVULESCU J	149	0.182
4	LI J	134	0.164
5	WANG Y	133	0.163
6	ANONYMOUS	122	0.149
7	SUGARMAN J	120	0.147
8	WEIJER C	118	0.144
9	WANG J	116	0.142
10	RACINE E	109	0.133
11	KNOPPERS BM	108	0.132
12	LIU Y	108	0.132
13	WANG L	106	0.13
14	LI Y	100	0.122
15	PARKER M	98	0.12
16	CHEN Y	94	0.115
17	ZHANG J	90	0.11
18	RESNIK DB	89	0.109
19	MILLER FG	86	0.105
20	LI L	82	0.1
21	LIU J	81	0.099
22	ILLES J	80	0.098
23	WILFOND BS	77	0.094
24	LI X	70	0.086
25	ZHANG L	70	0.086

Top Ranked Institutions on the Publications Productivity in the Field of Research Ethics

Table 4 analyzed the Top ranked institutions on the publications productivity in the field of Research Ethics. The study has been analyzed on the productivity of the publications were contributed by the researchers from the Institutions in the filed concerned. The study can notice that University of London was contributed 3477 (4.25%) of the publications have been witnessed in the field of Research Ethics. The institution was top ranked among twenty-five prolific institutions.

The study was observed that there another ranked second and ten in the series University of Toronto with 2030 (2.48%), University of Oxford 1870 (2.28%), University of Sydney 1771 (2.16%), University of California system 1727 (2.11), University of Melbourne 1690 (2.06), Harvard University 1577 (1.93%),

Figure 3. Network visualization of author collaboration

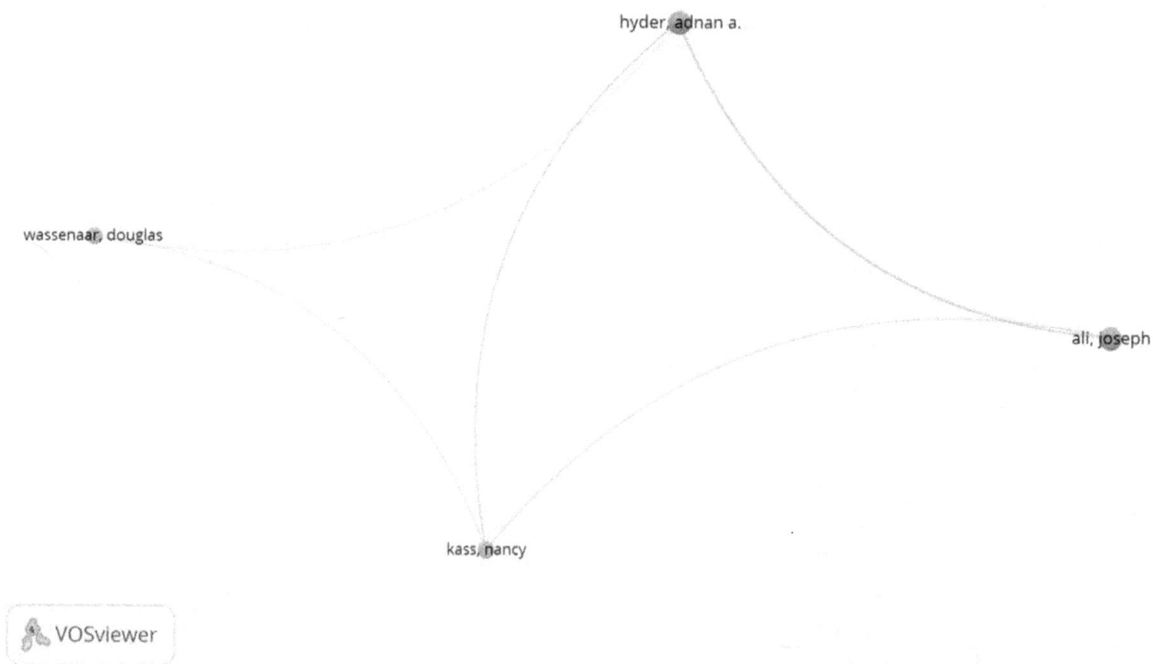

University College London 1389 (1.7%), Monash University1272 (1.55%), University of Queensland 1272 (1.55%).

Ranking of Research Areas in the Field of Research Ethics

Table 5 discussed about ranking of research areas wise distributions of the publications in the field of Research Ethics. The study was explored on the productivity of the publications were contributed by the researchers among twenty-five research areas have been given in the study. The top ranked research area was Social Science other topics was holding with 17100 (20.93%) of the publications distributed publications in the field concerned.

The study was analysed according to numbers publications as deceasing order out of twenty-five research areas are General internal Medicine as the second position with 11810 (1475%), Business Economics as third with 7451 (9.12%), Medical Ethics 6716 (8.22%), Biomedical Social Sciences 5953 (7.28%), Environmental Sciences Ecology 5492 (6.72%), Public Environmental Occupational Health 5492 (6.72%), Psychology 3984 (4.87%), Social Issues and Science Technology other Topics 3544 (4.38%) as the ranked order fourth, fifth, Sixth, Seventh, Eighth, Ninth and tenth respectively (Fig.2).

Ranking of Countries Wise Distribution of the Publications in the Field of Research Ethics

Table 6 discussed about ranking of countries accordingly productivity of Publications from respective country in the Research Ethics. The study was noticed that there are twenty-five countries are listed on

Table 4. Top ranked institutions on the publications productivity in the field of research ethics

RANK	NAME OF THE INSTITUTIONS	RECORD COUNT	%
1	UNIVERSITY OF LONDON	3477	4.256
2	UNIVERSITY OF TORONTO	2030	2.485
3	UNIVERSITY OF OXFORD	1870	2.289
4	UNIVERSITY OF SYDNEY	1771	2.168
5	UNIVERSITY OF CALIFORNIA SYSTEM	1727	2.114
6	UNIVERSITY OF MELBOURNE	1690	2.069
7	HARVARD UNIVERSITY	1577	1.931
8	UNIVERSITY COLLEGE LONDON	1389	1.7
9	MONASH UNIVERSITY	1272	1.557
10	UNIVERSITY OF QUEENSLAND	1272	1.557
11	UNIVERSITY TORONTO AFFILIATES	1269	1.553
12	UDICE FRENCH RESEARCH UNIVERSITIES	1177	1.441
13	UNIVERSITY OF BRITISH COLUMBIA	1075	1.316
14	UNIVERSITY OF NEW SOUTH WALES SYDNEY	1070	1.31
15	KING S COLLEGE LONDON	1059	1.296
16	MCGILL UNIVERSITY	1044	1.278
17	NATIONAL INSTITUTES OF HEALTH NIH USA	954	1.168
18	JOHNS HOPKINS UNIVERSITY	930	1.138
19	UNIVERSITY OF WASHINGTON	862	1.055
20	UNIVERSITY OF CAMBRIDGE	857	1.049
21	UNIVERSITY OF PENNSYLVANIA	852	1.043
22	PENNSYLVANIA COMMONWEALTH SYSTEM OF HIGHER EDUCATION PCSHE	830	1.016
23	UNIVERSITY OF WASHINGTON SEATTLE	829	1.015
24	UNIVERSITY OF NORTH CAROLINA	825	1.01
25	UNIVERSITY OF OTTAWA	824	1.009

the productivity of the publications. The top ranked country was USA 23355 (28.59%) of the publications as top ranked counties out of twenty-five. The study has been explored ranking of the counties in the order as second ranked was England with 13353 (16.34%), Australia 11777 (14.41%), Canada 8424 (10.31%), Peoples R china 5293 (6.48%), Germany 4512 (5.52%), Netherlands 3920 (4.79%), France 2733 (3.34%), Spain 2391 (2.92%) and Sweden 2294 (2.80%).

The study has been analysed that publications quantity was measured more than ten thousand publications contributed by USA, England, and Australia. Above twenty thousand of publications have witnessed only by USA. Above five thousands of publications from People R China and Canada published. Only five counties were contributed two thousands of publications, there are twelve countries are witnessed above thousands of Publications. The study was disclosed that Ireland only the country with less than thousand publications was brought out in the field of Research Ethics.

Table 5. Ranking of research areas in the field of research ethics

RANK	RESEARCH AREAS	Record Count	%
1	SOCIAL SCIENCES OTHER TOPICS	17100	20.933
2	GENERAL INTERNAL MEDICINE	11810	14.457
3	BUSINESS ECONOMICS	7451	9.121
4	MEDICAL ETHICS	6716	8.222
5	BIOMEDICAL SOCIAL SCIENCES	5953	7.287
6	ENVIRONMENTAL SCIENCES ECOLOGY	5492	6.723
7	PUBLIC ENVIRONMENTAL OCCUPATIONAL HEALTH	4786	5.859
8	PSYCHOLOGY	3984	4.877
9	SOCIAL ISSUES	3584	4.387
10	SCIENCE TECHNOLOGY OTHER TOPICS	3544	4.338
11	HEALTH CARE SCIENCES SERVICES	3081	3.772
12	NURSING	2814	3.445
13	PHILOSOPHY	2717	3.326
14	ENGINEERING	2538	3.107
15	EDUCATION EDUCATIONAL RESEARCH	2423	2.966
16	HISTORY PHILOSOPHY OF SCIENCE	2345	2.871
17	NEUROSCIENCES NEUROLOGY	2322	2.843
18	ZOOLOGY	2313	2.832
19	GOVERNMENT LAW	1928	2.36
20	PSYCHIATRY	1865	2.283
21	RESEARCH EXPERIMENTAL MEDICINE	1515	1.855
22	GENETICS HEREDITY	1425	1.744
23	MARINE FRESHWATER BIOLOGY	1340	1.64
24	SOCIOLOGY	1276	1.562
25	PEDIATRICS	1251	1.531

CONCLUSION

The study has discussed about number of publications based on how they researchers did in research ethics. In today's higher education environment, the study area was now a very important predominate factor in the publication of research papers. During the course of the study, compilations of bibliographic records were used to obtain the information from Web of Science regarding the availability of records. From 2000 to 2005, the researchers contributed less than a thousand publications that were analyzed as part of the study. Over thousand and fewer than 5,000 of distributions found during 2006-2016. The study demonstrated that the ratio of publications to EGR rises as publications rise, so by default, EGR rises as well. The review has examined that the distributions ranges somewhere in the range of 164 and 70 was contributed by Sparkle R and Zhang L separately. According to the findings, the University of Toronto ranked second with 2030 (2.48%), followed by the University of Oxford with 1870 (2.28%),

Figure 4. Ranking of research areas (network visualization)

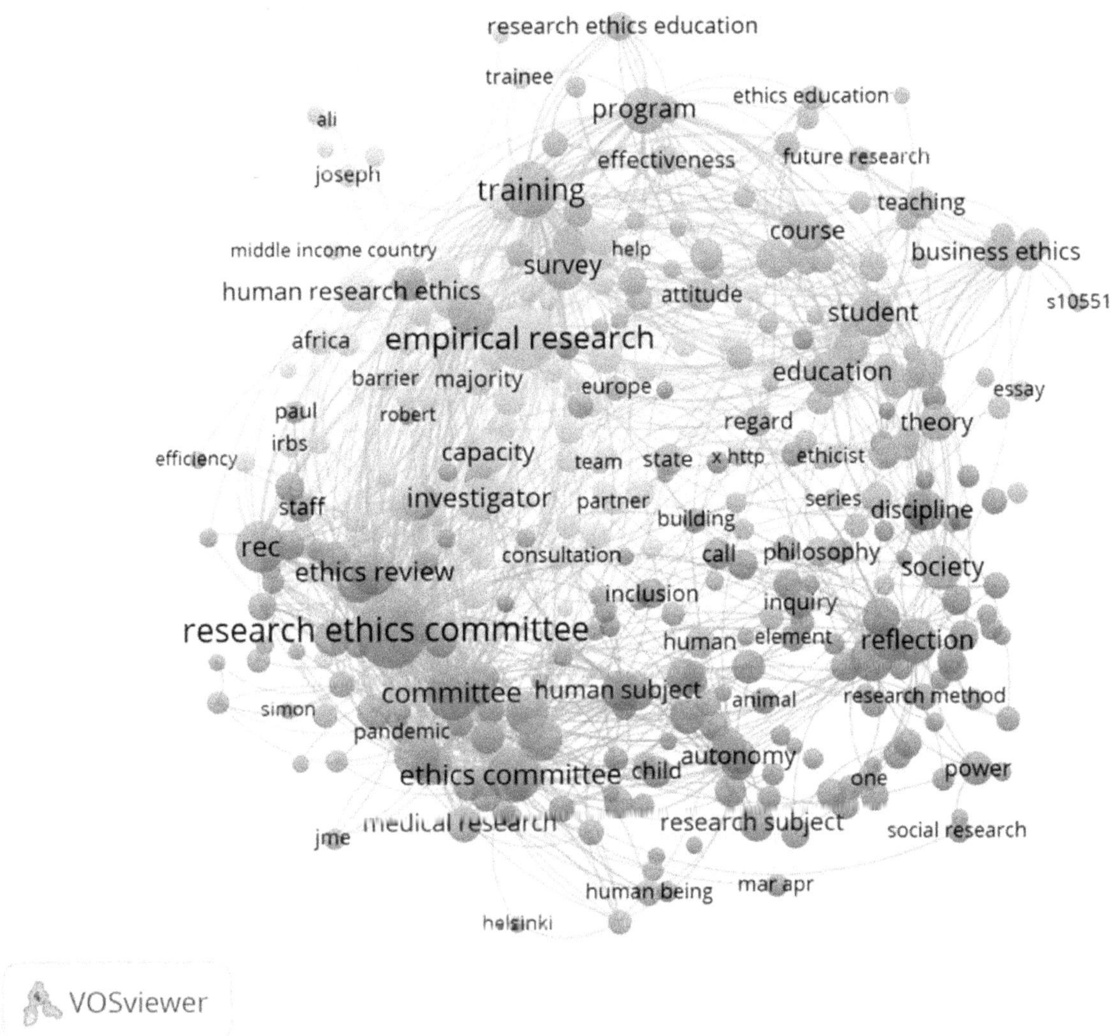

the University of Sydney with 1771 (2.16%), the University of California system with 1727 (2.11%), the University of Melbourne with 1690 (2.06), Harvard University with 1577 (1.93%), the University of London with 1389 (1.7%), Monash University with 1272 (1.55%), and the University of Queensland with 1272 (1.55%). With 17100 (20.93%) of the publications distributed in the subject matter, Social Science and Other Topics was the top-ranked research area. The review was seen that there are 25 nations are recorded on the efficiency of the distributions. The highest level nation was USA 23355 (28.59%) of the distributions as highest level provinces out of 25 for examined.

Table 6. Ranking of countries wise distribution of the publications in the field of research ethics

RANK	COUNTRIES/REGIONS	RECORD COUNT	%
1	USA	23355	28.59
2	ENGLAND	13353	16.346
3	AUSTRALIA	11777	14.417
4	CANADA	8424	10.312
5	PEOPLES R CHINA	5293	6.48
6	GERMANY	4512	5.523
7	NETHERLANDS	3920	4.799
8	FRANCE	2733	3.346
9	SPAIN	2391	2.927
10	SWEDEN	2294	2.808
11	SOUTH AFRICA	2120	2.595
12	SWITZERLAND	2031	2.486
13	SCOTLAND	1957	2.396
14	ITALY	1949	2.386
15	NEW ZEALAND	1880	2.301
16	BRAZIL	1760	2.155
17	BELGIUM	1749	2.141
18	NORWAY	1730	2.118
19	JAPAN	1563	1.913
20	DENMARK	1554	1.902
21	IRAN	1522	1.863
22	INDIA	1228	1.503
23	TURKEY	1070	1.31
24	FINLAND	1066	1.305
25	IRELAND	920	1.126

REFERENCES

Bakri, A., & Willet, P. (2008). The Malaysian Journal of Library and Information Science 2001- 2006: A Bibliometric study. *Malaysian Journal of Library and Information Science*, *13*(1), 1–13.

Baskaran, C. (2019). Scholarly information share through Social Networks (SNs) and Medias among social science scholars in selected State Universities in Tamilnadu. *International Journal of Library and Information Studies*, *9*(3), 83–92.

Baskaran, C. (2020). Research Patterns on the Social Networks and Media: A Scientometric Portrait. Handbook of Research on Emerging Trends and Technologies in Library and Information Science, pp.189-207. IGI Global. / doi:10.4018/978-1-5225-9825-1.ch014

Baskaran, C. (2021). An Exponential trend and Mapping of Research in Web 2.0, *Library Philosophy and Practice (e-journal)*, 5008. https://digitalcommons.unl.edu/libphilprac/5008

Brennan, N. (2017). Fifteen Years of Research Information Management in Ireland. *In Proceedings of the EuroCRIS Membership Meeting*, Dublin, Berlin.

DuBois, J. M., & Antes, A. L. (2018). Five dimensions of research ethics: A stakeholder framework for creating a climate of research integrity. *Academic Medicine*, *93*(4), 550–555. doi:10.1097/ACM.0000000000001966 PMID:29068823

Kagan, S. (1998). *Normative Ethics.* Westview Press. https://www.jstor.org/stable/2660152

Karpagam, R. (2014). Global research output of Nano-biotechnology research: A scientometrics study. *Current Science*, *106*(11), 1490–1499.

Klausen, M. H. (2016). *Even minor integrations can deliver great value—A case study.* In *Proceedings of the CRIS2016 Conference*, St Andrews, UK.

Martinson, B. C., Anderson, M. S., & De Vries, R. (2005). Scientists behaving badly. *Nature*, *435*(7043), 737–738. doi:10.1038/435737a PMID:15944677

Kalichman, M. (2010). *Introduction: What is Research Ethics? Resources for Research Ethics Education.* Online Ethics. https://onlineethics.org/cases/resources-research-ethics-education/introduction-what-research-ethics

Nalimov, V., & Kordon, V. a. Ya., & Korneeva. (1971). Geograficheskoe Raspredelenie Nauchnoi Informatsii. [Geographic Distribution of Scientific Information.] [V. V. Nalimov trans.] Informatsionnye Materialy Moscow:an SSSR Nauchnyi Sovet po Kompleksnoi Probleme Kibernetiki. [Moscow: Soviet Academy of Science, Scientific Council on Cybernetics.] 2: 3–37.

Nalimov, V. V., & Mulchenko, Z. M. (1969a). Eshche raz k voprosu o kontseptsii eksponentsial'nogo rosta.[A word to add on the exponential growth concept.] Nauchno-Tekhnicheskaya Informatsiya. *Seriya*, *2*(8), 12–14.

Nalimov, V. V., & Mulchenko, Z. M. (1969b). *Naukometriya. Izuchenie Razvitiya Nauki kak Informatsionnogo Protsessa. [Scientometrics. Study of the Development of Science as an Information Process].* Nauka.

Patra, S. K., Bhattacharya, P., & Verma, N. (2006). Bibliometric study of literature on bibliometrics. *DESIDOC Bulletin of Information Technology*, *26*(1), 27–32. doi:10.14429/dbit.26.1.3672

Singh, G., Mittal, R., & Ahmad, M. A. (2007). bibliometric study of literature on digital libraries. *The Electronic Library*, *25*(3), 342–348. doi:10.1108/02640470710754841

Tiew, W. S. (2000). Characteristics of self-citations in Journal of Natural Rubber Research 1988-1997: A ten-year bibliometric study. *Malaysian Journal of Library and Information Science*, *5*(1), 95–104.

Vancauwenbergh, S. (2017). The transition of FRIS 1.0 towards FRIS 2.0: Focus on semantic interoperability. In *Proceedings of the EuroCRIS Membership Meeting*, Dublin, Berlin.

Vancauwenbergh, S. (2017). Research data management at the Flemish universities: State-of-affairs. In *Proceedings of the CRIS2018 Conference*, Umeå, Sweden.

Chapter 6
Evolving Role of Instagram as a Digital Marketing Tool in Libraries

Sanmati Jinendran Jain
Central University of Tamil Nadu, Tiruvarur, India

T. Anil Kumar
https://orcid.org/0000-0002-0334-3549
Indian Institute of Corporate Affairs, Manesar, India

ABSTRACT

The role of social media in libraries has been gaining attention in recent years as libraries look for new ways to connect with users and promote their services. This chapter assesses the use of Instagram, a popular photo and video-sharing platform, by libraries as a means of outreach and engagement. Through a literature review and case studies of library Instagram accounts, the chapter explores how libraries use Instagram to showcase their collections, promote events and services, and create a sense of community. It also examines the benefits and challenges of using Instagram as a digital marketing tool and the implications for library practice. The chapter explains that Instagram can effectively reach and engage with users, particularly younger generations.

INTRODUCTION

Social media has become an integral part of the modern communication landscape, and libraries are no exception to this trend. Libraries have recognized the potential of social media platforms in reaching out to their users and stakeholders, promoting their services and resources, and engaging with their community. Amongst the various social media platforms, Instagram has emerged as a powerful marketing tool for libraries, owing to its visual appeal and the growing number of active users. Instagram is a mobile application available on iOS and Android devices that allows users to quickly transform their mobile photographs into visually captivating images, which can be shared with other users on the platform.

DOI: 10.4018/978-1-6684-8805-8.ch006

The photos can also be shared on social networks like WhatsApp, Twitter, Facebook, Tumblr, Flickr, and Pintrest. Instagram was launched in October 2010 and was acquired by Facebook in April 2012 (Salomon, 2013).

Instagram is a photo-sharing app that allows users to share images and videos with their followers. The platform's visual nature makes it an ideal tool for libraries to showcase their resources, events, and services. It also offers features such as stories, live videos, and hashtags, which enable libraries to engage with their users and promote their content. Using Instagram as a digital marketing tool, libraries can reach a wider audience and connect with potential users who might not be aware of their services. Moreover, it gives libraries insights into their audience's interests and behavior, which can help them tailor their marketing strategies to suit their users' needs better. Despite the potential benefits, using Instagram as a digital marketing tool for libraries also presents challenges, such as maintaining a consistent brand image, creating engaging content and managing user interactions.

LITERATURE REVIEW

Sobreira et al. (2020) investigated the extent to which libraries should explore social networks, focusing on themes that foster closer ties within society and thereby exhibit its ability to disseminate information beyond the physical boundaries of libraries. Using qualitative research, a comparative analysis was conducted on content produced by Brazilian library councils and published on their Instagram profiles in 2019 and 2020, before and during the Covid-19 pandemic. The content was classified according to its themes, and these categories were compared between the two time periods. The findings revealed a significant shift in the content analyzed, indicating a greater emphasis on topics of general interest to society.

Chan et al. (2020) employed a mixed-method approach to evaluate the efficacy of Facebook and Instagram usage at The University of Hong Kong Libraries (HKUL). The analysis involved reviewing the posts on the two social media platforms of HKUL and collecting user feedback through on-campus interviews. The content analysis indicated that both Facebook and Instagram received minimal user engagement, and the most frequently posted content was related to library operational news and events. The interviewees expressed their support for HKUL's use of Facebook and Instagram. Still, they suggested that the library improve its content style on social media and use the available features to grab users' attention. The study offered valuable insights for librarians regarding effective social media management and utilization, particularly in content design and management.

Rachman (2020) analyzed the Instagram posts of academic libraries to propose a classification system for grouping these posts. The study found that the largest category of posts, accounting for 29%, was related to library news and information. Baskaran (2020) analyzed that majority of the publications, 2,207 (18.48%), were published from communication area, on account of the specific nature of the research concentrating on Social networks. Computer communications information systems area scores second with 1,172 (9.81%) of the publications.

One of the most preferred topics in the research of social media, which is a group of Internet-based platforms that let users create, share, and exchange content as well as connect with other users' information on the same platform (Blank & Reisdorf, 2012; 2007 by Boyd and Ellison; Kaplan and Haenlein, 2010). According to Allen, Ryan, Gray, McInerney, & Waters (2014), there has been a rise in socializing, sharing personal content (such as pictures), and playing games online in recent years. Ryan, Chester, Reece, and Xenos, 2014), Kuss and Griffiths (2011) reasoned that the utilization of online entertainment has become exceptionally famous among teens and youthful grown-ups. Web-based entertainment likewise incorporates subtypes, like online journals (e.g., WordPress, Twitter), conversation gatherings (e.g., Reddit), computerized content sharing stages (e.g., Pinterest, Flickr, Instagram), and social organizing destinations (e.g., Facebook, LinkedIn) (Saravanakumar and Sugantha Lakshmi, 2012). Among these subtypes, the most often involved applications in the U.S. are Facebook (79%), Instagram (32%), Pinterest (31%), LinkedIn (29%), and Twitter (24%) (Greenwood and Perrin, 2016).Kuss, Griffiths, Karila, and Billieux's (2014) survey found that as innovations progress, different the new devices utilized to get to virtual entertainment additionally can possibly influence the general recurrence of Web fixation. Brenner and Smith (2013) say that, in comparison to older computing devices like desktop or laptop computers, modern mobile devices like smartphones and tablets make it easier for people to quickly access social media.

OBJECTIVES OF THE STUDY

- To examine how much libraries currently use Instagram as a digital marketing tool.
- To identify the potential benefits and challenges of using Instagram as a library marketing tool.
- To explore the strategies and best practices libraries can adopt to optimize their impact on Instagram.
- To investigate the impact of Instagram on user engagement and library promotion.
- To provide recommendations for libraries on effectively using Instagram as a digital marketing tool to promote their services and engage with their audience.

METHODOLOGY

This study is based on a qualitative method using content analysis techniques. This study highlights the applications of Instagram in library services and its possible strategies for marketing. By analyzing various library Instagram pages and consulting literature relating to library use, the role of Instagram as a digital marketing tool in libraries has been enumerated.

USE OF INSTAGRAM FOR THE PROMOTION OF LIBRARIES IN INDIA

The rapid proliferation of social media platforms has transformed how organizations connect with their audience and promote their services. Libraries in India are no exception to this trend, as they have increasingly adopted social media platforms to promote their collections, events, and services. Among these platforms, Instagram has emerged as a popular library choice due to its visual nature and broad user base.

Vikram Sarabhai Library, IIM Ahmedabad (@vsl_iima)

Figure 1. Instagram page of VSL Library, IIM Ahmedabad

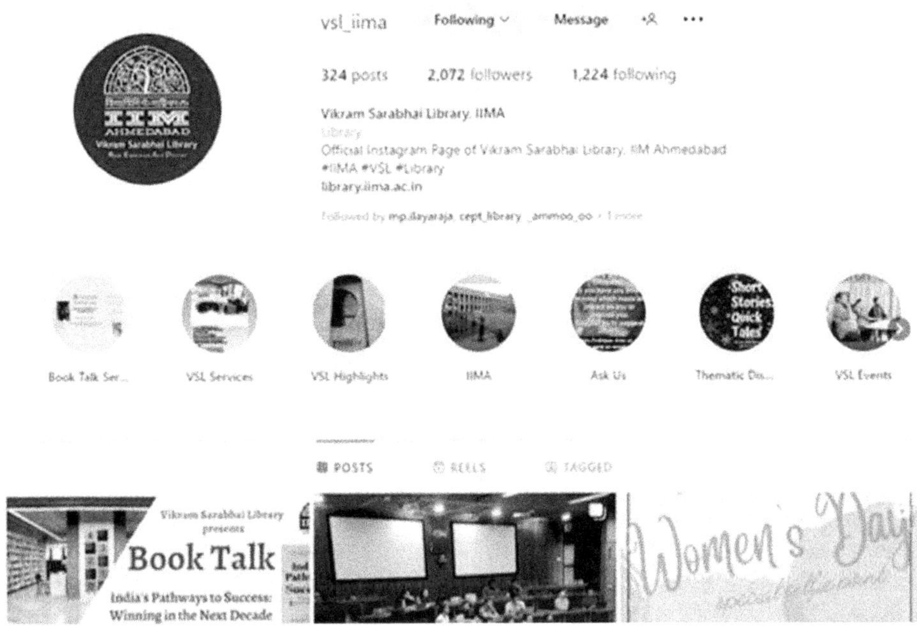

Library, IIM Udaipur (@iimu_library)

Figure 2. Instagram page of library, IIM Udaipur

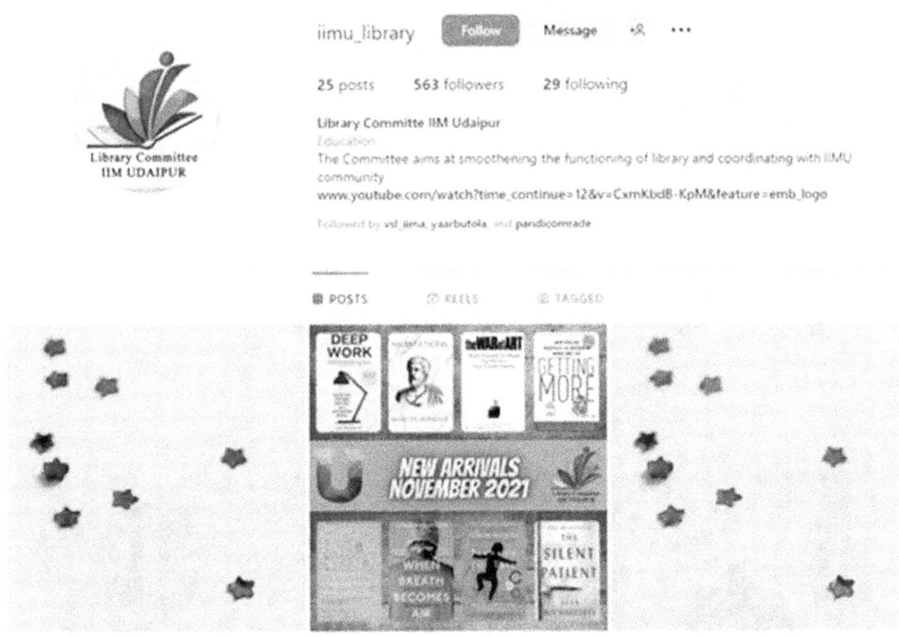

Library, IIT Gandhinagar (@iitgn_library)

Figure 3. Instagram page of library, IIT Gandhinagar

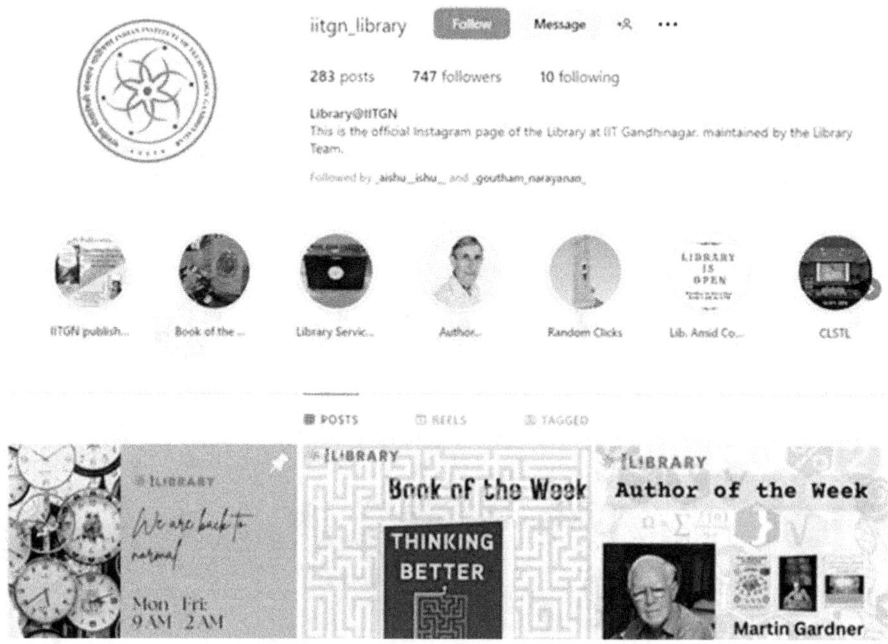

Library, CEPT University, Ahmedabad (@cept_library)

Figure 4. Instagram page of library, CEPT University

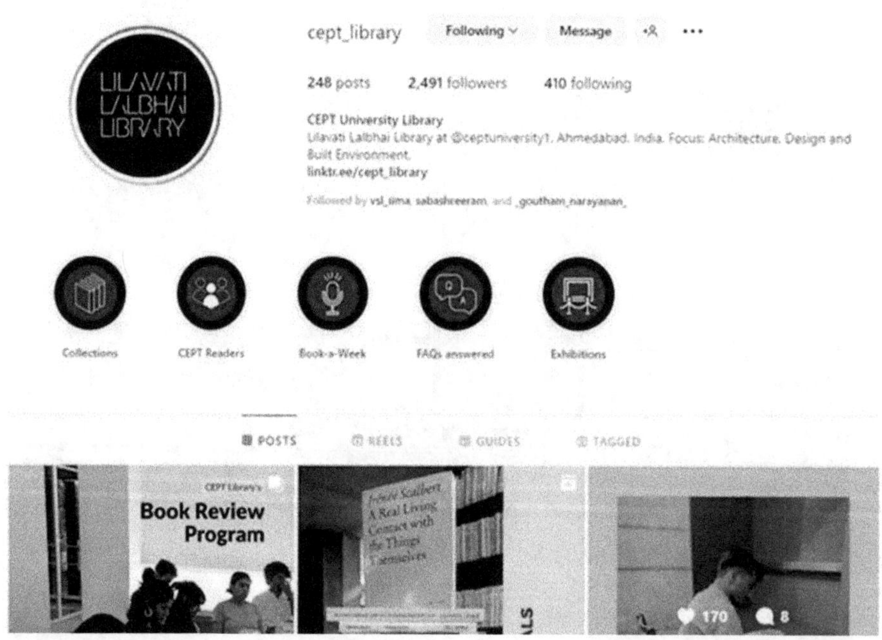

Delhi Public Library (@librarydpl)

Figure 5. Instagram page of Delhi Public Library

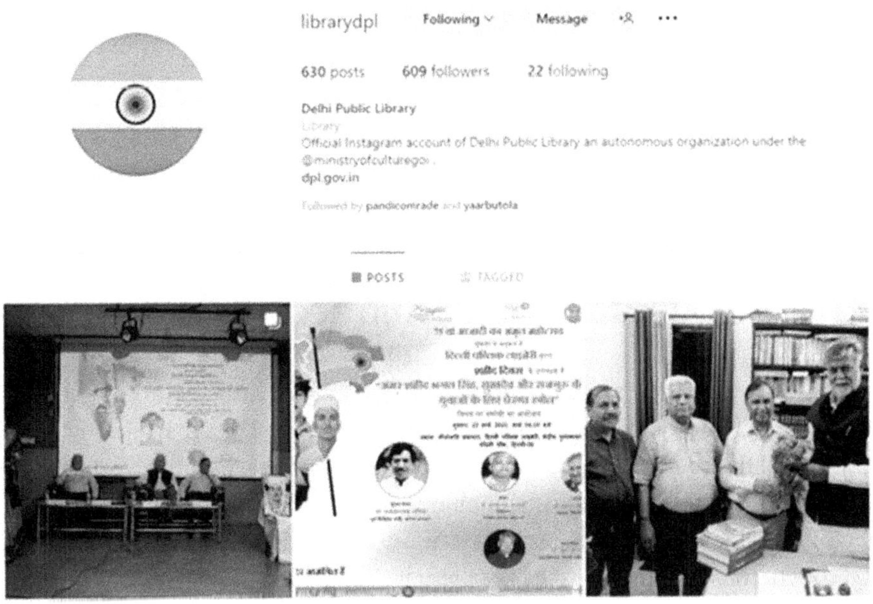

DEVELOPING INSTAGRAM ACCOUNT FOR THE LIBRARY

Developing an Instagram account for your library can be a great way to engage with your community and promote your services and resources. Here are some steps to get started:

- **Create an Instagram account**: If you don't already have an Instagram account for your library, create one. Choose a username that is easy to remember and reflects your library's name.
- **Determine your target audience:** Identify the target audience for your Instagram account, such as students, faculty, or the general public. Knowing your audience will help you tailor your content to their interests.
- **Define your goals:** Set goals for your Instagram account, such as increasing engagement, promoting events, or showcasing your resources. Having clear goals will help you measure your success.
- **Plan your content:** Plan your Instagram content to ensure you have a consistent and engaging feed. Consider creating a content calendar to help you stay organized.
- **Post regularly to keep your audience engaged:** Post at least once a day or a few times weekly.
- **Create a visually appealing profile:** Libraries should create a visually appealing Instagram profile that reflects their brand and showcases their resources and services. This could include high-quality images of library spaces, book displays, and events.
- **Use hashtags:** Use hashtags to help your content reach a wider audience. Use relevant hashtags, such as #librarylife, #bookstagram, or #research.

- **Engage with your audience:** Engage with your audience by responding to comments and direct messages. Encourage your audience to share their experiences with your library and tag your account in their posts.
- **Collaborate with other libraries:** Collaborate with other libraries on Instagram to increase your reach and engage with a broader audience.
- **Measure your success:** Measure your success by tracking your engagement and follower growth. Use this data to adjust your content and strategy as needed.

Following these steps, you can develop a successful Instagram account for your library and engage with your community in new and exciting ways.

MARKETING OF LIBRARY RESOURCES AND SERVICES USING INSTAGRAM

To maintain user engagement, Instagram continuously introduces new features within its application. In 2018, Instagram launched several new tools, including Nametag, third-party direct publishing, video chatting via IG Direct, and topic channels (Alfonzo, 2019). It can be a highly effective digital marketing tool for libraries, offering a range of benefits, including:

Visual storytelling: Instagram's emphasis on visual content makes it an ideal platform for libraries to tell their story through images and videos. Libraries can showcase their collections, events, and services and create an emotional connection with their patrons.

Increased engagement: Instagram's highly engaged user base allows libraries to reach and connect with their target audience. By creating compelling content and using hashtags, libraries can increase their reach and engage with patrons meaningfully.

Promoting events and services: Instagram can be an effective way for libraries to promote their events and services, such as author talks, book clubs, and workshops. By posting visually appealing content about upcoming events, libraries can attract more attendees and generate interest in their services.

Raising awareness: Instagram can help libraries raise awareness of their resources and services among a wider audience. By sharing posts about library resources and services, libraries can increase visibility and attract new patrons.

Building community: Instagram can help libraries build community among patrons. By responding to comments and direct messages, libraries can foster meaningful relationships with their patrons and create a loyal following.

CONCLUSION

The role of Instagram as a digital marketing tool in libraries is becoming increasingly important in today's digital age. It offers libraries a unique opportunity to engage with patrons, promote resources and services, and raise awareness of library events. However, there are also challenges that libraries may face in implementing an Instagram marketing strategy, such as maintaining a consistent content schedule or measuring the success of marketing campaigns. Therefore, libraries must develop a clear plan and goals to guide their Instagram marketing efforts. In the future, Instagram will likely continue evolving as a digital marketing tool for libraries. As new features and best practices emerge, libraries must stay up-to-date with the latest trends to maximize this digital marketing tool.

REFERENCES

Alfonzo, P. (2019). Chapter 4. Instagram in the Library. *Library Technology Reports*, *55*(2). https://journals.ala.org/index.php/ltr/article/view/6960/9415

Baskaran, C. (2020). Research Patterns on the Social Networks and Media: A Scientometric Portrait. Handbook of Research on Emerging Trends and Technologies in Library and Information Science, pp.189-207. IGI Global. / doi:10.4018/978-1-5225-9825-1.ch014

Blank, G., & Reisdorf, B. C. (2012). The participatory web: A user perspective on Web 2.0. *Information Communication and Society*, *15*(4), 537–554. doi:10.1080/1369118X.2012.665935

Boyd, D. M., & Ellison, N. B. (2007). Social network sites: Definition, history, and scholarship. *Journal of Computer-Mediated Communication*, *13*(1), 210–230. doi:10.1111/j.1083-6101.2007.00393.x

Brenner, J., & Smith, A. (2013). *72% of online adults are social networking site users.* Pew Research. http://pewInternet.org/Reports/2013/social-networking-sites.aspx

Chan, T. T. W., Lam, A. H. C., & Chiu, D. K. W. (2020). From Facebook to Instagram: Exploring user engagement in an academic library. *Journal of Academic Librarianship*, *46*(6), 102229. doi:10.1016/j.acalib.2020.102229 PMID:34173399

Kuss, D. J., Griffiths, M. D., Karila, K., & Billieux, J. (2014). Internet addiction: A systematic review of epidemiological research for the last decade. *Current Pharmaceutical Design*, *20*(25), 4026–4052. doi:10.2174/13816128113199990617 PMID:24001297

Rachman, Y. B. (2020). Examining academic libraries' use of instagram. *College & Undergraduate Libraries*, *27*(2–4), 462–473. doi:10.1080/10691316.2021.1915221

Ryan, T., Chester, A., Reece, J., & Xenos, S. (2014). The uses and abuses of Facebook: A review of Facebook addiction. *Journal of Behavioral Addictions*, *3*(3), 133–148. doi:10.1556/JBA.3.2014.016 PMID:25317337

Salomon, D. (2013). Moving on from Facebook: Using Instagram to connect with undergraduates and engage in teaching and learning. *College & Research Libraries News*, *74*(8), 408–412. doi:10.5860/crln.74.8.8991

Saravanakumar, M., & SuganthaLakshmi, T. (2012). Social media marketing. *Life Science Journal*, *9*, 4444–4451. doi:10.7537/marslsj090412.670

Sobreira, D., Santos de Oliveira, D., & García-Peñalvo, F. J. (2020). The Use of Instagram as a Digital Marketing Tool by the Brazilian Library Councils in times of Covid-19. *Eighth International Conference on Technological Ecosystems for Enhancing Multiculturality*, (pp. 582–587). ACM. 10.1145/3434780.3436599

Chapter 7
Research Impact of the Scholarly Literature on Google Scholar:
A Scientometric Study

M. Chandrasekaran
Alagappa University, Karaikudi, India

ABSTRACT

The study analyses the research impact of scholarly literature on Google scholar; a Scientometric study during 2002 -2022. A total number of 25760 records were retrieved from Web of Science database for the period of study. Out of 25760, the highest 5017(19.47%) of the publications have been witnessed in the year 2021. Following is 4903, then 3835 contributed in the field of Google Scholar during 2022 and 2020 respectively. The top ranking of authors is found to be a Maffulli N with 113 (0.4%) of the publications on Google Scholar. The review articles hold the maximum of publications with 19132 (74.27%). Pharmacology pharmacy was top ranked in subject categories among the listed subjects and holds 2590 (10.05%) of the publications. Tehran University of Medical Sciences is among the most prolific institutions in the listed twenty-five of them. The institute holds 706(2.74%) of the publications in Google Scholar during period of study.

INTRODUCTION

Nalimov and Mulchenko (1969) coined the scientometrics to describe the study of science in terms of structure, growth, linkages, and production. In the 1960s and 1970s, historians, philosophers of science, and social scientists like Derek J. de Solla Price published a sizable collection of research papers on the theoretical underpinnings of the discipline of quantitative science studies, resulting in a "full-fledged research programme" (Price 1976). Scientometrics plays a significant role in the sociology of scientific knowledge and significantly influences the creation of scientific policies (Hess 1997). Its primary interests are the creation and use of methods, tools, and quantitative indicators to study the relationship between

DOI: 10.4018/978-1-6684-8805-8.ch007

scientific and technical domains, the processes of improvement in the design of information systems for knowledge acquisition, analysis, and distribution; social variables; and the study of cognitive and socio-organizational structures in scientific fields (van Raan 1997). In a recent review of the literature (Mingers and Leydesdorff 2015), the conceptual and empirical foundations of this problem-oriented field's disciplinary traditions and key turning points were examined.

Google Scholar

Google Scholar Profiles offer authors an easy method to highlight their scholarly output. You may look up who is citing your work, chart citation trends over time, and calculate a variety of citation metrics. You can also choose to make your profile public, which will allow it to show up in Google Scholar search results when someone types in your name, such as Richard Feynman. The best part is that it's quick and easy to set up, even if you've produced hundreds of articles and your name is used by several different academics. Your citation metrics are generated and updated automatically as Google Scholar discovers new citations to your work on the web, and you can add groups of relevant articles rather than simply one article at a time. You have a choice can choose whether to manually update your articles at any time, to have your list of articles updated automatically, to check the updates yourself.

REVIEW OF LITERATURE

Recent Scientometric studies have also been concentrated in conference series related to CHI that take place in particular nations or regions, including but not limited to the UK (Padilla et al. 2014), Korea (Lee et al. 2014), India (Gupta 2015), New Zealand (Nichols and Cunningham 2015), Brazil (Barbosa et al. 2016), and Australia (Mubin et al. 2017). These research used methods such subject modelling, trend analysis, author gender analysis, institutional-level analysis, co-authorship networks, citation analysis, and writing frequency (Mubin et al. 2017). The majority of the time, scientometric analyses a quantitative emphasis on authorship and citation analysis. It is also important to note that compared to other sorts of publications (like system design), similar studies in the field of HCI have poor citation rates. Specificity standard relies primarily on the Horn et al.'s (2004) categorization, which acknowledged ACM CSCW, GROUP, and ECSCW as devoted venues in the field of CSCW. The authors also recognized JCSCW as a CSCW-related outlet, while According to Holsapple and Luo (2003), JCSCW articles are extremely specialized and have a significant impact on shaping research. This selection criterion was also discussed in other studies, which considered the included sources to be the leading venues in the field for the regular publication of CSCW authors from North America and Europe (Jacovi et al.). 2006; 2012, Grudin and Poltrock).

Horn et al. on the other hand (2004) thought about CHI and the ACM Conference on UI Programming and Innovation (UIST) as non-CSCW distribution outlets. Because we wanted to include only documents that had been reviewed by other people, we didn't include book chapters in order to keep the data analysis consistent. Due to the particular nature of the research focusing on social networks, the majority of the publications, 2,207 (18.48 percent), were published in the communication field. The area of computer communications and information systems comes in second, with 1,172 publications (9.81 percent). Researchers from the University of California systems contributed to the majority of 134 (or 1.96 percent) of the publications. Zhang Y was the top creator was contributed 16(0.23%) of the distributions in the field of Web 2.0. Baskaran (2019) looked at how researchers at a few Tamil Nadu universities used social media and other media. To most noteworthy 87 (31.9%) of respondents Annamalai College, (16.8%) and (15%) of them Male and Female individually out of 273 all out respondents answered the review. Baskaran (2020) looked at the data and found that the majority of the publications, 2,207 (18.48 percent), were in the communication field due to the particular nature of the research that focused on social networks. The area of computer communications and information systems comes in second, with 1,172 publications (9.81 percent). Baskaran (2021) investigated the greater part 134 (1.96%) of the distributions was contributed by the analysts from the College of California frameworks. Zhang Y was the most prolific author, contributing 16 (0.23 percent) of the publications in the Web 2.0 field.

METHODOLOGY

The present study aims at the distribution of research output on the basis of research publications contributed by Scientists in the field of Google Scholar. The required data was collected from Web of Science (WoS) database for the period 2000-2022. A total of 57388 records were retrieved from the database using the search strings "Google Scholar", analysed by using MS-Excel software package as per the objectives of the study and data has been presented in tabular as well as in graphical form. This study explores the Relative Growth Rate (RGR) of publications and citation patterns, author productivity and collaboration, most productive journals, Top ranked authors, Institutions and Network Visualization portrait presented through VoS viewer.

RESULTS AND DISCUSSION

Year wise Distribution of Publications

The distribution of year-wise publications was measured growth of publications in the field of Google Scholar. A total of 25760 publications were found on Google Scholar during 2002-2022. Table 1 shows the maximum number 5017 (19.47%) of publications contributed by the researchers in the year 2021. It followed 49 (19.03%), 3838 (14.88%), 2707 (10.48%) of the publications found in the year 2022, 2020 and 2019 respectively. The overall growth trend on the publication in the field of Google scholar has been slowly increased during 2002-2022, excluded the year 2022. The study was analyzed less than 500 publications are contributed by the researchers during 2000-2012 (Fig.1).

Relative Growth Rate (RGR) and Doubling Time (Dt) in Google Scholar Research

The Relative Growth Rate (RGR) is the increase in number of articles or pages per unit of time. The mean RGR over the specific period of interval can be calculated from the following equation.

Mean RGR = 1-2R = logeW2 – logeW1/T2-T1

Table 1. Year wise distribution of the publications on Google Scholar

YEAR	NO. OF PAPERS	%
2002	1	0.004
2004	1	0.004
2005	19	0.074
2006	29	0.113
2007	66	0.256
2008	109	0.423
2009	160	0.621
2010	235	0.912
2011	300	1.165
2012	456	1.77
2013	602	2.337
2014	928	3.602
2015	1180	4.581
2016	1422	5.52
2017	1765	6.852
2018	1967	7.636
2019	2701	10.485
2020	3835	14.887
2021	5017	19.476
2022	4903	19.033
	25760	

Figure 1. Year wise distribution of the publications on Google Scholar

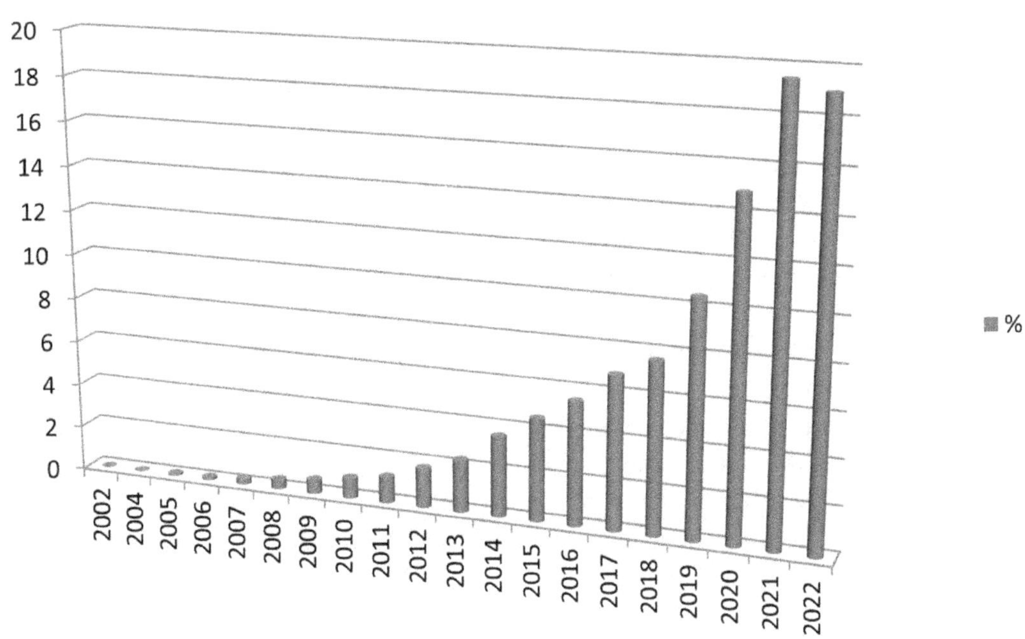

Table 2. Relative growth rate and doubling time of the publications

YEAR	PUBLICATIONS	W1	W2	RGR (W2-W1)	DT
2002	1	0	6.470	0	0
2004	1	6.470	6.459	-0.011	0.0158
2005	19	6.459	6.378	-0.081	0.116
2006	29	6.378	6.532	0.154	0.221
2007	66	6.532	6.496	-0.036	0.322
2008	109	6.496	6.804	0.308	0.443
2009	160	6.804	6.961	0.157	0.226
2010	235	6.961	8.759	1.798	2.590
2011	300	8.759	7.232	-1.527	-2.200
2012	456	7.232	7.416	0.184	0.265
2013	602	7.416	7.488	0.072	0.103
2014	928	7.488	7.717	0.229	0.329
2015	1180	7.717	7.765	0.048	0.069
2016	1422	7.765	7.870	0.105	0.151
2017	1765	7.870	7.852	-0.018	0.025
2018	1967	7.852	8.015	0.163	0.234
2019	2701	8.015	8.062	0.047	0.067
2020	3835	8.062	8.153	0.091	0.131
2021	5017	8.153	8.220	0.067	0.096
2022	4903	8.220	8.266	0.046	0.066
	25760	8.226	8.374	0.148	0.213

where, 1-2R is mean relative development rate over the particular time of stretch; The log of the first number of articles is W1; logeW2 is log of conclusive number of articles after a particular time of span and; T2-T1 is the unit contrast between the underlying time and the last time. Here, the year can be used as a measurement of time.

As a result, the mean relative growth rate per unit of articles per unit of year over a particular interval can be represented by 1-2R (aa-1 year-1).

According to the calculation, Table 2 can be used to analyze the RGR as an increasing and decreasing trend as seen that fluctuate more has been performed throughout the study period. Doubling time (DT) = 0.693/R Doubling time for articles (DT) = 0.693/1-2R (aa-1 year-1) As uncover that greatest worth of RGR was seen as 0.308 and most minimal RGR was 0.011 in the year 2003 and 2000 separately. Then again, the multiplying time pattern is likewise show up as vacillate over the review period. Further, it has investigated that a normal RGR and Dt viewed as 0.09 and 0.158 separately (Fig.2).

Table 3 analyzed the Prolific Authors are found from the Web of Science contributing by the research in the field of Google Scholar. There were listed top 25 authors based on their research contribution in Google Scholar research. The prolific author series, Maffulli N was the top author has witnessed his contribution 113 (0.43%) of the publications on the core area of Google Scholar. It followed by Szbebkar as the second ranked authors with 104 (0.40%) of the publications. Mohaammali M 72(0.28%), Wang

Figure 2. Relative growth rate (RGR) and Doubling time (Dt)

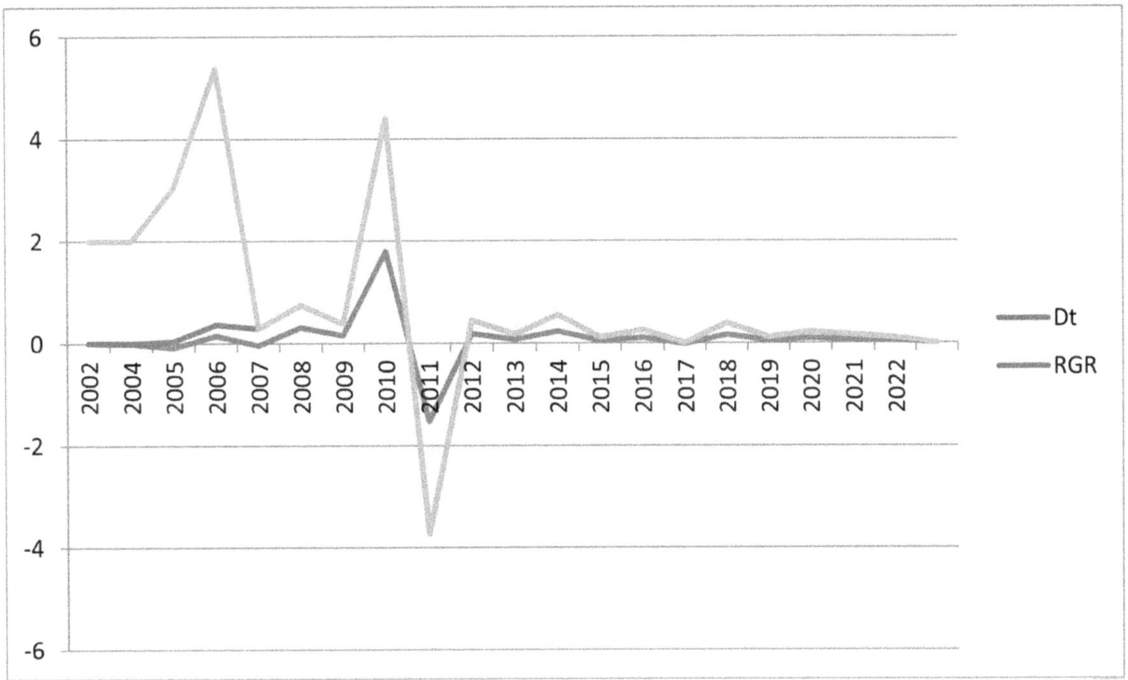

Y 72 (0.28%), Zhang Y 71 (0.27%), Kumar A 65 (0.25%), Li J 64 (0.24%), Denaro V 62 (0.24%), Pergialiotis 62 (0.24%) and Lio Y 48 (0.18%) of the publications was ranked as Third, Fourth, Fifth, Sixth, Seventh, Eighth, Ninth and Tenth respectively. The study can be noticed that the publications activity between 58 and 41 was being ranked as 11 and 25.

Sources Wise Distribution of the Publications in the Field of Google Scholar

Table 4 analyzed out of 17 sources are listed on the contribution made by the authors in the field of Google Scholar. It is analyzed that Review article was the hold majority 19132 (74.24%) of the publications are witnessed in the field of Google Scholar. Article with 6448 (25.03%), Early Access 824 (3.19%), Editorial Materials 136 (0.52%), Proceedings papers 126 (0.48%) of the publications. The study found that the rank of sources between 7 and 17, the sources have less than publications was hundred.

Subject Wise Distribution of the Publications in the Field of Google Scholar

Table 5 analyzed the Subject wise distribution of the publications in the field of Google scholar. The result find that Pharmacology Pharmacy was found that top ranked subject based on the publications with 2590 (10.05%) of the publications. The study found ranked of the subjects in the series Medicine General Internal 2574 (9.99%), Public Environmental Occupational Health 2030(7.80%), Surgery 1697 (6.58%), Clinical Neurology 1134 (4.40%), Oncology 934 (3.62%), Integrative Complementary Medicine 892 (3.46%), Chemistry Medicinal 881 (3.42%), Psychiatry 859 (3.30) and Health Care Sciences

Table 3. Prolific authors in the field of Google Scholar

RANK	AUTHORS	NO. OF PAPERS	%
1	MAFFULLI N	113	0.439
2	SAHEBKAR A	104	0.404
3	MOHAMMADI M	72	0.28
4	WANG Y	72	0.28
5	ZHANG Y	71	0.276
6	KUMAR A	65	0.252
7	LI J	64	0.248
8	DENARO V	62	0.241
9	PERGIALIOTIS V	62	0.241
10	LIU Y	60	0.233
11	BELLOS I	58	0.225
12	WANG J	58	0.225
13	ESMAILLZADEH A	54	0.21
14	CHEN Y	51	0.198
15	ZHANG L	50	0.194
16	LI Y	48	0.186
17	ABDOLLAHI M	47	0.182
18	KUMAR S	47	0.182
19	MIGLIORINI F	47	0.182
20	WANG L	45	0.175
21	BRAMER WM	44	0.171
22	HADI A	42	0.163
23	SALEHI-ABARGOUEI A	42	0.163
24	ZHANG J	41	0.159
25	LARIJANI B	40	0.155

Services 848 (3.29%) of the publications. The study witnessed that rest of the subjects ranked as 11 and 25 with range of publications 793 and 554.

Institutions Wise Distribution of the Publications in the Field of Google Scholar

Table 6 has witnessed that Institutions wise distribution of the publications in the field of Google scholar during 2002-2022. The study was analyzed top 25 institutions are listed in the study, among 10 leading institutions provided ranked publications in the field of study. The study can noticed Tehran University of Medical Sciences was the top ranked institutions with 706 (2.74%), University of London 688(2.67%), Shahid beheshti 465 (1.80%) University Medical Ssciences 449 (1.74%), Harvard University 406 (1.57%), University of California system 389 (1.51%), Isfahan University Medical Science 340(1.32%), University

Table 4. Sources wise distribution of the publications in the field of Google Scholar

RANK	DOCUMENT TYPE	NO. OF PAPERS	%
1	REVIEW ARTICLE	19132	74.27
2	ARTICLE	6448	25.031
3	EARLY ACCESS	824	3.199
4	EDITORIAL MATERIAL	136	0.528
5	PROCEEDING PAPER	126	0.489
6	LETTER	22	0.085
7	BOOK CHAPTERS	12	0.047
8	RETRACTED PUBLICATION	12	0.047
9	BOOK REVIEW	5	0.019
10	MEETING ABSTRACT	5	0.019
11	CORRECTION	4	0.016
12	DATA PAPER	4	0.016
13	NEWS ITEM	4	0.016
14	BIOGRAPHICAL-ITEM	1	0.004
15	DATABASE REVIEW	1	0.004
16	HARDWARE REVIEW	1	0.004
17	SOFTWARE REVIEW	1	0.004

of Toronto 325 (1.3%) and University Rotterdam 325 (1.26%) of the publications have contributed by the scientists in the field of Google Scholar.

Table 7 analyzed the publications was indexed by the databases in the category of Web of Science in the field of Google scholar. The study is found that Science Citation Index Expanded (SCI-EXPANDED) was the top indexed with 23954 (92.98%) of the publication was top ranked database Indexing system in the study. It followed Social Sciences Citation Index (SSCI) with 7389 (26.68%), Conference Proceedings Citation Index – Science (CPCI-S) 112 (0.43%), Arts & Humanities Citation Index (A&HCI) 93 (0.36%), Conference Proceedings Citation Index – Social Science & Humanities (CPCI-SSH 42 (0.163%), Book Citation Index – Science (BKCI-S)11 (0.004%) and Book Citation Index – Social Sciences & Humanities (BKCI-SSH) with 1 (0.004%) of the publications was distributed in the above mentioned databases were Indexed the publications in the field of Google Scholar.

CONCLUSION

The study has analysed based on the bibliographical records of web of Science in the field of Google researcher during 2000-2022. Over time, the publication trend increased gradually. Between 2002 and 2022, excluding the year 2022, the overall publication growth trend on Google Scholar has increased slowly. The researchers contributed less than 500 publications to the study between 2000 and 2012. As uncover that greatest worth of RGR was seen as 0.308 and most minimal RGR was 0.011 in the year 2003 and 2000 separately. Then again, the multiplying time pattern is likewise show up as vacillate over

Table 5. Subject wise distribution of the publications in the field of Google Scholar

RANK	SUBJECT CATEGORIES	NO. OF PAPERS	%
1	PHARMACOLOGY PHARMACY	2590	10.054
2	MEDICINE GENERAL INTERNAL	2574	9.992
3	PUBLIC ENVIRONMENTAL OCCUPATIONAL HEALTH	2030	7.88
4	SURGERY	1697	6.588
5	CLINICAL NEUROLOGY	1134	4.402
6	ONCOLOGY	934	3.626
7	INTEGRATIVE COMPLEMENTARY MEDICINE	892	3.463
8	CHEMISTRY MEDICINAL	881	3.42
9	PSYCHIATRY	859	3.335
10	HEALTH CARE SCIENCES SERVICES	848	3.292
11	ENVIRONMENTAL SCIENCES	793	3.078
12	MEDICINE RESEARCH EXPERIMENTAL	788	3.059
13	MULTIDISCIPLINARY SCIENCES	763	2.962
14	PLANT SCIENCES	762	2.958
15	OBSTETRICS GYNAECOLOGY	733	2.845
16	NURSING	729	2.83
17	INFORMATION SCIENCE LIBRARY SCIENCE	717	2.783
18	NUTRITION DIETETICS	712	2.764
19	NEUROSCIENCES	668	2.593
20	ORTHOPAEDICS	649	2.519
21	BIOCHEMISTRY MOLECULAR BIOLOGY	630	2.446
22	PAEDIATRICS	614	2.384
23	DENTISTRY ORAL SURGERY MEDICINE	604	2.345
24	INFECTIOUS DISEASES	555	2.155
25	CARDIAC CARDIOVASCULAR SYSTEMS	554	2.151

the review period. The average RGR and DT are 0.09 and 0.158, respectively, as shown by the result. Google Scholar research ranked the top 25 authors according to their contribution to the field. Maffulli N was the most prolific author in the series, contributing 113 (0.43%) of the core area of Google Scholar publications. According to the findings of the study, sources with fewer than 100 publications had ranks between 7 and 17. Based on the number of publications, Pharmacology Pharmacy ranked first, with 2590 (or 0.05%) of the publications.

Table 6. Institutions wise distribution of the publications in the field of Google Scholar

RANK	NAME OF INSTITUTION	NO. OF PAPERS	%
1	TEHRAN UNIVERSITY OF MEDICAL SCIENCES	706	2.741
2	UNIVERSITY OF LONDON	688	2.671
3	SHAHID BEHESHTI UNIVERSITY MEDICAL SCIENCES	465	1.805
4	HARVARD UNIVERSITY	449	1.743
5	UNIVERSITY OF CALIFORNIA SYSTEM	406	1.576
6	ISFAHAN UNIVERSITY MEDICAL SCIENCE	389	1.51
7	UNIVERSITY OF TORONTO	386	1.498
8	ERASMUS UNIVERSITY ROTTERDAM	340	1.32
9	UDICE FRENCH RESEARCH UNIVERSITIES	335	1.3
10	ERASMUS MC	325	1.262
11	MASHHAD UNIVERSITY MEDICAL SCIENCE	324	1.258
12	IRAN UNIVERSITY OF MEDICAL SCIENCES	320	1.242
13	UNIVERSITY OF SYDNEY	305	1.184
14	TABRIZ UNIVERSITY OF MEDICAL SCIENCE	271	1.052
15	UNIVERSITY COLLEGE LONDON	262	1.017
16	EGYPTIAN KNOWLEDGE BANK EKB	249	0.967
17	UNIVERSITY TORONTO AFFILIATES	249	0.967
18	HARVARD MEDICAL SCHOOL	248	0.963
19	UNIVERSITY OF TEXAS SYSTEM	244	0.947
20	UNIVERSITY OF KWAZULU NATAL	243	0.943
21	IMPERIAL COLLEGE LONDON	242	0.939
22	JOHNS HOPKINS UNIVERSITY	241	0.936
23	MONASH UNIVERSITY	237	0.92
24	STATE UNIVERSITY SYSTEM OF FLORIDA	237	0.92
25	NATIONAL KAPODISTRIAN UNIVERSITY OF ATHENS	224	0.87

Table 7. Indexed databases wise distribution of the publications in the field of Google Scholar

S.No	NAME OF THE INDEXED	NO. OF PAPERS	%
1	Science Citation Index Expanded (SCI-EXPANDED)	23954	92.989
2	Social Sciences Citation Index (SSCI)	7389	28.684
3	Conference Proceedings Citation Index – Science (CPCI-S)	112	0.435
4	Arts & Humanities Citation Index (A&HCI)	93	0.361
5	Conference Proceedings Citation Index – Social Science & Humanities (CPCI-SSH)	42	0.163
6	Book Citation Index – Science (BKCI-S)	11	0.043
7	Book Citation Index – Social Sciences & Humanities (BKCI-SSH)	1	0.004

REFERENCES

Barbosa, S. D. J., Silveira, M. S., & Gasparini, I. (2016). What publications metadata tell us about the evolution of a scientific community: The case of the Brazilian Human–Computer Interaction conference series. *Scientometrics, 110*(1), 275–300. doi:10.100711192-016-2162-4

Baskaran, C. (2019). Scholarly information share through Social Networks (SNs) and medias among social science scholars in selected State Universities in Tamilnadu. *International Journal of Library and Information Studies*, *9*(3), 83–92.

Baskaran, C. (2020). Research Patterns on the Social Networks and Media: A Scientometric Portrait. Handbook of Research on Emerging Trends and Technologies in Library and Information Science, pp.189-207. IGI Global. doi:10.4018/978-1-5225-9825-1.ch014

Baskaran, C. (2021). An Exponential trend and Mapping of Research in Web 2.0. *Library Philosophy and Practice (e-journal),* 5008. https://digitalcommons.unl.edu/libphilprac/5008

Gupta, A. (2015). Five years of India HCI: A scientometric analysis. In *Proceedings of the 7th international conference on HCI (India HCI),* (pp. 56–61). ACM.

Jacovi, M., Soroka, V., Gilboa-Freedman, G., Ur, S., Shahar, E., & Marmasse, N. (2006). The chasms of CSCW: A citation graph analysis of the CSCW conference. *In Proceedings of the 20th anniversary ACM conference on computer supported cooperative work,* (pp. 289–298). ACM.

Lee, H. E., Park, J. H., & Song, Y. (2014). *Research collaboration networks of prolific institutions in the HCI field in Korea: An analysis of the HCI Korea conference proceedings.* In Proceedings of HCI Korea, (pp. 434–441). ACM.

Mingers, J., & Leydesdorff, L. (2015). A review of theory and practice in scientometrics. *European Journal of Operational Research, 246*(1), 1–19. doi:10.1016/j.ejor.2015.04.002

Mubin, O., Al Mahmud, A., & Ahmad, M. (2017). HCI down under: Reflecting on a decade of the OzCHI conference. *Scientometrics, 112*(1), 367–382. doi:10.100711192-017-2293-2

Nalimov, V. V., & Mulchenko, B. M. (1969). *Scientometrics. Studies of science as a process of information.* Science.

Nichols, D. M., & Cunningham, S. J. (2015). A scientometric analysis of 15 years of CHINZ conferences. *Proceedings of the 15th New Zealand conference on human–computer interaction*, pp. 73–80. 10.1145/2808047.2808060

Padilla, S., Methven, T. S., & Chantler, M. J. (2014). Is British HCI important? A topic-based comparison with CHI. In *Proceedings of the 28th international BCS human computer interaction conference HCI*, pp. 365–370. 10.14236/ewic/HCI2014.2

Price, D. S. (1963). *Little science, big science.* Columbia University Press. doi:10.7312/pric91844

Van Raan, A. (1997). Scientometrics: State-of-the-art. *Scientometrics, 38*(1), 205–218. doi:10.1007/BF02461131

Chapter 8
A Study on Dissemination of Information for Teaching and Research Competence Among Faculty Members in the Deemed Universities in the Coimbatore Region

P. Ramesh
Sri Ramakrishna Polytechnic College, Coimbatore, India

ABSTRACT

The chapter examines the dissemination of information for teaching and research competence among the faculty members in the Deemed Universities in Coimbatore region by a questionnaire-based survey. The result finds out of 364 respondents, a majority of 26.92% reported from Karunya University. The Boolean searches for access to e-resources by the respondents in the Deemed Universities. The majority 147 (48.8%) of respondents preferred Boolean search technique to very large extent while 74 (20.32%) respondents preferred the large extent. The majority with 132 (36.26%) of the respondents have noticed preferred e-books among the deemed Universities in the Coimbatore region as Excellent towards preferred e-book.

INTRODUCTION

Electronic format materials are bring a new delivery concept of LIS professionals. It is an electronic delivery for information requisites. Electronic resources removed several divides such as (a) reduce the space, (b) save the timing (both professional and user), (c) Geographical boundaries and (d) equity (Rural and urban) information access, multiple usage etc. E-sources are redefining access to information and its usage habits; now-a-days E- resources comes to several formats, they can be easily downloaded. Accesses

DOI: 10.4018/978-1-6684-8805-8.ch008

to these Electronic resources are increasing day by day, so user's confidence is also growing. E-resource makes to greatest revolution in knowledge communication. The emerging information and communication technologies (ICT), especially the interactive digital devices will drive the information future. It is a strong force, but it is not an independent agent. It helps in providing efficient and effective services to the users. Now, it is a transition period for the Libraries, since they are going from traditional printed materials collections to the Libraries with Electronic resources. With the research and advancement of new technology (whether it may be Computers, Networking, Communication or Internet, shortly ICT) and by adopting the same, the publishing industry and the Libraries have entered a fascinating period.

The collections are in the electronic form and the information is retrieved over networks. All types of traditional resources, too much extent, are giving way for resources in electronic form. The Libraries are striving towards giving large-scale, better services to their patrons. With the effort of the Government, UGC and other higher academic bodies, it is now possible for the academic institutions to have access for Electronic resources, especially E-Journals, for the development of their academic fraternity, which would help them in expanding their knowledge. Electronic resources are nothing but the traditional resources materials which are in the electronic form. These resources are stored on magnetic or optical media such as CDs or DVDs. Significant proportion of these Electronic resources are available in machine-readable format (as opposed to that of print), accessible by means of computers. The digital content may be locally held or accessed remotely via computer networks or also through the web. The Libraries which are maintaining only these Electronic resources are called "digital Libraries". However, the Libraries which are maintaining both Traditional and Digital Resources are called as "Hybrid Libraries".

REVIEW OF LITERATURE

Ajuwon, (2003) conduct a study of uptake of ICTs by health science students at the University College Hospital, Ibadan. He found that 57% of students sampled could not use a computer and that the use of the database was poor due to lack of awareness and lack of ICT skills. Ali (2005) study taken on online searches of scientific information in science and technology libraries in Delhi reveals that a sizeable number of users (almost 60%) are facing numerous problems while browsing electronic information, such as lack of knowledge about the resources, lack of trained staff and inadequate terminals. Dastforoush and Venkatesha (2011) conduct a study to determine dependency on electronic and print journals by research scholars and faculty in Iran. Major find in of the study are: (i) 43.9% of the respondents did not find e-journals to be equivalent to print journals (ii) 57.9% of the respondents preferred electronic versions over printed ones. Tahir, Mahmood, and Shafique (2010) have surveyed the use of electronic information resources and facilities by humanities scholars at the University of the Punjab, Lahore, Pakistan. Major findings of the study are: (i) 69% of respondents preferred both print and electronic resources. Dilek-Kayaoglu (2013) examined the use of electronic journals by faculty at Istanbul University, Turkey. Major findings of the study are: (i) 89% of the respondents stated that one of the benefits of e-journals was that there was no need to visit the library. Baskaran (2019) analyses the majority (83.5%) of respondents using "Whaatsapp" out of respondents. The respondents reported (77.7%) and (73.6%) using devices tools "Laptop" and Smartphone/Mobiles are respectively. Tlakula (2017) in the same University is also seen in in the literature.

A recent Indian study by Narendara Kumar and Lalita (2017) with the University of Delhi is noteworthy. A collaborative study by Bhat and Ganaie (2016) described the use of e-resources by users of

Dr. Y.S. Parmar University of Horticulture and Forestry, Solan, Himachal Pradesh. A similar joint study by Adeloke and Emeahar (2016) of the University of Ibadan, Nigeria described the use of e-resources by students. A study of e-journal resource done by Katabalwa (2016) at the University of Dar es Salaam was seen in the literature. In their study, Kumar and Reddy (2016) found that round the clock (24*7) availability of e-resources and ease to use them encourage the research scholars and faculty members to use the e-resources especially electronic journals for their research work.

Baskaran (2019) analyzed that use of social network and Medias by the Research scholars in Selected Universities of Tamilnadu. To highest 87 (31.9%) of respondents Annamalai University, (16.8%) and (15%) of them Male and Female respectively out of 273 total respondents responded to the study. Baskaran (2020) analyzed that majority of the publications, 2,207 (18.48%), were published from communication area, on account of the specific nature of the research concentrating on Social networks. Computer communications information systems area scores second with 1,172 (9.81%) of the publications. Baskaran (2021) explored the majority 134 (1.96%) of the publications was contributed by the researchers from the University of California systems. Zhang Y was the top author was contributed 16(0.23%) of the publications in the field of Web 2.0.

METHODOLOGY

The studies apply to undertake the survey the sample for collecting under questionnaire based method. The Self-designed structured questionnaires were circulated among the faculty members in the Selected Deemed Universities in Coimbatore to Collect the data for the study. To collect the data on awareness and usage of e-resources and services, type of e-resources and services used, issues and challenges in use of e-resources and services and user satisfaction on e-resources and services provided by the library. Total no. of 436 questionnaires was distributed to the respondents in the deemed universities in the Coimbatore region and 364 filled up questionnaire received for the analysis. The Data was analyzed by using a simple method of calculation for using simple percentage technique in the study.

ANALYSIS AND INTERPRETATION

Table 1 analyses the institutions wise respondents in the Coimbatore region for the study. The results of the study has been witnessed among four deemed Universities in the Coimbatore Region out of 364 respondents majority 26.92% of the respondents reported from Karunya University, it followed by

Table 1. Institutions-wise respondents in the Coimbatore region

S.No	Name of the University	No. Of Respondents	%
1.	Karunya University	98	26.92
2	Amritha University	89	24.45
3	Karpagam Academy of Higher Education	85	23.35
4	Avinashilingam University	92	25.27
		364	

Avinashilingam University (25.27%), Amritha University (24.45%), Karpagam Academy of Higher Education 23.35%) of the respondents reported to the study (Figure 1).

Table 2 witnessed subject wise distribution of the respondents found on the analysis made to the study. There are four subjects listed out those belong to the area of discipline of the respondents, among them majority 57.99% of the respondents from Engineering Technology. The study has been found the respondents of remaining areas with Management, Architecture and Arts/Science/other was reported by 20.87%, 16.20% and 05.68% of the respondents respectively.

Searching Method of Access E- Resources

(a) Boolean

Table 1, Boolean search was select for accessing e-resources by the respondents in the Deemed Universities. The majority 147 (48.8%) respondents preferred Boolean search technique to Very Large Eextent while 74 (20.32%) respondents to Large extent. However, 38 (10.45%) respondents with some extent and followed by 56 (15.38%) of them reported No comment.

(b) Truncation

The maximum number of 117 (32.14%) respondents preferred this searching technique to access E-resources was stated as Very Large Extent. 98 (26.92%) respondents say large extent. 67 (18.40%) of the

Figure 1. Institutions-wise respondents in the Coimbatore region

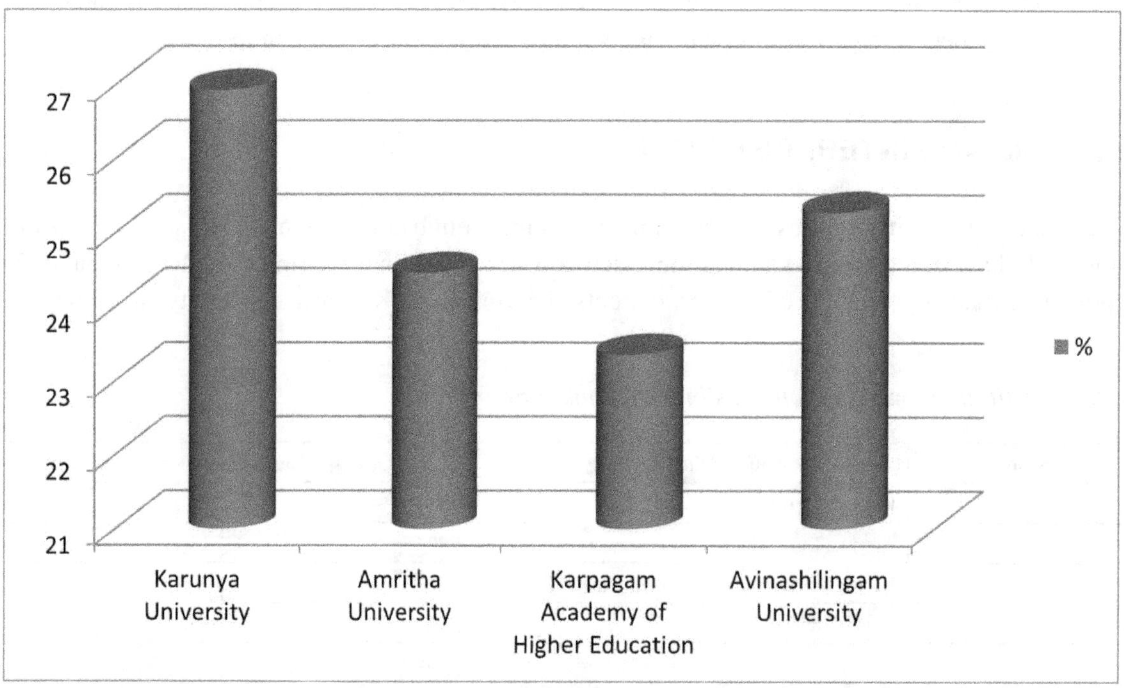

Table 2. Subject wise distribution of the respondents

S.No	Name of the University	No. Of Respondents	%
1.	Engineering and Technology	214	57.99
2	Management	76	20.87
3	Architecture	59	16.20
4	Arts/Science/others	15	05.68
		364	

Table 3. Searching method of access E-resources

Sl.No	Methods	Very Large Extent	Large Extent	Some Extent	Less Extent	No Comments	Total
1.	Boolean	147(40.38)	74(20.32)	38(10.43)	49(13.46)	56(15.38)	364
2.	Truncation	117(32.14)	98(26.92)	67(18.40)	54(14.83)	28(07.69)	364
3.	Content/Text	134(36.81)	81(22.25)	49(13.46)	38(10.43)	58(15.93)	364
4.	Advanced search	128(35.16)	76(20.87)	59(16.20)	45(12.36)	56(15.38)	364

respondents preferred as Some Extent. 54 (14.83%) respondents have mentioned Less extent followed by 28 (07.69%) respondents said No comment.

(c) Content/Text

The Majority of 134 (36.81%) respondents have preferred Content/Text as Very Large Extent, 81 (22.25%) respondents to Large extent. 67(18.40%) of them have noticed as Some Extent, 54 (14.83%) and 28(15.93%) of them have been noticed as Less Extent and No Comment respectively.

(d) Advanced Search

The majority of 128 (35.16%) of the respondents have preferred the Advance search as Very Large Extent. It followed by 76 (20.89%), 59 (16.20%), 45 (12.36%) and 56 (15.28%) of them preferred the e-resources by the respondents from the Deemed Universities in Coimbatore region as Large Extent, Some Extent, Less Extent and No Comment receptively.

Use on the Effectiveness of E-Resources

(a) E-Books

Table 4 analyses majority of 132 (36.26%) of the respondents have noticed preferred e-books among the deemed Universities in the Coimbatore region as Excellent. It followed by 85 (25.75%), 82 (22.52%), 76 (20.81%) and 25 (06.86%) of the respondents have reported as Very Good, Poor and No Comment respectively.

Table 4. Use on the effectiveness of E-resources

Sl.No	Resources	Excellent	Very Good	Good	Poor	No Comment	Total
1.	E-Books	132(36.26)	85(23.35)	76(20.87)	82(22.52)	25(06.86)	364
2.	E-Journals	142(43.84)	71(19.50)	64(17.58)	56(15.38)	54(14.83)	364
3.	E- Databases	128(35.16)	63(17.30)	57(15.65)	49(13.46)	67(18.40)	364
4.	E-Newsletter	119(32.69)	57(15.65)	43(11.81)	39(10.71)	77(21.15)	364
5.	Others	127(34.89)	41(11.26)	39(10.71)	57(15.65)	35(09.61)	364

(b) E- Journals

The large number of 142 (43.84%) respondents rated that the information sought from E-Journals is Excellent, it followed by 71 (19.50%) of them rated as very good and 56 (15.28%) respondents have stated rated as Good.

(c) E- Database

The majority of 128 (35.16%) respondents have suggested that the information find from E-Database is Excellent, 71 (19.50%) rated as Very Good. It has also been noticed that 49(13.46%) respondents rated as good, 67(18.40%) of them reported as No Comment.

(d) E- Newsletter

119 (32.69%) respondents have preferred the E-Newsletter Information among the respondents from Deemed Universities in the Coimbatore Region. It followed by 79 (21.15%) of them dislike on used as No Comment. 57 (15.65%), 43 (11.81%), 39 (10.91%) of them preferred that Very Good, Good and Poor respectively.

(f) Other E-Resources

The majority of 127 (34.89%) respondents have preferred other type of e-resource as Excellent among the respondents in the Deemed Universities from Coimbatore Region. It followed by 57 (15.65%), 41 (11.26%), 35(09.68%) and 29 (10.71%) of them preferred as Very Good, No comment, Good and Poor respectively.

Satisfaction of Source of Information on e-resources

(a) Bibliographical Information

Table 5 shows that majority of 168 (46.15%) of the respondents have opted bibliographical information was very useful, 89 (24.45%) of them opted very useful, 66(18.13%) of them noticed as No Comment and 41 (11.26%) of them stated as Not useful about satisfaction access through e-resources among the respondents.

Table 5. Satisfaction of source of information on e-resources

Sl.No	Source of Information	Very useful	Useful	Not useful	No comment	Total
1.	Bibliographical information	168(46.15)	89(24.45)	41(11.26)	66(18.132)	364
2.	Academic Information	162(44.50)	77(21.15)	64(17.58)	61(16.75)	364
3.	Research Information	159(43.68)	81(21.95)	57(15.65)	67(18.40)	364
4.	Latest Information	149(13.46)	92(25.27)	69(18.95)	54(14.83)	364
5.	Social Information	132(36.26)	87(23.90)	74(20.32)	71(19.50)	364

(b) Academic Information

The data analysis indicates that a maximum number of 162 (44.50%) respondents rated as very useful, It followed by 77 (21.15%) of the respondents rated as useful, 64 (17.28%) respondents have rated as not useful about Academic Information.

(C) Research Information

The Majority 159 (43.68%) of them have noticed as Very Useful about Research Information access by e-resources, It followed by 77 (21.15%) of them stated useful, 64 (17.18%) of them with Not Useful and 61 (18.40%) of them have opined on No Comment.

(d) Latest Information

The study analyses majority 149 (13.46%) of the respondents have suggested Latest information access by e-resources as Very Useful. It followed by, 92 (25.27%) of them have suggested as Useful, 69 (18.90%) of them have noticed as Not Useful and 54 (14.83%) of them have witnessed as No Comment.

(e) Social Information

The majority 132 (36.26%) of the respondents have reported about usage of e-resource for purpose of access Social Information as Very Useful. It followed by 87 (23.90%) of them noticed as useful, 74 (20.32%) of them have reported as Not useful and 71(19.50%) of the have stated as No Comment.

Uniqueness of Access the E-Resources among the respondents

(a) Extending of Knowledge

Table 6 analyses a majority of 157 (41.48%) of the respondents have stated as Strongly Agree use of e-resource towards Extending of Knowledge. It followed by 84 (23.01%) of the respondents have stated as Agree, 68 (17.85%) of them with Disagree, 38 (40.43%) of them have sated as Strongly Disagree and 26 (07.14%) of them have noticed as No Comment.

(b) Find Relevant Information

Table 6. Uniqueness of access the e-resources among the respondents

Sl.No	Uniqueness s	Strongly Agree	Agree	Disagree	Strongly disagree	No comment	Total
1.	Extending of Knowledge	151(41.48)	84(23.07)	65(17.85)	38(10.43)	26(07.14)	364
2.	Find relevant information	139(38.18)	67(18.40)	49(13.46)	31(08.51)	78(21.42)	364
3.	Teaching and Research Information	132(36.26)	54(14.83)	41(11.26)	39(10.71)	98(26.92)	364
4.	Research and Development Information	138(37.91)	49(13.46)	38(10.43)	51(14.01)	88(24.17)	364

The Majority 139 (38.18%) of the respondents have noticed as Strongly Agree, 67 (18.4) of them with Agree, 49 (13.46%) of them with Disagree, 31 (8.51%) of them with Strongly Disagree and 78 (21.42%) of them have opined as No Comment of use of e-resources for find relevant Information among respondents.

(c) Teaching and Research Information

The maximum number of 132 (36.26%) of the respondents Strongly agree that they obtain teaching and research information from e-resources while a fairly proportionate number of 54 (14.83%) respondents agree and 98 (26.92%) of them have stated as No Comment on use of e-resources for Teaching and Research Information among the respondents.

(d) Research and Development Information

The data notices majority 138 (31.91%) of the respondents have reported on use of e-resource towards Research and Development information as Strongly agree. It followed by 88(24.17%) of them have stated as No Comment, 51 (14.01%) of them have witnessed as Strongly disagree on uniqueness to access e-resource among respondents.

CONCLUSION

The study was analyzes the importance of the activities of the Library Consortium recognizes that the number of participating libraries in India is increasing day by day. The efforts of the UGC- INFLIBNET will be considerable and will definitely strengthen India's higher education system. Free and / or heavily subsidized access to scientific resources helps educational institutions actually fulfil their mission. In the long run, the consortium approach is much more popular in the user community, and it wasn't long before the consortium approach expanded its national information base. We at suggest that in order to effectively use the subscribed resources, professionals need to implement more awareness programs in order to sell new online services at the facility.

The result of the study explores there are four subjects listed out those belong to the area of discipline of the respondents, among them majority 57.99% of the respondents from Engineering Technology. The maximum number of 117 (32.14%) respondents preferred this searching technique to access E-resources was stated as Very Large Extent. 98 (26.92%) respondents said as Large extent on truncation. The majority of 128 (35.16%) of the respondents have preferred the Advance search as Very Large Extent on preferred by the respondents with Advanced Search. The large number of 142 (43.84%) respondents rated that the information sought from E-Journals is Excellent. The majority of 168 (46.15%) of the respondents have opted bibliographical information was very useful, 89 (24.45%) of them opted very useful, 66(18.13%) of them noticed as No Comment. The study analyses majority 149 (13.46%) of the respondents have suggested Latest information access by e-resources as Very Useful. majority of 157 (41.48%) of the respondents have stated as Strongly Agree use of e-resource towards Extending of Knowledge. It followed by 84 (23.01%) of the respondents have stated as Agree. The majority 138 (31.91%) of the respondents have reported on use of e-resource towards Research and Development information as Strongly agree.

REFERENCES

Adeleke, D. S., & Emeahara, E. N. (2016). Relationship between information literacy and use of electronic information resources by postgraduate students of the University of Ibadan. *Library Philosophy & Practice*. http://eds.b.ebscohost.com/eds/detail/detail?vid=0&sid=9a7caa11-b75

Akussah, M., Asante, E., & Adu-Sarkodee, R. (2015). Impact of Electronic Resources and Usage in Academic Libraries in Ghana: Evidence from Koforidua Polytechnic & All Nations University College, Ghana. *Journal of Education and Practice*, *6*(33), 33–38. https://eric.ed.gov/?id=EJ1083497

Amankwah, P. B. (2014). *Use of Electronic Resources by undergraduate students of the Ghana Institute of Management and Public Administration (GIMPA)*. [Doctoral dissertation, University of Ghana, Legon]. http://197.255.68.203/handle/123456789/6989

Ameyaw, S., Kwakye, J. O., & Asante, E. (2016). Electronic Resources Usage in PrivateUniversities: Evidence from Valley View University, Ghana. *ADRR Journal of Arts and Social Science, 14*(4), 1-15. https://journals.adrri.org/index.php/adrrijass/article/view/266

Baskaran, C. (2019). Scholarly information share through Social Networks (SNs) and medias among social science scholars in selected State Universities in Tamilnadu. *International Journal of Library and Information Studies*, *9*(3), 83–92.

Baskaran, C. (2020). Research Patterns on the Social Networks and Media: A Scientometric Portrait. Handbook of Research on Emerging Trends and Technologies in Library and Information Science. ACM. http://doi.org/ doi:10.4018/978-1-5225-9825-1.ch014

Baskaran, C. (2021). An Exponential trend and Mapping of Research in Web 2.0. *Library Philosophy and Practice (e-journal)*, 5008. https://digitalcommons.unl.edu/libphilprac/5008

Bhat, N. A., & Ganaie, S. A. (2016). E-resources: Use and search strategies adopted by users of Dr YS Parmar University of Horticulture and Forestry. *Collection Building*, *35*(1), 16–21. doi:10.1108/CB-08-2015-0015

Dastforoush, M. T., & Venkatesha, Y. (2011). Dependency on electronic and print journals: A case study. *SRELS Journal of Information Management*, *48*(4), 441–448.

Dilek-Kayaoglu, H. (2008). Use of electronic journals by faculty at Istanbul University, Turkey: The results of a survey. *Journal of Academic Librarianship*, *34*(3), 239–247. doi:10.1016/j.acalib.2008.03.007

Katabalwa, A. S. (2016). Use of electronic journal resources by postgraduate students at the University of Dar es Salaam. *Library Review*, *65*(6/7), 445–460. doi:10.1108/LR-11-2015-0108

Kumar, N., & Lalita, L. (2017). Usage of electronic resources at University of Delhi: A case study. *Collection Building*, *36*(3), 96–107. doi:10.1108/CB-11-2016-0031

Panneerdas, C. (2022). Utilization of E- Resources by Faculty Members and Students in Management Institutions, Coimbatore: A Study, *Library Philosophy and Practice (e-journal)*, 6802. https://digitalcommons.unl.edu/libphilprac/6802

Tahir, M., Mahmood, K., & Shafique, F. (2010). Use of electronic information resources and facilities by humanities scholars. *The Electronic Library*, *28*(1), 122–136. doi:10.1108/02640471011023423

Chapter 9

A Study on Disseminating Scholarly Information Through Online Resources in Universities: A Questionnaire-Based Study

P. C. Binu
St. Paul's College, Kalamassery, India

ABSTRACT

The chapter analyzes the dissemination of scholarly information through online resource use by the respondents in university. The study found that, out of 460 respondents, 234 (50.86%) belong to engineering /architecture—this is the top one category among the categories of disciplines of the study. The majority 435 (94.6%) of the respondents are aware of e-resources. 413 (89.78%), they prefer to make use of e-journals. E-Books are widely used by 426 (92.60%) respondents, followed by the use of e-magazines by 212 (46.08%) respondents. The result shows 185 (40.21%) of the respondents preferred general books. The majority 197 (42.82%) of the respondents preferred to use electronic resources for publishing papers in journals. The majority 142(30.86%) of them have preferred the electronic resources for preparing articles to seminars and conference; the respondents have opined as satisfied.

INTRODUCTION

Libraries have become digital and virtual libraries, books, periodicals, and magazines have become e-books, e-magazines and e-journals. This facilitated the global dissemination of information (Abinew and Vuda, 2013). Electronic resources such as e-magazines, e-books, e-databases, and website resources such as e-series are readily available in the surrounding area. Jone (2008) agreed that printed sources are digitized as electronic resources solve storage problems and control information overload. The rapid growth of new technologies has changed the communication process and lowered the cost of commu-

DOI: 10.4018/978-1-6684-8805-8.ch009

nication for individuals. Electronic information resources can be defined as electronic presentations of information that can be accessed through electronic systems and computer networks (Johnson, Evensen, Gelfand, Lammers, Sipe, and Zilper, 2012). They prove that more and more electronic data sources are considered state-of-the-art information technology and are available in various formats such as e-books, digital libraries, online journals, e-learning tutors, and online tests. These electronic resources are sourced with effective use of multimedia tools. Electronic journals offer many solutions to the problems facing today's information management centers.

First, it saves space because storage is more efficient than shelf space storing multiple volumes (Bandyopadhay, 1999; Cook, 1992: 33; and Sweeney, 1997). Electronic diaries speed up communication by providing up-to-date information on newly published material, allowing rapid transmission of research results and communication of scientific results (Sweeney 1997; Rao 1998 and Mountifield and Van Brakel 1994).). Chan (1999) agrees that electronic submission of articles saves valuable time, especially in the peer review process, and creates network communication between authors, editors, and reviewers. Third, e-journals provide powerful search tools. Words and terms in database entries can be searched and combined using Boolean operators (AND, NOT, OR). You can also broaden or narrow your search using specific time periods or years (Fisher 1995). Multiple databases can be searched simultaneously, so large collections of material can be searched instantly. Fourth, it provides immediate access to required information (Fisher 1995, Sweeney 1997, and Bandyopadhay 1999). Rao (1998) confirms that electronic journals provide accurate access to data archives.

How to Find Articles and Magazines Wait times for printed magazines are longer, but can range from minutes to seconds (Chan 1999). Finally, electronic journals are cheaper when printed on paper and mailed to subscribers (Rao 1998). Academic and public institutions face budget cuts and have been forced to unsubscribe from multiple sites. Information published in e-journals is usually up-to-date due to delays in printing and delivery. Rao (1998) emphasizes that published electronic papers are available for preview within 48 hours after his team's approval.

REVIEW OF LITERATURE

Namdar (2004) discussed the traditional forms of libraries and their changing appearance in the modern digital environment. He explained how new technologies enable libraries to collect up-to-date information. He focused on training library professionals in library automation and information processing. He suggested that libraries should assess their services and information literacy to meet their users' needs. Kaur and Verma (2006) conducted a study in libraries in Punjab to investigate user perceptions of library services and electronic resources. We also investigated user attitudes toward library resources and the Internet, particularly those who wanted to use INFONET, a CD-ROM database available in libraries. Sharma (2008) discusses various aspects of automation and its application in libraries. It also discusses the importance of electronic resources and how users access them over the Internet. The Internet has been a boon to libraries and networks that provide library services anywhere. The author described academic libraries and their contributions to the academic community. Hariharan (2006) reviewed a research paper entitled 'Enhancing the Use of Library Resources in Structural Engineering and Research Center Libraries with Special Reference to Electronic Resources'. Parameshwar (2009) analyzed electronic resources and their use by faculty and researchers at Gulbarga University. Data is collected by questionnaire to survey the information needs of faculty and researchers in various PG departments of

Gulbarga University. Prabakaran (2013) studied 'Usage of electronic resources among faculty members of a college of technology in Cadalore district'. Data from a total of 198 respondents were analyzed to assess various factors such as the organizational distribution of respondents using electronic resources. Montgomery and King (2002) focused on library services and cost effectiveness. They explored the impact of a library's preference for e-journals on both staffing and costs. They conducted a comparative analysis of the library (the Drexel library) and developed a usage model.

Natarajan (2010) popularized the concept of modern library and e-resource growth, posing the library challenges of acquiring, storing and distributing e-resources in a digitized environment. The authors also discussed long-term storage and access of information. Jaiswal (2011) describes e-resources as a very important information resource for the modern community of the world, as e-resources have become an increasingly preferred source of information among researchers. says there is. Modern libraries prefer electronic resources that can be accessed from all over the world. Sharma and Gupta (2013) described library resources and their storage and retrieval systems. They defined goals and guidelines. New management thinking, scientific management, various functions of the library were explained. Nishi and Naushad Ali (2013) described the use of e-journals by the Delhi Institute of Technology Library and the Delhi University Library System. The author conducted a survey on the usage of e-journals in both libraries. Based on a sample survey of 300 users, 66.66% of users are aware of e-journals and use them to keep their knowledge up-to-date and collect materials related to their studies and research. It turns out that there is Baskaran (2019) analyzed social her networks and media use by researchers from selected universities in Kerala. In this study, up to 87 (31.9%) of the 273 Annamalai University respondents were male (16.8%) and female (15%). Baskaran (2020) analyzed that the majority of his papers, 2,207 (18.48%), were published from the communication field, as the research focused on social his networks. Computer Communication and Information Systems was second with 1,172 (9.81%) publications.

METHODOLOGY

Data were collected in a questionnaire targeting faculty respondents from selected universities in Kerala. The questionnaire was personally distributed to faculty members of Kerala University, Cochin University of Science and Technology, Mahatma Gandhi University and Kannur University. Of the 470 questionnaires, respondents were distributed and 460 were returned by faculty. Data were analyzed with the Software Package for Social Science (SPSS). This study investigated how selected faculty members access and search electronic information resources. This study includes various parameters of statistical tools with percentile analysis.

RESULTS AND DISCUSSION

Table 1 has been analyzed about Disciplines wise of the respondents. It is understood that out of 460 respondents among three Universities in Kerala, 234 (50.86%) of the respondents belong to Engineering /Architecture; this is the top one category among the category of Disciplines of the study. It follows, 105 (22.82%), 82 (20.19%), 39 (08.47%) of the respondents under Arts/Science, Management, and others respectively (Figure 1).

Table 1. Discipline-wise respondents of the study

Sl.No	Qualification	Frequency	%
1.	Arts/Science	105	22.82
2.	Engineering /Architecture	234	50.86
3.	Management	82	20.19
4.	Others	39	08.47
Total		460	100

Figure 1. Discipline-wise respondents of the study

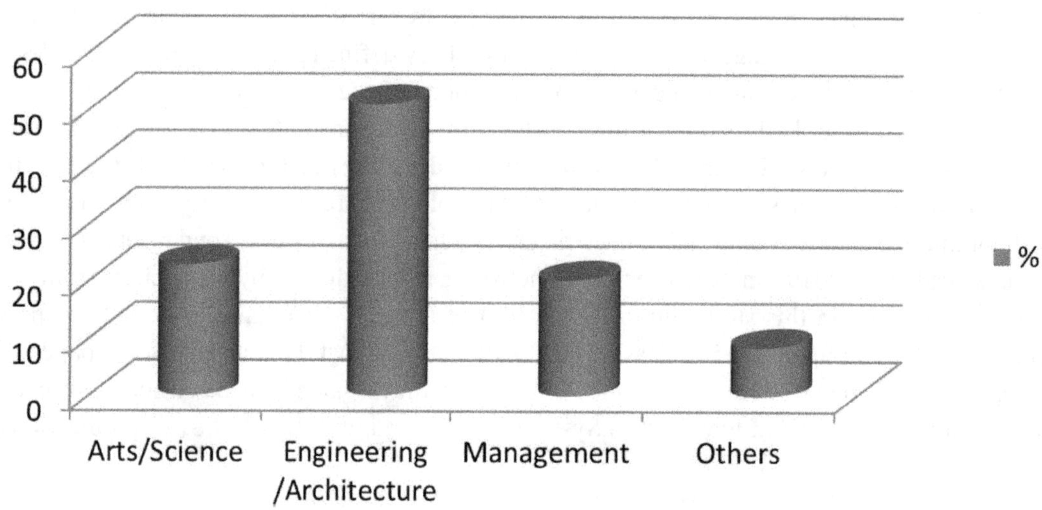

Table 2 was analyzed that awareness on electronic resources by the respondents in the Selected Universities in Kerala. The result finds majority 435 (94.6%) of the respondents aware on e-resources. Whereas only 25(05.43%) of the respondents are not aware with electronic resources by the respondents in the Selected Universities in Kerala (Figure 2).

- **Multiple Responses**

Table 3 discloses that highest number of respondents 413 (89.78%), they prefer to make use of E-Journals. E-Books are widely used by 426 (92.60%) respondents followed by the use of E-Magazines

Table 2. Awareness of electronic resources by the respondents

Sl.No	Response	Frequency	%
1.	Yes	435	94.6
2.	No	25	05.43
Total		460	100

Figure 2. Awareness of electronic resources by the respondents

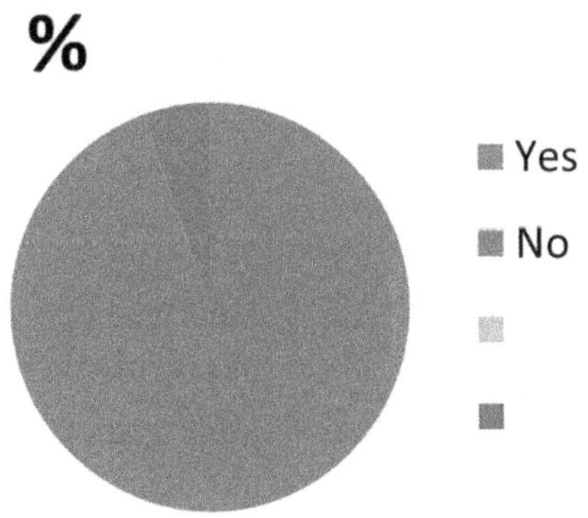

Table 3. The respondents preferred E-resources

Sl.No	Electronic resources	Frequencies	%
1.	E-Journals	413	89.78
2.	E-Books	426	92.60
3.	E-magazines	212	46.08
4.	E-databases	135	29.34
5.	E-Theses and Dissertations	385	83.69

by 212 (46.08%) respondents. The respondents of the study were used E-Databases is reported by 135 (29.34%) of them in the study, The study has been reported that 385 (83.69%) of the respondents used e-theses and dissertations.

- **Multiple Responses**

Table 4 analyzed that the respondents need information to collect from various sources for academic and research in the Selected Universities in Kerala. The result shows 185 (40.21%) of the respondents

Table 4. Type of information resource needed for academic and research

Sl.No	Resources	Frequency	%
1.	General books	185	40.21
2.	Text Books	237	51.52
3.	Reference books	241	52.39
4.	Periodicals/Magazines	292	63.47
5.	Thesis & Dissertation	91	19.78
6.	CD/ROM databases	85	18.47
7.	Electronic/ web resources	232	50.43

preferred General Books. The respondents have preferred other sources with 237 (51.52%), 241 (52.39%), 292 (63.47%), 91 (19.78%), 85 (18.47%) and 232 (50.43%) of the respondents they have preferred Text books, Reference books, Periodicals/Magazines, Theses and Dissertations, CD/ROM database and Electronic /Web resources.

(a) Teaching Materials

Table 5 analyzes the Majority 187 (40.65%) respondents are highly satisfied with Teaching materials, it ffollowed by 132 (28.69%), 80(17.39%), 40 (08.69%) and 21 (04.56%) of respondents have reported Satisfied, Less Satisfied, Not Satisfied and No Comments respectively among the respondents in the Selected Universities in Kerala.

(b) Publish Paper in Journals

The majority 197 (42.82%) of the respondents preferred as Satisfied to use electronic resources for using publish the papers in journals. It followed 138(30%), 67 (14.56%),39 (08.47%) and 19 (04.13%) of the respondents preferred as Highly Satisfied, Less Satisfied, Not Satisfied and No Comment respectively.

(c) Preparing Articles for Seminar/Conference

The majority 142(30.86%) of them have preferred the electronic resources for preparing articles to seminars and conference, the respondents have opined as satisfied, it followed 128 (27.82%), 66(14.34%), 97 (21.08%) and 27(05.86%) of the respondents have reported Highly Satisfied, Less Satisfied, Not Satisfies and No comment on preferred the Electronic resources.

(d) Research and Development

The maximum 147(31.95%) of the respondents have used electronic resources for research and development as Highly Satisfied. The study has been witnessed 129 (28.04%), 76 (16.52%), 97 (21.08%)

Table 5. Level of satisfaction of using electronic resources

Sl.No	Using Electronic Resources	Highly Satisfied	Satisfied	Less Satisfied	Not Satisfied	No Comments
1.	Teaching materials	187 (40.65)	132 (28.69)	80 (17.39)	40 (08.69)	21 (04.56)
2.	Publishing paper in journals	138 (30.0)	197 (42.82)	67 (14.56)	39 (08.47)	19 (04.13)
3.	Preparing articles for seminar/ conference	128 (27.82)	142 (30.86)	66 (14.34)	97 (21.08)	27 (05.86)
4	Research and development	147 (31.95)	129 (28.04)	76 (16.52)	97 (21.08)	11 (02.39)
5.	Project works	175 (38.04)	138 (30.0)	91 (19.78)	30 (6.52)	25 (5.43)

and 11 (02.39%) of the respondents given their opinion about use of electronic resources as Highly Satisfied, Less Satisfied, Not satisfied and No Comment.

(e) Project Work

The study has explored about use of electronic resources for them develop project work, the majority 17(38.04%) of the stated as Highly satisfied, 138 (30%) of them with Satisfied, 91 (19.78%) of them with Less Satisfied, 30 (6.52%) of them with Not Satisfied and 25 (5.43%) of them have suggested as No Comments on the use the electronic resources for Project work.

(a) Google

Table 6 explicit a large number of 212 (46.08%%) respondents are highly satisfied in using Google as their search engine while 81 (17.60%) respondents are satisfied, 97(21.08%) with Less satisfied, 32 (6.95%) with Not Satisfied and 38 (8.26%) of them preferred as No Comment said the respondents in the Selected Universities in Kerala.

(b) Yahoo

The majority of 191 (41.52%) respondents are Highly satisfied in using yahoo as against 79 (17.17%) respondents who are satisfied. 82 (17.82%) of respondents are less satisfied in the use of yahoo search engines.

(c) Alta Vista

The large number of 142 (30.86%) respondents are Highly satisfied as against 138 (30%) respondents Less satisfied. It also been noticed that 54 (11.73%) respondents are not satisfied followed by 97(21.08%) respondents with No Comment.

Table 6. Level of satisfaction for use of search engines on internet and e-resources

Sl.No	Search Engines	Highly Satisfied	Satisfied	Less Satisfied	Not Satisfied	No Comments
1.	Google	212 (46.08)	81 (17.60)	97 (21.08)	32 (6.95)	38 (8.26)
2.	Yahoo	191 (41.52)	79 (17.17)	82 (17.82)	29 (6.30)	79 (17.17)
3.	Alta Vista	164 (35.65)	69 (15.0)	76 (16.52)	54 (11.73)	97 (21.08)
4.	Mozilla firebox	138 (30.0)	142 (30.86)	59 (12.82)	43 (9.34)	78 (16.95)
5.	Ask	112 (24.34)	123 (26.73)	89 (19.34)	72 (15.65)	64 (13.91)
6.	My Web Search	136 (29.56)	118 (25.65)	76 (16.52)	68 (14.78)	62 (13.47)
7.	hotmail	174 (37.82)	135 (29.34)	68 (14.78)	74 (16.08)	09 (01.9)

(d) Mozilla firebox

The majority 142 (30.86%) respondents have suggested as satisfied, 138 (30%) respondents with highly satisfied. It has also been observed that 58 (12.82%) respondents are Less satisfied followed by 78 (16.95%) respondents with No Comment.

(e) Ask

The data shows that a large number of 123 (26.73%) respondents are satisfied as against 112 (24.34%) respondents who are highly satisfied with Ask. It has also been noticed that 89 (19.30%) respondents are Less satisfied followed by 64 (13.91%) respondents who did not comments.

(f) My Web Search

The data elucidates that as many as 136 (29.56%) respondents are Highly satisfied in 'my web search' and 118 (25.65%) respondents are satisfied. There were 68 (14.78%) respondents with not satisfied, it followed by 62 (13.47%) respondents have suggested as No comment.

(g) hotmail

The majority most of the respondents 174 (74.4%) have reported as Highly satisfied. On the other hand, 135 (29.34%) respondents are satisfied followed by 74 (16.08%) respondents are not satisfied and 09 (01.9%) of them have sated No Comment on preferred Hotmail.

Method of Preferences While Retrieving the Information

(a) Download as PDF

Table 7 analyzes that majority 198 (42.82%) of the respondents prefer to retrieve the information could download as PDF on the electronic information by Always. It followed 168 (36.52%) with some times, 55 (11.95%), 12 (02.60%) with Not required and 28 (46.66%) with No Comment of using electronic resources among the respondents in the Selected Universities in Kerala.

(b) Save through External Drive

Table 7. Method of preferences while retrieving the information

Sl.No	Method	Always	Some times	Rarely	Not required	No Comment
.1.	Download as PDF	197 (42.82)	168 (36.52)	55 (11.95)	12 (02.60)	28 (46.66)
2.	Save through external Drive	174 (37.82)	149 (32.39)	38 (8.26)	19 (4.13)	80 (17.39)
3.	Save by Desktop	161 (35.0)	128 (27.82)	62 (13.47)	49 (10.65)	60 (13.04)

The majority of 174 (37.82%) respondents have been preferred electronic resources save by external drive as always. It followed by the respondents have preferred electronic resources among 149(32.39%) with Some times, 38 (8.26%) with Rarely, 19 (4.13%) with not required and 80 (17.39%) with No Comment of preferring electronic resources among the respondents save through external Drive.

(c) Save by Desktop

The study explores majority of 161 (35%) respondents have preferred the electronic information access and shared as always information save by Desktop. The study has been witnessed, the series of analysis finds 128 (27.82%) with Some times, 62 (13.17%) with rarely, 49 (10.65%) with Not required and 60 (13.04%) with No comment on preferring the respondents save the information by Desktop.

CONCLUSION

With the Advancement of computer technology and information technology, libraries and information centers have been blessed with electronic information resources, so libraries are gradually switching to electronic mode. As a result, today's libraries are increasingly involved in the creation and procurement of electronic resources. The availability of IT-based e-resources is increasing the pressure on libraries, and there is no doubt that e-resources are growing rapidly. A majority of respondents, 191 (41.52%), are very satisfied with their use of Yahoo, while 79 (17.17%) are satisfied. 82 (17.82%) respondents are not very satisfied with using Yahoo search engine. A majority of the respondents, 174 (74.4%), said they were very satisfied. On the other hand, 135 people (29.34%) were satisfied with her, followed by 74 people (16.08%) who were not satisfied. 09 people (01.9%). A majority of respondents, 198 (42.82%), preferred to download information in her PDF format on the Always electronic information platform. Regarding the use of electronic resources, 168 people (36.52%) "Several times", 55 people (11.95%) "Unnecessary", 12 people (02.60%) "Unnecessary" and 28 people (46.66%) "No comment".

To meet the ever-increasing demands of faculty, staff, and students for a digital environment, libraries need access to materials in electronic form, much as they have shared and lent print resources among libraries for more than 100 years. Should be developed to manage and share effectively. When it comes to engineering, technology libraries need more funding to maintain modern infrastructure and subscribe to online resources, especially databases. This instrument enhances the research culture more spontaneously. Faculty should be more involved in pursuing the latest technological advances and research innovations.

REFERENCES

Abinew, A. A., & Vuda, S. (2013). A case study of acceptance and use of electronic library services in universities based on SO-UTAUT model. *International Journal of Innovative Research in Computer and Communication Engineering*, *1*(4), 903–910. http://ijircce.com/upload/2013/june/17_A%20Case%20Study.pdf

Ajuwon, G. A. (2003). Computer and internet use by first year clinical and nursing Students in a Nigerian teaching hospital. *BMC Medical Informatics and Decision Making*, *3*(1), 10. doi:10.1186/1472-6947-3-10 PMID:14498997

Ali, N. (2005). The use of electronic resources at IIT Delhi Library: A study of search behaviors. *The Electronic Library*, *23*(6), 691–700. doi:10.1108/02640470510635773

Bandyopadhay, A. (1999). Accessing SCI-tech literature: Commercial document delivery services and online full-text databases. *Collection Building*, *18*(1), 10–15. doi:10.1108/01604959910256481

Baskaran, C. (2019). Scholarly information share through Social Networks (SNs) and Medias among social science scholars in selected State Universities in Kerala. *International Journal of Library and Information Studies*, *9*(3), 83–92.

Baskaran, C. (2020). Research Patterns on the Social Networks and Media: A Scientometric Portrait. Handbook of Research on Emerging Trends and Technologies in Library and Information Science, pp.189-207. IGI Global. doi:10.4018/978-1-5225-9825-1.ch014

Baskaran, C. (2021). An Exponential trend and Mapping of Research in Web 2.0. *Library Philosophy and Practice (e-journal)*. 5008. https://digitalcommons.unl.edu/libphilprac/5008

Chan, L. (1999). Electronic journals and academic libraries. *Library Hi Tech*, *17*(1), 10–16. doi:10.1108/07378839910267145

Fisher, S. (1995). Access to information. *Library Management*, *16*(5).

Hariharan, A., Chitra, H. M., Mymoon, M., Abdunnasar, A., & Suryakala, A. (2006). Enhancing the usage of library resources at SERC library with particular reference to e- resource. In Harish Chandra (Ed). Initiatives in Libraries and Information Centers in the Digital Era, (13-17), Chennai: Society for the Advancement of Library and Information science.

Inamdar, S. A. N. (2004). Information Handling in Libraries. *Keralan Library and Society*, *17*(2), 64–70.

Jaiswal, S. (2011). E-resources, their types. *National Seminar on digital resources*. Hyderabad.

Johnson, S., Evensen, O. G., Gelfand, J., Lammers, G., Sipe, L., & Zilper, N. (2012). *Key issues for e-resource collection development; A guide for libraries*. International Federation of Library Associations and Institutions.

Jones, P. H. (2008). Turning the page: Learning about the future of electronic books from students today. Toronto: University of Toronto Libraries, 38.

Kaur, B., & Verma, R. (2006.) Use of electronic resources at Tiete library, Patiala: A case study. *ILA Bulletin*, *12*(3), 18-20.

Mentegomery, C. H. (2002). Comparing Library and User related cost of print and electronic journal collections. *D-Lib Magazine : the Magazine of the Digital Library Forum*, *8*(10), 1–15. doi:10.1045/october2002-montgomery

Mountifield, H. M., & van Brakel, P. A. (1994). Network-based electronic journals: A new source of information. *South African Journal of Library and Information Science*, *62*(1), 28–33.

Natrajan, K, K. (2010). Use and user perception of electronic resources. *Annals of Lib. & Inf. Studies, 57*(3), 59-64.

Nishi, F. & Ali, N. (2013). Witness and use of e-journals by IIT Delhi and Delhi University Library System. *International Journal of Digital Library Services, 32*(2), 57-64. http://www.ijodls.in/uploads/3/6/0/3/3603729/23-42.pdf

Parameshwar, S., & Patil, D. B. (2009). Use of electronic resources in university libraries of Karnataka: A case study of Gulbarga University Library. *Keralan Journal of Information Science and Services, 3*(1), 13-21. 5.

Prabakaran, T. (2013). Use of E-Resources Among Faculty Members of Engineering Colleges in Cuddalore District. *Journal of Advances in Library and Information Science, 2 (2), pp-71-75.*

Rao, S. S. (1998). Trends in publishing and delivery of electronic information. *New Library World, 99*(1145).

Sharma, H.P. (2008). Moving beyond library automation: Role of e-resources in academic libraries. *University news, 46*(34), 6-10.

Sharma, J.N. & Gupta, S. (2013). Management of libraries in modern era: A view. *Lib. & Inf.Tech, 9*(1), 7–20.

Sweeney, L. (1997). The future of academic journals: Considering the current situation in academic libraries. *New Library World, 98*(1132), 4–15. doi:10.1108/03074809710155569

Chapter 10
Accessing E-Content on Excel:
Information Literacy Among Faculty Members in the Deemed Universities in Chennai, India

M. Prasad
Kumaraguru College of Technology, Coimbatore, India

ABSTRACT

The study analyses the information literacy skills with access e- contents among faculty members in the deemed universities in Chennai, India. The chapter has shown that a majority of 97.38% of the respondents have been aware of the information literacy skill about access of e-documents. The result shows SRM University and Sathiyabama University have respondents of each 130 (34.21%), 85 (22.36%) respondents are belonging to St. Peters University, 35(9.22%) respondents are from Crescent University. Out of 380 respondents, it is seen that 210 (55.26) respondents are extremely satisfied on OPAC/Web OPAC. 205(53.90) respondents are extremely satisfied with e-databases, 192(50.52) respondents are extremely satisfied on automated circulation services. The majority of respondents are extremely satisfied with the OPAC/Web OPAC (55.26%) provided by university library followed by e-databases (53.90%).

INTRODUCTION

The conventional way of education is another name for the traditional higher education system. The fundamental goal of this traditional approach is to preserve societal norms, manners, and values for the next generation. Students study the society's traditions and customs under this framework. The amount of written and practical work is quite little, and it helps to improve oral reaction. There may be no specific written or practical tests, merely oral tests. Science and technology are pretty far from the traditional higher education structure. It might cover information about beliefs, traditions, and practises. The higher education system of today is entirely different from the higher education system of the past. Writing and practical skills are improved as a result of contemporary higher education, which is also used in science and technology. It encompasses writing, visualisation, critical thinking, and creative thinking. It aids

DOI: 10.4018/978-1-6684-8805-8.ch010

individuals in examining the right method of instruction through written and practical tests. Modern higher education systems may employ interactive teaching methods. The higher education system of today is an imitation of the older system.

According to Gui (2007), information literacy includes navigation skills, selection and assessment skills and the ability to use information. These skills enable a person scope with the changing content of computer and information resources and know where and how to find resources. Therefore, the concept of information literacy requires that an the information user knows when information is needed, seeks, evaluates, uses and distributes information especially in an electronic environment to solve certain research problems goals Recently, college students have faced the problem of multiple use forms of information resources effectively (Dalvi, 2010). Also amazing the growth of electronic resources has changed the entire scenario of education and information industry. Therefore, students need to quickly acquire information literacy because it ensures effective use of electronic information resources. Students don't just have to have knowledge not only about the technology, but also about the application domain and the skills needed to configure it what they need and how they use it. Librarians must understand that students acquire information reading skills through various user training programs. That is why libraries are in demand expanding its user training programs. In addition, librarians need to rethink their role this new information age characterized by the electronic resources that students need information literacy, working with faculty to protect and support integration of information literacy courses into curricula.

Traditional and contemporary higher education is similar but different from one another. But it is clear that the conventional higher education system has been upgraded by modern higher education. The two systems are contrasted in the following. The educational landscape now is very different from that of the past. Modern higher education was not valued at the time, much as traditional education is not valued today. The traditional higher educational system is required because of education. It has abruptly converted to a modern higher education system, which is progressively winning over the public. People used to teach their kids how to meet their educational needs in the past, but only human needs have changed. As time went on and human needs increased, so too did the demand for knowledge. It would be challenging to meet needs in the modern world without an evolution in education. All of the modern conveniences we take for granted, such as washing machines, lights, fans, cars, and refrigerators, would not be around today if education had not been changed. Therefore, higher education must evolve if knowledge is to advance in modern times. Considered Universities in Chennai, India, are doing an outstanding pivotal.

REVIEW OF LITERATURE

Print medium is increasingly giving way to the electronic form of materials (Sharma, 2009) Hence it has become a great challenge for the electronic resource producers and providers to understand the variety of users' demands in order to improve the efficiency and scholarly value of the electronic resources. This needs to be taken care so that the libraries can attract more potential users and enhance the service quality and customer satisfaction (Liyi Zhang et al. 2010). Blogs can be used to facilitate group working, a number of commentators argue that blogs encourage reflective learning (Chen et al. 2005) they can be used as a diary to keep a record of achievement and of interesting material. The way that blogs are typically organised allows the reader/viewer to re-contextualize posts according to categories and dates as well as allowing the user to use key words to search for posts on any topic. Blogs can decrease feelings of isolation for distance students (Dickey, 2004) and promote the shift from surface to deep learning

(Bartlett-Bragg, 2003). Ferdig and Trammel (2004) argue that the immediacy and commentary based systems of blogging lead to reflection and analysis and contextualisation of learning via hyperlinks. They further contend that blogs are more successful in promoting interactivity that is conversational as opposed to other online discussion. Research also suggests that blogs allow a more creative response from students (Oravec, 2005). Ajayi, Shorunke & Aboyade (2014) conducted a survey to find the Influence of Electronic Resources Use on Students' Reading Culture in Nigerian universities.

Felicitas Ciabere Ratanya (2017) revealed that majority of the academic staff at the Egerton University are still not aware of the existence of the IR. Staff also faced challenges in accessing and using the content available. JosepLladós et al. (2017) have revealed that a combination of personal and contextual factors such as cultural, social and subjective factors affects the use of Wikipedia as a tool for faculty in higher education institutions. Faculty perceptions and attitudes in two large Spanish Deemed Universities in Chennai were analysed. Franklina Adjoa Yebowaah and Florence Dedzoe Dzokotoe Plockey (2017) stated the objective of this study was to examine the use of electronic resources by lecturers of the University for Development Studies, Wa campus. This suggests that awareness is high but utilization is low. The study recommends that efforts to improve the use of e-resources. Roaganachi Mercy A and Izuagbe Roland (2018) examined the role of access to online databases as the basis for faculty research output in six Deemed Universities in Chennai (comprising two each of federal, state and private) in two South-western states in Nigeria. Baskaran (2019) analyzed that use of social network and Medias by the Research scholars in Selected Universities of Tamilnadu. To highest 87 (31.9%) of respondents Annamalai University, (16.8%) and (15%) of them Male and Female respectively out of 273 total respondents responded to the study. Baskaran (2020) analyzed that majority of the publications, 2,207 (18.48%), were published from communication area, on account of the specific nature of the research concentrating on Social networks. Computer communications information systems area scores second with 1,172 (9.81%) of the publications. Baskaran (2021) explored the majority 134 (1.96%) of the publications was contributed by the researchers from the University of California systems. Zhang Y was the top author was contributed 16(0.23%) of the publications in the field of Web 2.0. Ananda et. al. (2017) The article emphasis the use and awareness of Electronic Information Resources (EIR) among UG & PG students of T John College, the study identifies the awareness of electronic information resources among students i.e. 81% opined that they are aware of Electronic Information Resources (EIR) and 19% opined that they are not aware of Electronic information Resources. Akuffo, Maxwell, N &Budu, Stephen (2019) Their study revealed high awareness levels, adequate computer competencies, derivation of multiple benefits, use of electronic resources for academic purposes and inadequate search skills of most respondents because of the dearth of training. Boud (2001) comments that journal writing for learning can be used to "capture an experience, record an event, explore our feelings, or make sense of what we know."

OBJECTIVES OF THE STUDY

The following are the objectives of this study.

1. To study the awareness Information Literacy skill on accesses the e-documents among the faculty members in four Deemed Universities in Chennai in Tamil Nadu.
2. To find out the training provided by the University Libraries on accessing E-Resources by the faculty members of four Deemed Universities in Chennai in Tamil Nadu.

3. To find out the level of satisfaction on Library E-Resources and Services by the faculty members of four Deemed Universities in Chennai in Tamil Nadu.
4. To examine the frequency of access of E-Resources by the Faculty members among four Deemed Universities in Chennai, Tamil Nadu.

METHODOLOGY

The population of the present study comprises of the teaching faculty members, in four Deemed Universities in Chennai in Tamil Nadu. The Deemed Universities are there in Chennai namely are St. Peters University, SRM University, Sathiyabama University, and Crescent University, were chosen the study. A total of 450 questionnaires distributed to the respondents in the four Deemed Universities in Chennai in Tamil Nadu. Out of these 450 questionnaires, the 400 questionnaires (88.88%) were received back. Among 400 questionnaires 20 could not be used for the final analysis as they were incomplete. Further, the analysis was made by SPSS 6.0 for quantifying the data wherever necessary results of study incorporate with percentile analysis, Chi-Square test, ANOVA and more.

ANALYSIS AND DISCUSSION

In research presentation, Data Analysis is a very important stage. It involves with the data calculation and its result which leads to the finding and conclusion. These processes are calculation oriented and the output is the result oriented to get a conclusion to the selected problem. Here the variables are played a vital role to give its end result.

Table 1 examines the Awareness of Information Literacy skill for access e-documents among the faculty members in the Deemed Universities in Chennai in Tamilnadu. Figure 1 analysed majority of 97.38% of the respondents have witnessed that they have been aware about Information Literacy skill about access of e-documents. Whereas only 1.84% of the respondents have witnessed as No about the Information Literacy skill about access of e-documents among respondents in the Deemed Universities are in Chennai.

Gender Wise Respondents of the Study

Data presented in table 2 represents the Gender wise respondents of the selected state Deemed Universities are in Chennai. Out of 380 respondents, the male respondents are found 263(69.21%) which is followed by female respondents 117 (30.79%) It is very clear from the above discussion that the male

Table 1. Awareness of information literacy skill for access e-documents

Awareness	No. of Respondents	%
Yes	373	97.38
No	07	1.84
Total	**380**	

Figure 1. Awareness of information literacy skill for access e-documents

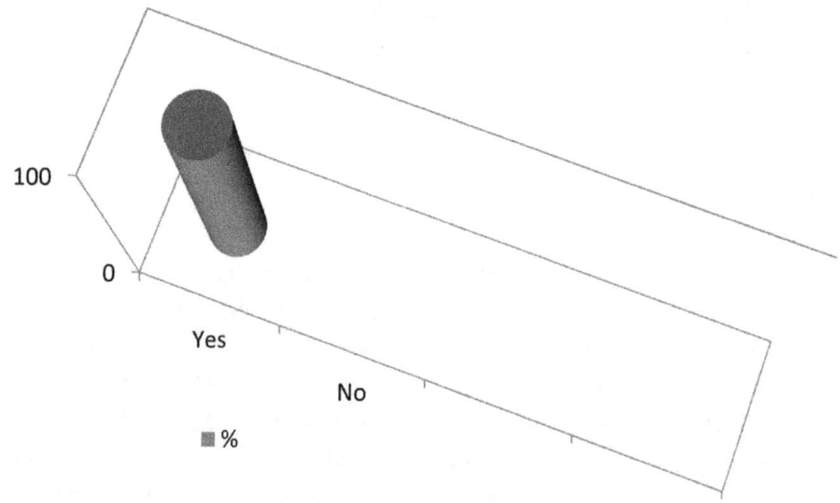

Table 2. Gender wise distribution of respondents

Gender	No. of Respondents	Percentage
Male	263	69.21
Female	117	30.79
Total	**380**	

*Source Computed

Figure 2. Gender wise distribution of respondents

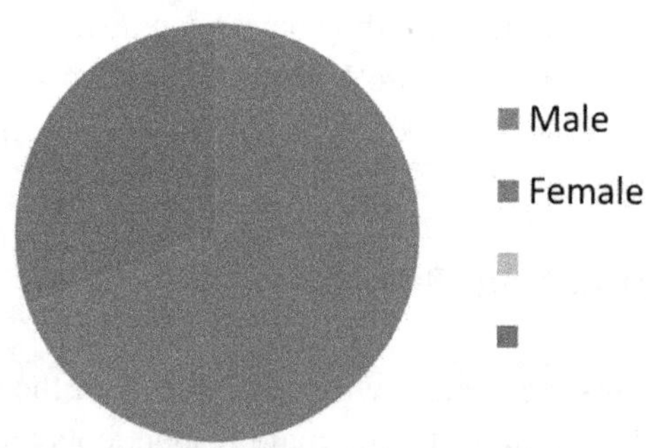

respondents are more in numbers and the female respondents are lesser in numbers than the male respondents (Figure 2).

Age Wise Respondents of the Study

The data in table 3 indicate the distribution of respondents according to Age. Out of 380 respondents, it is found that 80(21.06%) respondents are belonging to 41 to 45 age group. 72(18.94%) respondents are

Table 3. Age wise distribution of respondents

Age Group	No. of Respondents	Percentage
Up to 25	10	2.63
26 to 30	35	9.22
31 to 35	68	17.89
36 to 40	72	18.94
41 to 45	80	21.06
46 to 50	64	16.84
Above 50	51	13.42
Total	**380**	

*Source Computed

belonging to 36 to 40 age group and 68 (17.89%) respondents are belonging to 31 to 35 age group. It is found that 64 (16.84%) respondents are belonging to 46 to 50 age group and 51(13.42%) respondents are belonging to above 50 age group. It is found that 35(9.22%) respondents are belonging to 26 to 30 age group and 10 (2.63%) respondents are belonging to up to 25 age group. (Figure 3)

Educational Qualification Wise Respondents

Data presented in table 4 represents the Educational Qualification wise respondents of the selected Deemed Universities are in Chennai. Out of 380 respondents, the Ph.D. qualified respondents are found 285(75%) which is followed by M.Phil. with NET qualified respondents 51 (13.42%) and PG with NET qualified respondents 44 (11.57%).

University Wise Distribution of the Respondents

Table 5 observed that the University wise respondents are accounted for this study. Out of 380 respondents, it is found that SRM University and Sathiyabama Universityhave respondents of each 130 (34.21%), 85 (22.36%) respondents are belonging to St. Peters University, 35(9.22%) respondents are from Crescent University (Figure 3).

Table 4. Educational qualification wise distribution of respondents

Educational Qualifications	No. of Respondents	Percentage
PG with NET	44	11.57
M.Phil with NET	51	13.42
Ph.D.	285	75.0
Total	**380**	

*Source Computed

Table 5. University wise distribution of respondents

Sl.No.	Name of the University	No. of the Respondents	Percentage
1	St. Peters University	85	22.36
2	SRM University	130	34.21
3	Sathiyabama University	130	34.21
4	Crescent University	35	9.22
	Total	**380**	

*Source Computed

* * Figures in Parentheses denoted Percentages

Figure 3. University wise distribution of respondents

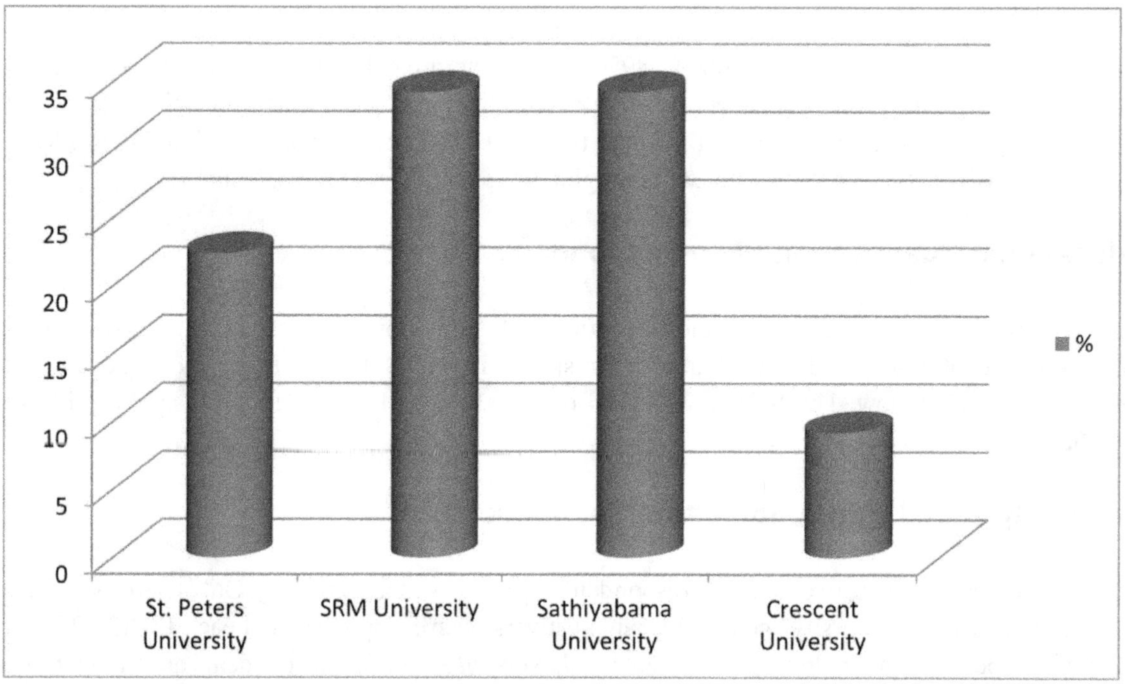

Designation Wise Respondents

Table 5 indicates the distribution wise respondents according to Designation wise. Out of 380 respondents, Assistant Professor Respondents are found 265 (69.74%) which is followed by Professor Respondents 65(17.11%) and Associate Professor Respondents 50 (13.15%).

Faculty Wise Respondents Acquired Training

Table 7 indicates the distribution of respondents according to Faculty wise training provided for accessing the E-Resources. It has seen that out of 104 (27.40%) respondents from Faculty of Arts the majority of 100 (96.20%) respondents got training from University Libraries and only 4 (3.80%) respondents are

Table 6. Designation wise distribution of the respondents

Designation	No. of Respondents	Percentage
Assistant Professor	265	(69.74)
Associate Professor	50	(13.15)
Professor	65	(17.11)
Total	380	(100)

*Source Computed

* * Figures in Parentheses denoted Percentages

Table 7. Faculty wise respondents acquired training

Sl. No.	Faculty Wise	Training provided by University Library		Total
		Yes	No	
	Arts	100 (96.20)	4 (3.80)	104 (27.40)
	Science	177 (100)	0 (0)	177 (46.60)
	Management	49 (94.20)	3 (5.80)	52 (13.70)
	Education	45 (95.70)	2 (4.30)	47 (12.40)
	Total	371 (97.60)	9 (2.40)	380 (100)

*Source Computed

not getting the Training. It is found that out of 177 (46.60%) respondents from Faculty of Science, all the 177 (100%) respondents are got training.

It is observed that out of 52 (13.70%) respondents from Faculty of Management the majority of 49 (94.20%) respondents got training and only 3 ((5.80) respondents are not getting the training. It is also seen that out of 47(12.40%) respondents from Faculty of Education the majority of 45 (95.70%) respondents are getting training and 2 (4.30%) respondents are not getting training. It is very clear from the above discussion that the majority of 177 (100%) respondents are got training from Faculty of Science.

Satisfaction of Respondents for Their Information Needs for Accessing the Resources

Table 8 observed the distribution of respondents according to the level of Satisfaction for their information needs by accessing various E-Resources and services provided by University Libraries. Out of 380 respondents it is seen that 210 (55.26%) respondents are extremely satisfied on OPAC/Web OPAC. 205(53.90%) respondent are extremely satisfied on E-Databases, 192(50.52%) respondents are extremely satisfied on Automated circulation services, 140(36.84%) respondents are extremely satisfied on E-Journals, 122(32.10%) respondents are extremely satisfied on E-Theses and Dissertations, 116(30.52%) respondents are extremely satisfied on E-Books for their information needs by accessing the E-Resources and services provided by the respondents four selected Deemed University Libraries in Chennai.

Further, It is discussed that 94(24.73%), 92(24.21%), 72(18.94%) respondents are extremely satisfied on E-Question Bank, Email alert services, other E-Resources/E-services respectively provided by the University Libraries. It is found that 140(36.84%) respondents are very satisfied with E-Theses and

Table 8. Satisfaction of respondents for their information needs for accessing the resources

Library E- Resources/Services	Extremely Satisfied	Very Satisfied	Moderately Satisfied	Slightly Satisfied	No Comment/ Not Used
CDs/DVDs	90(23.68)	110(28.94)	82(21.57)	72(18.94)	26(6.84)
E – Books	116(30.52)	127(33.42)	95(25)	25(6.57)	17(4.47)
E Journals	140(36.84)	131(34.47)	79(20.78)	18(4.73)	12(3.15)
E- Databases	205(53.90)	108(28.42)	42(11.05)	11(2.89)	14(3.68)
E-Theses and Dissertations	122(32.10)	140(36.84)	58(15.26)	32(8.42)	28(7.36)
E-Question Bank	94(24.73)	102(26.84)	106(27.89)	46(12.10)	32(8.42)
Email alert service	92(24.21)	112(29.47)	88(23.15)	52(13.68)	36(9.47)
OPAC/Web OPAC	210(55.26)	101(26.57)	47(12.36)	10(2.63)	12(3.15)
Automated Circulation Services	192(50.52)	125(32.89)	32(8.42)	18(4.73)	13(3.42)
Other-Resources/E-Services	72(18.94)	126(33.15)	75(19.73)	65(17.10)	42(11.05)

*Source Computed

* * Figures in Parentheses denoted Percentages

Dissertations and 106(27.89%) respondents are moderately satisfied with E-Question Bank. It is also observed that 72(18.94) respondents are slightly satisfied with CD/DVDs and 42(11.05) respondents are given no comment/not used option on other E-resources/E-services provided by the University Library. It is very clear from the above discussion that the majority of respondents are extremely satisfied on OPAC/Web OPAC (55.26%) provided by University Library followed by E-Databases (53.90%) of the respondents four selected Deemed University Libraries in Chennai.

Major Findings

1. Out of 380 respondents, the male respondents are found 263(69.21%) which is followed by female respondents 117 (30.79%).
2. 80(21.06%) respondents are belonging to 41 to 45 age group. 72(18.94%) respondents are belonging to 36 to 40 age group and 68 (17.89%) respondents are belonging to 31 to 35 age group.
3. Out of 380 respondents, the Ph.D. qualified respondents are found 285(75%) which is followed by M.Phil.
4. Out of 380 respondents, it is found that SRM University and Sathiyabama Universityhave respondents of each 130 (34.21%), 85 (22.36%).
5. 380 respondents, it is seen that 181 (47.66%) respondents are accessing the scholarly information via E-Resources on daily basis, 109 (28.68%) respondents are using once in two days.
6. 104 (27.40%) respondents from Faculty of Arts the majority of 100 (96.20%) respondents got training from University Libraries and only 4 (3.80%) respondents are not getting the Training.
7. 210 (55.26) respondents are extremely satisfied on OPAC/Web OPAC. 205(53.90) respondents are extremely satisfied on E-Databases, 192(50.52) respondents are extremely satisfied on Automated circulation services.

CONCLUSION

Since the 1970s, libraries have provided access to records such as censuses and surveyed data. During the microcomputer revolution of the 1980s, libraries began to acquire software and data on floppy disk and CD ROMs, including the full text. The library provided web-based catalogs, bibliography and full-text databases, e-journals, and ultimately e-books over the Internet. The library was more efficient and easier to access and search, eliminating the need for users to go to the library to do most of their research. In the mid-1960s, all electronic resources began to dramatically change the way users can access library resources.

One of the benefits of using information and communication technology in the classroom is that the novelty of the activity and the interest in working with computers will improve the quality of education. The use of information and communication technology is related to enabling a combination of material visualization, its "activation", the ability to make visual journeys, the ability to visualize the phenomena that cannot be demonstrated otherwise, and control. Open lessons learned and training procedures. According to Lawal (2012), each stage of information seeking is the process model shows the progressive development that would lead the student to achieve a a sense of ownership in the area of expertise, which is an important part information literacy and lifelong learning. Hence the ultimate goal of information seeking the process model is to get students who know how to expand their knowledge and skills have information literacy when using different information sources both inside and outside the school (Kuhlthau et al., 2007). The above results show that information literacy is a basic requirement for using electronic information resources and the reflection of information literate student performance indicators.

REFERENCES

Ajayi, S. A. Oludare, Shorunke, A &Modupe, Aboyade, A. (2014) The Influence of electronic resources use on students' reading culture in Nigerian Universities: A case study of AdelekeUniversite Ede, Osun State, *Library Philosophy and Practice (e-journal),* http://digitalcommons.unl.edu/ libphilprac /1182

Akuffo, M., & Budu, S. (2019, May). Use of Electronic Resources by Students in a Premier Post-graduate Theological University in Ghana. *South African Journal of Information Management*, *21*(1). doi:10.4102ajim.v21i1.1026

Ananda, S. K. (2017, January-March). Use and Awareness of Electronic Information Resources among UG & PG students of T John College, Bangalore: A Study. *International Journal of Library and Information Studies*, *7*(1), 20–25.

Bala, S., & Lal, P. (2016, July). Usage of Electronic Resources and their Impact on Reading Culture: Study of Punjab Agricultural University. *International Journal of Digital Library Services*, *6*(3), 59–66.

Baskaran, C. (2019). Scholarly information share through Social Networks (SNs) and medias among social science scholars in selected State Universities in Tamilnadu. *International Journal of Library and Information Studies*, *9*(3), 83–92.

Baskaran, C. (2020). Research Patterns on the Social Networks and Media: A Scientometric Portrait. Handbook of Research on Emerging Trends and Technologies in Library and Information Science, pp.189-207. IGI Global. doi:10.4018/978-1-5225-9825-1.ch014

Baskaran, C. (2021). An Exponential trend and Mapping of Research in Web 2.0, *Library Philosophy and Practice (e-journal)*. 5008. https://digitalcommons.unl.edu/libphilprac/5008

Boud, D. (2001). Using journal writing to enhance reflective practice. In L. M. English & M. A. Gillen (Eds.), *Promoting Journal Writing in Adult Education. New Directions in Adult and Continuing Education. No. 90* (pp. 9–18). Jossey-Bass. doi:10.1002/ace.16

Chen. (2005). Using wikis and weblogs to support reflective learning in an introductory engineering design course. *Proceeding of the 2005 American Society for Engineering Education Annual Conference & Exposition.* https://citeseerx.ist.psu.edu/viewdoc/summary?doi=10.1.1.127.5481

Dalvi, M. (2010). In Information literacy for effective use of e-resources: A study of selected colleges in Mumbai Tiruchirappalli. *International Conference on e-resources in higher education: Issues, Developments, Opportunities and Challenges*, 82-120. Bharathidasan University.

Ferdig, R. E., & Trammell, K. D. (2004). Content delivery in the Blogosphere. The academic staff at Egerton University, Kenya. *Library Management, 384*(5), 276–284. doi:10.1108/LM-02-2017-0018

Yebowaah., F., Dedzoe., F., & Plockey, D. (2017). Awareness and Use of Electronic Resources in University Libraries: A Case Study of University for Development Studies Library. *Library Philosophy and Practice (e-journal).* 1562, 1-32.

Gui, M. (2007). *Formal and Substantial Internet Information Skills: The Role of Socio-Demographic Differences on the Possession of Different Components of Digital Literacy,* 12, 9–3. First Monday. https://www.firstmonday.org/htbin/cgiwrap/bin/ojs/index.php/fm/article/view/2009/1884

Iroaganachi, M. A., & Izuagbe, R. (2018). Access to Online Databases: Predicate for Faculty Research Output, *Library Philosophy and Practice (e-journal).* https://www.researchgate.net/publication/330243888_Access_to_Online_Databases_Predicate_for_Faculty_Research_Output

Kuhlthau, C. C., Maniotes, L. K., & Caspari, A. K. (2007). *Guided Inquiry: Learning in the 21stcentury.* Libraries Unlimited. https://iasl-online.org/resources/Documents/IASL2009_2_KN_KUHLTHAU.pdf

Lawal, V. (2012). A contextual study of the information literacy of aspirant barristers in Nigeria. *South African Journal of Library and Information Science, 78*(2), 101–111.

Lladós, J. (2017). Explaining teaching uses of Wikipedia through faculty personal and contextual features. *Online Information Review, 41*(5), 728–743. doi:10.1108/OIR-10-2016-0298

Sharma, C. (2009). Use and Impact of E-Resources at Guru Gobind Singh Indraprastha University (India): A Case Study. *E-JASL 1999-2009,* 1-10, 123. https://digitalcommons.unl.edu/ejasljournal/123

Zhang, L., Ye, P., & Liu, Q. (2011). A survey of the use of electronic resources at seven universities in Wuhan, China. *Program, 45*(1), 67–77. doi:10.1108/00330331111107402

Chapter 11
A Quantitative Study on Usage Measure of Social Media Platforms in Bangladesh

M. Prasad
Kumaraguru College of Technology, Coimbatore, India

Muhammad Mezbah-ul Islam
University of Dhaka, Dhaka, Bangladesh

ABSTRACT

The study analyzes the Impact and usage statistics of social medias platforms in Bangladesh. As the country races to become a mini-Silicon Valley, changes from the traditional way of life to AI-dominated living is inevitable. Investments from global venture capitalists and equity investors are tapping the potential of 166 million Bangladeshis with its big population of youths embracing the future. The study analyzes the majority of the people in Bangladesh preferred all type of social medias platforms; among listed social medias is FaceBook, with percent between 89.69 and 93.79 in August and January in the year 2022. YouTube usage rate follows, at between 2.58 and 5.91, Linked In (0.71 and 1.63), Twitter (0.38 and 1.77), Printerest (0.26 and 0.58), Instagram (0.3 and 1.01), Reddit (0.2 and 0.31), Tumblr (0.01 and 0.04).

INTRODUCTION

The computer-based technology that makes it easier to share ideas, thoughts, and information through virtual communities and networks is referred to as social media. Users can quickly share personal information, documents, videos, and photos electronically through social media, which are internet-based. Through web-based software or applications, users engage with social media on a computer, tablet, or smartphone. While web-based entertainment is pervasive in America and Europe, Asian nations like Indonesia lead the rundown of online entertainment use. (As of October 2021, more than 4.5 billion individuals use social media (Digital Reportal, 2021). The most popular social media sites are Facebook,

DOI: 10.4018/978-1-6684-8805-8.ch011

Twitter, Instagram, and YouTube. These social media sites play a significant role in the lives of students and are frequently accessed multiple times per day (Junco et al., 2010). Social media have changed how individuals, communities, and associations create, share, and seek information from one another over the past few years. The majority of students are also so accustomed to social media that the various forms of online engagement, including communication, self-expression, community building, and others, are now part of their only reality.

As a result, it's critical to step back and consider how we got here and what our most recent social media "advances" might mean for us in the future. This is on the grounds that the advanced web-based entertainment varies from the conventional media (e.g., paper and TV) in the particulars of their availability, intelligence, convenience and pervasiveness. In 2017, web clients spent over 2 hours on normal for each day on interpersonal organizations and informing administrations (30 minutes every day longer than five years sooner), which estimated to around 33% of their entire day of day to day PC time (Mander, 2017). One of the best inventions of our time, social media allows people to communicate and interact with one another around the world, enabling them to live in a global village. Although mobile devices are not required to access social media, their effectiveness is enhanced by mobile technologies because of the increased frequency of use. More than 95% of understudies approach some kind of cell phone that interfaces with the web, implying that involving portable innovation for instructive reason conceivably fewer than 5% of understudies. Be that as it may, PC labs are accessible nearby at most foundations. Some of them also give students the option to temporarily borrow a laptop or tablet, which can encourage students to search for information and investigate the accessibility issue. One of the primary objectives for students is to learn how to socialize with others and build social networks that will support them throughout their education and beyond.

Users typically do not question the short- and long-term implications and potential risks of their choices, despite the fact that executives of social media platforms have acknowledged that their platforms have negative effects. According to Bennett, Maton, & Kervin (2008), a generation of digital natives who have grown up with social media are now making up a significant portion of the workforce as well as the customers of many businesses. The majority of adult users have also become so accustomed to social media that the various forms of communication, self-expression, community building, and other forms of online engagement are now part of the only reality they are aware of. Therefore, taking a step back to consider how we got here and what our most recent social media "advances" might mean for us in the future is crucial. This article highlights the dual nature of social media: Despite the many positive aspects of social media, there are also negative aspects that need to be looked into so that we can make better decisions and become more aware of the risks they pose. We start by explaining what we mean by the clouded side of virtual entertainment, and why it is a worry for society. We then present a system for understanding the clouded side of the center functionalities of various online entertainment stages. We then outline a number of important research opportunities that could assist in facilitating a healthier use of the "media" by better understanding their related negative impact on the "social" fabric of society. We do this by using the ideas and issues that this framework highlights.

Social Medias Platforms Usage in Bangladesh

With recent development and infrastructure, bandwidth expansion, software hub creation, start-up funding for digital and technology-based businesses, and continuous price reductions in Mbps costs, Bangladesh's digital transformation has finally caught up with the rest of the world. Credit must be given to the government, which has been building the foundation over the past decade. National ID and passport applications, tender submissions, registrations, car licenses, utility bill payments, and even COVID-19 vaccinations all continue to progress toward digitization. Regular daily existence is going advanced and on the web. The Sadai app was recently launched by the government of Bangladesh, making it possible to sell and buy agricultural and farming products online. Connectivity, app and online use, the accumulation of knowledge, and financial independence have all been facilitated by falling smartphone prices. According to the statistics released by the BTRC in June 2021, approximately 120.95 million people in Bangladesh are currently connected to the internet, with 110.90 million of the 166 million people in the country using their mobile phones to access it.

Peer-to-Peer Digital Connectivity

The greatest effect of digitisation was felt on coordinated correspondence - voice approaches cell phones are quick declining and progressively moving towards applications like FB Courier, WhatsApp, imo, Viber, LinkedIn and others with the expectation of complimentary voice connection, free informing and individual virtual cooperation and amusement extravagance. The usage patterns of Bangladeshi users of imo, for instance, indicate the communication app industry's future potential. As per its true delivery, in 2020, 96bn messages, 30bn worldwide messages, 26bn sound and video calls, and 15bn global calls were made by Bangladeshis utilizing imo. On imo platforms, during the pandemic in 2020, there was an 8.2% increase in messages and a 7.8% increase in video and audio calls compared to 2019.

News and Entertainment Thrive Online

More people read news online than in newspapers or on television. According to Alexa analytics, online local news portals receive approximately 3.5 million unique page visits per day, with an estimated average time spent of more than six minutes. All newspapers and televisions have been forced to focus more on their online and streaming versions as a result of this shift toward online news consumption.

- Facebook used to be a big source of breaking news and news consumption. However, the recent rise in unverified sensationalist content and fake news is causing readers to turn to traditional news sources and platforms.
- Niche urban geographies are experiencing an increase in the popularity of international news platforms like the Wall Street Journal and Huffington Post.
- Throughout the country, apps like TikTok and Likee have gained a lot of popularity with the new alpha generation. However, the quality of their content, including the promotion of violent and unacceptable lifestyles, has drawn regulatory scrutiny and social criticism.

Virtual entertainment is as yet overwhelmed by Facebook, Instagram and LinkedIn. NapoleanCat and media reports claim that as of June 2021, Facebook had over 47.2 million active users in Bangladesh, or approximately 28% of the population; 21.2 million users are between the ages of 18 and 24, with 69.1% being male and 30.9% being female. This clearly demonstrates Generation Z's love for social media, whereas Generation Alpha prefers apps like Instagram, which has 4.7 million users in the country. In May 2021, LinkedIn had 4.1 million clients and Twitter is still in its early stage. The delivery of entertainment has also undergone significant shifts since the advent of digital life. With an average of 19 minutes spent per day, YouTube has become the alternative to television for all demographics, and over the past two years, both local and international streaming apps and OTT platforms have increasingly entered urban homes. Netflix had an estimated user base of over 200,000 as of 2020 and was rapidly expanding. The battleground for content consumption is being heated up by additional video content streaming apps like Amazon Prime, Hoichoi, Zee5, Bongo, and a few other local apps.

People between the ages of 18 and 34 are the primary consumers of streaming content. The majority of the 134 publications, or 1.96% was written by researchers from the University of California systems, according to Baskaran (2021). Zhang Y was the top creator was contributed 16(0.23%) of the distributions in the field of Web 2.0. Spotify's new send off in the market is set to impel music utilization to another level, including to music utilization VAS administrations by cell phone administrators and YouTube. By partnering with a local operator, the audio and video content distribution platform Stingray, which is based in Canada, is also aiming for Bangladesh's large music consumption population. International online entertainment brands have a chance to enter the music, fashion, and movie industries thanks to the young population, rapid growth in internet use, and the introduction of 5G. Be that as it may, achievement will possibly occur on the off chance that there is solid comprehension of nearby requirements and worldwide substance is adjusted to the neighbourhood taste and language.

THE E-COMMERCE BOOM AND GOING CASHLESS

The transfer in how people shop online has been a big part of the shift to digital life. While it was developing at a guaranteed pace over the most recent couple of years at around 25%, Coronavirus has sped up it to help web based business organizations and F-trade administrators. Everything is moving online, from necessities and last-minute purchases to food delivery and fashion, with the estimated e-commerce market in Bangladesh expected to reach US$3 billion by 2023. Global giants like Alibaba already have a strong presence in the country with investments in various start-ups and local businesses, and Amazon is anticipated to arrive in the near future. In 2020, the pandemic sparked growth of more than 150%.

Every other month, new local e-commerce businesses open, and while these entire look promising, worrying signs include recent trends of delivery and refund failures, the absence of consumer rights protection, and unhealthy under-pricing. The public authority as of late sent off new orders to safeguard buyers from misdirection and make the business more supportable and productive. The expanded utilization of credit/charge cards and online monetary instalment passages has been a major impetus for development. Local and international players alike will be able to enter the market as long as Bangladeshis continue to embrace online shopping. Finding the right balance between inventory, pricing, delivery, and users is the key to profitability. At the end of 2020, more than 1.62 million people will have credit cards and 19.99 million will have debit cards, according to data from the Bangladesh Bank. Then again, portable monetary administrations (MFS) month to month exchange in April 2021 remained at BDT63, 478 crore (1 crore = 10 million) or US$7,381 million.

USAGE OF DIGITAL PLATFORMS

Another sector on the cusp of transformation due to the digital surge is the advertising scene. In a country like Bangladesh – where non-digital platforms like television have always been the leader, followed by print and radio – digital advertising has already overtaken print and radio to be the second biggest media platform in terms of consumption and reach.

- Five years ago, the top 20 brands allocated 3–5% of their annual ad spend to digital.
- Today, digital has risen to 10–15% of the annual advertising budget and industry analysts predict that in the next two years, digital ad spend will rise to more than 25%.
- The biggest bulk of ad spend is on Facebook, accounting for more than 60%, followed by Google, YouTube, local online news and entertainment platforms.
- Programmatic advertising, which was largely Google-based, has started to spread across multiple operators, both local and international.

Digital advertising is a necessity today for engagement and interaction with fans. Any brand entering the Bangladesh consumer landscape can use a plethora of measurable social media and online advertising platforms to penetrate the market at less than a quarter of the cost of traditional advertising modes. The revolution in digital life is being triggered and pushed forward by the continued emergence of online and app-based start-ups across multiple sectors, transforming the urban and rural lives of Bangladeshis. From B2C and B2B e-commerce, to motherhood care-giving and medical and health services, skills development and education, agriculture and enhancing farmers' livelihoods, women empowerment and entrepreneurial skill shaping, financial and online insurance and logistics and advertising, the list goes on. AI and data are the two core components driving these start-ups. Most recently, nine Bangladeshi starts up were featured in Forbes 30 under 30 Asia 2021 listing, clearly signalling the strong innovation-driven business culture in Bangladesh.

DIGITAL MODE IS BUSINESS

As the country races to become a mini-Silicon Valley, change – from the traditional way of life to AI-dominated living – is inevitable. Investments from global venture capitalists and equity investors are tapping the potential of 166 million Bangladeshis with its big population of youths embracing the future. The government plans to introduce 5G by the end of 2021 and this will surely disrupt internet-based services across all sectors by improving the current internet speed benchmarks, which have been among the lowest in the world. According to Ookla, Bangladesh's business and investment landscape – with its healthy GDP growth in the last decade making it one of the fastest-growing economies in the world – is ready for business in the digital age (Rabeth Khan D').

Face Book Usage by the People

Bangladesh is a nation of low Web insight albeit a major internet based local area is step by step creating. The socioeconomic and political issues are actively addressed by this online community. In today's society, social media have both positive and negative effects. People in Bangladesh are increasingly incorporating social media into their daily lives. Its resembles lose situation. There are many uses for social media. Nowadays, urban residents, particularly those in the young community, are required to have a Facebook account. Facebook is used as a daily opinion poll by young people. It is used for a variety of purposes by adults as well as young people who have been educated. Rural residents do not lag behind. s race is being run by them. Different people use social media for different reasons. The United States made an effort to develop a new communication technology that would replace fax, telephone, and telegraph in order to communicate with its overseas troops and security experts. They created the internet during the Cold War. The wealthy nations of Europe and Asia gradually gained access to it. In contrast, Bangladesh receives it a few years after its creation.

In 1996, Bangladeshis were given the opportunity to use the Internet for the first time. In June of 1996, the Internet went live. The results of national polls were made available to internet users via the World Wide Web shortly after the introduction of the internet. In Bangladesh, this was the very first usage event of its kind. The effort was lauded by many, particularly by Bangladeshis living abroad. Continuously Bangladeshi individuals become presented with various uses web; however countless individuals don't approach web till today. Today, social networking sites are used by more Bangladeshi Internet users than any other application. Facebook is the single most significant and widely used social networking site in our nation. News and data stream rapidly through Facebook, at times which astoundingly affect individuals of Bangladesh. Facebook launched in 2004, and after a few years, Bangladesh became its first country. In 2010, Facebook's user base surpassed 500 million, just six years after its launch. This number now exceeds 800 million. The total number of internet users in Bangladesh is 99,5560, with 83% of those users using social media. (Md. Monirul Islam, 2022) There is a large and active community that uses a variety of social media platforms.

Table 1 observed that majority of the people in Bangladesh preferred all type of Social Medias Platforms, among listed Social Medias, the Face Book with % between 89.69 and 93.79 in August and January in the year 2022. It follows YouTube usage rate between 2.58 and 5.91, Linked In (0.71 and 1.63), Twitter (0.38 and 1.77), Prnterest (0.26 and 0.58), Instagram (0.3 and 1.01), Reddit (0.2 and 0.31), Tumblr (0.01 and 0.04), Vimeo (0.01 and 0.05). Further it is discussed that the People of preferred the Social Medias rest of above all Social Medias platforms, the study categorize Type of other, it has usage % was very low with only 0.01 during January and Decembers in the Year 2022. Further, the study

Table 1. Social medias using in all platforms in Bangladesh

Year/Month	Facebook	YouTube	LinkedIn	Twitter	Pinterest	Instagram	reddit	Tumblr	Vimeo	Other
2022-01	93.79	3.65	1.02	0.43	0.48	0.36	0.18	0.04	0.05	0
2022-02	92.64	4.8	0.96	0.4	0.49	0.42	0.21	0.02	0.04	0
2022-03	92.09	4.87	1.1	0.58	0.48	0.5	0.31	0.02	0.04	0
2022-04	92.2	4.72	1.23	0.53	0.58	0.47	0.2	0.03	0.03	0.01
2022-05	93.05	3.99	1.18	0.54	0.62	0.39	0.18	0.02	0.01	0.01
2022-06	95.27	2.58	0.86	0.59	0.42	0.18	0.09	0.01	0	0.01
2022-07	93.43	3.95	1.07	0.57	0.4	0.3	0.26	0.02	0	0.01
2022-08	89.69	7.6	1.09	0.58	0.47	0.45	0.1	0.01	0	0.01
2022-09	89.87	5.91	1.63	0.66	0.77	1.01	0.11	0.03	0	0.01
2022-10	95.95	2.16	0.69	0.38	0.33	0.44	0.03	0.01	0	0.01
2022-11	92.46	4.39	1.02	1.08	0.33	0.59	0.08	0.03	0	0.01
2022-12	91.67	5.03	0.71	1.77	0.26	0.48	0.05	0.01	0.01	0.01
Average Usage	93.42	4.47	1.04	0.67	0.49	0.46	0.15	0.2	0.01	0.007

Figure 1. Average usage of social medias among the public in Bangladesh (Jan-Dec 2023)

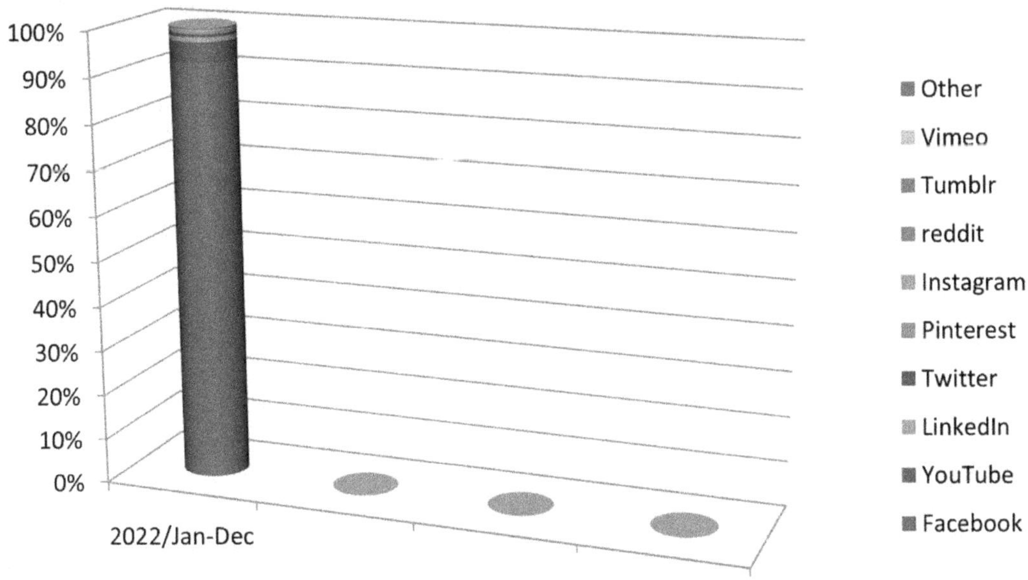

has been witnessed about average usage of Social Medias, Facebook (93.42), YouTube (4.47), Linked In (1.04), Twitter (0.67), Pinterest (0.49), Instagram (0.46), reddit(0.15), Tumblr (0.2), Vimeo (0.01) and Other (0.007).

Table 2 analyzed the that majority of the people in Bangladesh preferred all type of Social Medias Platforms, among listed Social Medias, the Face Book with % between 89.69 and 93.79 in August and January in the year 2022. It follows YouTube usage rate between 2.58 and 5.91, Linked In (0.71 and 1.63), Twitter (0.38 and 1.77), Prnterest (0.26 and 0.58), Instagram (0.3 and 1.01), Reddit (0.2 and 0.31),

Table 2. Desktop social media stats Bangladesh (Jan-Dec 2022)

Date	Facebook	YouTube	LinkedIn	Pinterest	Twitter	Instagram	reddit	Tumblr	Vimeo	VKontakte	Other
2022-01	69.33	14.52	7.44	3.69	1.83	1.33	1.16	0.35	0.33	0.01	0.01
2022-02	68.38	15.76	7.09	4.13	1.5	1.3	1.35	0.17	0.29	0.01	0.02
2022-03	60.22	18.6	9.07	4.41	2.96	2.16	2.09	0.17	0.3	0.02	0.01
2022-04	62.2	18.15	9.05	5.07	1.89	1.84	1.4	0.18	0.18	0.02	0.01
2022-05	63.75	17	9.53	5.22	2.33	1.01	0.93	0.14	0.06	0.01	0.01
2022-06	65.82	15.26	9.35	5.07	2.96	0.79	0.57	0.13	0	0.02	0.01
2022-07	55.68	21.04	12.41	4.89	3.17	1.15	1.39	0.21	0.01	0.03	0.01
2022-08	49.7	28.26	11.82	5.18	3.06	1.18	0.63	0.1	0	0.04	0.01
2022-09	54	21.76	13.45	6.68	1.97	1.24	0.62	0.24	0	0.02	0.01
2022-10	68.73	14.1	9.43	4.61	1.61	0.91	0.4	0.14	0	0.02	0.05
2022-11	61.29	19.41	10.72	3.55	2.82	1.24	0.68	0.2	0	0.03	0.04
2022-12	61.88	21.22	8.44	3.03	3.82	0.97	0.41	0.14	0.01	0.01	0.06

Tumblr (0.01 and 0.04), Vimeo (0.01 and 0.05). Further it is discussed that the People of preferred the Social Medias rest of above all Social Medias platforms, the study categorize Type of other, it has usage % was very low with only 0.01 during January and Decembers in the Year 2022.

Table 3 observed that majority of the people in Bangladesh preferred all type of Social Medias Platforms, among listed Social Medias, the Face Book with % between 89.69 and 93.79 in August and January in the year 2022. It follows YouTube usage rate between 2.58 and 5.91, Linked In (0.71 and 1.63), Twitter (0.38 and 1.77), Prnterest (0.26 and 0.58), Instagram (0.3 and 1.01), Reddit (0.2 and 0.31), Tumblr (0.01 and 0.04), Vimeo (0.01 and 0.05). Further it is discussed that the People of preferred the Social Medias rest of above all Social Medias platforms. The study categorize the type of other, it has usage % was very low with only 0.01 during January and Decembers in the Year 2022.Figure 2 was witnessed the average usage of Mobile Social Media usage statistics, Facebook (95.01), YouTube (3.39), Twitter (0.53), LinkedIn(0.39), Instagram (0.4), Pinterest (0.15), reddit (0.08) and Other (0.01).

CONCLUSION

Web-based entertainment began as a method for collaborating with loved ones however was subsequently embraced by organizations that needed to exploit a famous new specialized strategy to contact clients. The ability to connect and share information with anyone on Earth or with a large number of people simultaneously is a key feature of social media's power. More than 3.8 billion people worldwide use social media. Virtual entertainment is a steadily changing and consistently developing field, with new applications, for example, TikTok and Clubhouse coming out apparently consistently, joining the positions of laid out informal communities like Facebook, YouTube, Twitter, and Instagram.

Table 3. Mobile social media stats Bangladesh

Date	Facebook	YouTube	Twitter	LinkedIn	Instagram	Pinterest	reddit	Other
2022-01	96.09	2.62	0.29	0.43	0.27	0.18	0.08	0.03
2022-02	94.96	3.76	0.29	0.39	0.32	0.16	0.1	0.03
2022-03	94.83	3.7	0.36	0.45	0.34	0.15	0.14	0.03
2022-04	94.89	3.52	0.4	0.54	0.34	0.18	0.09	0.03
2022-05	95.67	2.82	0.38	0.45	0.34	0.21	0.12	0.01
2022-06	97.12	1.78	0.44	0.33	0.14	0.12	0.06	0.01
2022-07	95.75	2.9	0.41	0.38	0.25	0.13	0.19	0.01
2022-08	92.38	6.22	0.41	0.37	0.4	0.15	0.06	0.01
2022-09	93.01	4.52	0.54	0.61	0.99	0.25	0.07	0.01
2022-10	97.37	1.54	0.31	0.24	0.42	0.1	0.01	0.01
2022-11	94.63	3.34	0.96	0.35	0.55	0.11	0.04	0.02
2022-12	93.49	4.04	1.66	0.23	0.45	0.09	0.03	0.01
Average Usage	95.01	3.39	0.53	0.39	0.4	0.15	0.08	0.01

Stat Counter Global statistics https://gs.statcounter.com/social-media-stats

Figure 2. Average usage mobile social media in Bangladesh

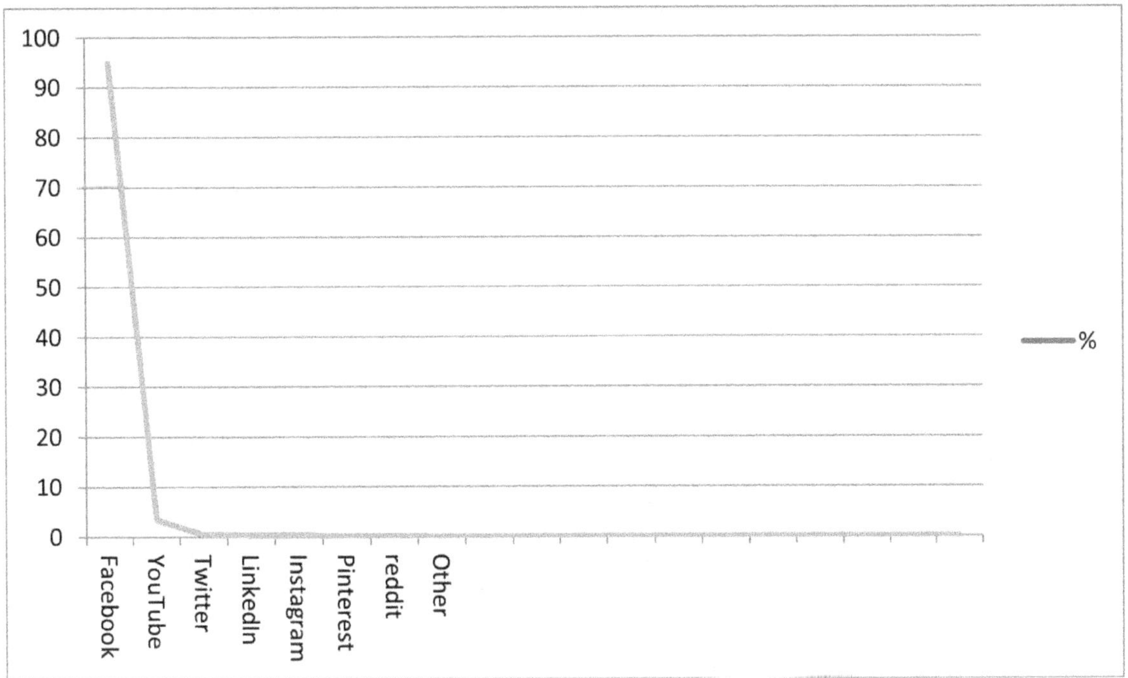

It is anticipated that the number of people using social media in the United States will reach approximately 257 million by 2023. The way we interact with one another online has changed as a result of social media. It enables us to connect with one another, keep in touch with long-distance friends, and access an infinite amount of information at our fingertips, as well as to learn about what's going on in the world in real time. In many ways, social media has made the world seem more approachable and helped many people online find common ground. As per an overview by Seat Exploration center, the utilization of web-based entertainment is corresponded with having more companions and more different individual organizations, particularly inside arising economies. some youngsters, fellowships can begin essentially, with 57% of teenagers meeting a companion on the web.

REFERENCES

Baskaran, C. (2020). Research Patterns on the Social Networks and Media: A Scientometric Portrait. Handbook of Research on Emerging Trends and Technologies in Library and Information Science, pp.189-207. IGI Global. doi:10.4018/978-1-5225-9825-1.ch014

Bennett, S., Maton, K., & Kervin, L. (2008). The 'digital natives' debate: A critical review of the evidence. *British Journal of Educational Technology*, *39*(5), 775–786. doi:10.1111/j.1467-8535.2007.00793.x

Data Reportal. (2021). *Global Social Media Stats October 2021*. Data Reportal. https://datareportal.com/reports/digital-2021-october-global-statshot

Digital Marketing Institute. (2021). *Social Media: What Countries Use It Most & What Are They Using?* Digital Marketing Institute. https://digitalmarketinginstitute.com/blog/social-media-what-countries-use-it-most-and-what-are-they-using

Junco, R., Heiberger, G., & Loken, E. (2010). The effect of Twitter on college student engagement and grades. *Journal of Computer Assisted Learning*, *27*(2), 119–132. doi:10.1111/j.1365-2729.2010.00387.x

Mander, J. (2017). *Daily time spent on social networks rises to over 2 hours*. Global Web Index. https://blog.globalwebindex.com/chart-of-the-day/daily-time-spent-on-social-networks

Monirul Islam, Md. (2022). *Social Media Adaptation and Its Impact in Bangladesh*. Research Gate. https://www.researchgate.net/publication/358470840_Social_Media_Adaptation_and_Its_Impact_in_Bangladesh

Pew Research Center. (2021). *Social media users more likely to interact with people who are different from them*. Pew Research. https://www.pewresearch.org/internet/2019/08/22/social-media-users-more-likely-to-interact-with-people-who-are-different-from-them/

Rabeth, K. (2021). *A peek into the digital emergence of Bangladesh*. WARC. https://www.warc.com/newsandopinion/opinion/a-peek-into-the-digital-emergence-of-bangladesh/en-gb/4397

Chapter 12
Information Literacy Competence on Social Media Use Against Scholarly Communication:
A Case Study Among Faculty Members and Scholars in the Faculty of Management of Alagappa University

C. Baskaran
Alagappa University, Karaikudi, India

ABSTRACT

The chapter examines information literacy competence on social media use against scholarly communication. A case study among faculty members and scholars in the faculty of management of Alagappa University, India. 340 respondents participated from eight departments, with the majority of them 51(15%) of them from Alagappa Institute of Management. sex and Age wise respondents of study, the higher potion of population being male with 238 (70%), less than half in the respondents 102 (30%) were female. The study witnesses the multiple responses about usage the different social Medias listed in the table. Majority of 324(95.29%) of them preferred face book. The respondents 11.47%, 10%, 0.35%, 0.41%, 0.35%, 0.32% and 0.55% of them have witnessed as ''Not Satisfied'' with Academia, LinkedIn, Research Gate, Mendeley, Google Scholar, Web of Science and Scopus respectively.

INTRODUCTION

Social media can serve a useful and direct purpose for businesses looking for ways to sell their products and services and promote their brands. Of course, there are other ways to take advantage of social media. Most people use social media to socialize, communicate with family and former colleagues, and keep

DOI: 10.4018/978-1-6684-8805-8.ch012

in touch with old classmates. Facebook grew out of the idea of a technological way to keep in touch with people. Facebook adds 500,000 new accounts every day. This equates to 6 new profiles per second (Smith, 2019). So who exactly is on your Facebook friends list? Who is subscribed to your channel on YouTube? Who is retweeting your tweets?, If you want to extend your influence beyond your area of expertise, you need to establish a strong connection base within your network. The average person has 5.54 social media accounts (Smith, 2019). Whatever their purpose, will these accounts be projected across multiple platforms? Who is engaged? These are all important questions about your potential reach as an individual. The point is that if you want to be involved in community activities at work, you must first be active in that particular community.

Social media and Research

Research Communication relies on scientists and faculty contributing research papers, chapters, and proceedings. Now let's turn around and focus on another way scientists are using social media to advance their research. All social media channels have the potential for massive reach, but it all depends on who you connect with and who you hire. However, while the rapid and active aspect of social media may deter skeptical researchers from publications, good response and rapid responses may increase the researchers' credibility. Research is about generating new information, and social media offers a unique opportunity to showcase new content. As scientists, once we publish our research, we want to share it with as many colleagues and people as possible to read new insights. I want to share my hard work with many people. Nearly all researchers send emails to colleagues and individuals within their institution. This is essentially the first step in promoting your research. What if you could take it one step further and reach out to the scientific community through social media? Perhaps you already have social media accounts and have accounts that are relevant to the scientific community. Research Gate is a popular academic social media portal. Research Gate is a website that provides a forum for scientists to share and discuss research and find collaborators. Share your research results on your personal account so only those you contact can see them. However, if she is active in various communities related to her field or on sites of special interest and participates in regular discussions with other researchers on those sites, potentially wider reach. Again, it all depends on what you want to achieve with your research.

Social media can be useful and direct purpose for businesses that sell a product or service and are searching for ways to advertise their brand. Of course, there other ways to use social media. Most people use social media to be, well, social, and communicate with family, former colleagues, or keep in touch with old classmates. The idea of a technological way of staying in touch with people is how Facebook was created. Facebook adds 500,000 new accounts each day, which equates to 6 new profiles every second (Smith, 2019). So who exactly is in your friends list on Facebook? Who is subscribing to your channel on YouTube? Who is re-tweeting your tweets? If you want to broaden your impact beyond your discipline, you need to have a strong base of connections in your network. The average person has 5.54 social media accounts (Smith, 2019). Of those accounts, whatever one's goal is, is it being projected across multiple platforms? Exactly who is engaged? These are all important questions that deal with your potential reach as an individual. The bottom line is if you seek community engagement in what you are doing, you must first be active in that particular community.

REVIEW OF LITERATURE

The Surface learning is characterized by the learner's approach of completing only the minimum amount of content necessary to meet the assessment requirements, whether just learning the content presented in the test or participating in and completing the activity (Rose, 2000). Mortensen and Walker (2002) argued that the informal nature of blogging provided an opportunity to write, share and solicit comments from others about raw ideas before proceeding to formal publication of scholarly research. Ahmad, A. (2011) states that the use of collaborative technologies such as blogs and his SNS leads to instant his online his community where people can communicate with each other quickly and conveniently, and also states that "the individuals used and their It provides a wealth of information about the network." It can be used for various business purposes. Bicen and Cavus (2010) investigated student Internet use. Department of Computer Education, Department of Education Technology.

Coat&& Bansode (2017) shows that at Shivaji University in Kolhapur, 98.94% of respondents use Social He network technology for research purposes.1.06% of respondents do not use social networking his technology for research purposes. Perryman (2011) pointed out that social networks are now ubiquitous and used by people of all ages for a variety of purposes. Hadagali (2016) investigated scientists' attitudes toward social network use. Mohammad and Smithha (2011) investigated that the majority of research students use social media for friendly communication. Ruleman (2012) stated that both teachers and students regularly used the social networks for academic purposes. Bhatt and Kumar (2014) found that the majority (98%) of students have Facebook accounts. Madhusudhan (2012) investigated the use of social networks by academics at the University of Delhi. He found that social media helps researchers collaborate with colleagues to create a learning environment. The majority of his publications, 2,207 (18.48%), were published from the communications field. This is due to the special nature of research focused on social networks. In publications Computer Communication Information Systems ranked him second with 1,172 (9.81%). Baskaran (2020) analyzed that the majority of papers, 2,207 (18.48%), were published from the communication field, as the research focused on social networks. Computer Communication Information Systems ranked him second with 1,172 (9.81%) publications.

METHODOLOGY

The study was conducted based on a questionnaire method targeting respondents from the Faculty of Management in the Alagappa University. The Respondents are faculty and research Scholars. Out of 360 questionnaires distributed to the respondents by the Faculty of Management of the Alagappa University. In total, the complete filled up data, a total of 340 questionnaires, were received from the respondents for analysis in selected departments of the Faculty of Management of the Alagappa University. The collected data were tabulated and analyzed using statistical tools. Therefore, the results of this study are based solely on the questionnaire responses returned by the respondents.

RESULTS AND DISCUSSIONS

Table 1 analyses the respondents of the study was undertaken by the respondents towards share the scholarly communication by social Medias to get the experience of Information Literacy skill based on data collected for the study. The result finds the respondents participated out of 340 for eight departments, majority of them 51(15%) of them from Alagappa Institute of Management, it follows, International Business with 49 (14.41%), Corporate Secretaryship with 47 (14.41%), Banking Management with 45 (13.23%), commerce with 43 (12.64%), Logistics Management with 42 (12.35%) Hotel Management and Catering Technology with 42(12.35%) in the respondents were responded to the study (Figure 1).

The study discloses the respondents atop the information literacy skill where share scholarly communication through the Social media among the respondents in the faculty of management in Alagappa University. Table 2 analyses the sex and Age wise respondents of study, the higher potion of population being male with 238 (70%), less than half in the respondents 102 (30%) were female. The study has been also discussed more than 67% in the respondents were in the age group between 20 and 25. The study finds to reveal upon the small portion of population as less than 1% in the respondents above age of 25.

The states the respondents type of Social Media use by the Faculty members and Research Scholars in the faculty of Management in Alagappa University. Table 3 analyzes the respondents prefer the

Table 1. Department wise respondents of faculty members and research scholars of management

S.No	Name of Department	Frequency (n=340)	Percentage (n=340)
1	Alagappa Institute of Management	51	15
2	Corporate Secretaryship	47	13.82
3	International Business	49	14.41
4	Commerce	43	12.64
5	Banking Management	45	13.23
6	Logistics Management	42	12.35
7	Disaster Management	21	6.17
8	Hotel Management and Catering Technology	42	12.35

Figure 1. Department wise respondents of faculty members and research scholars of management

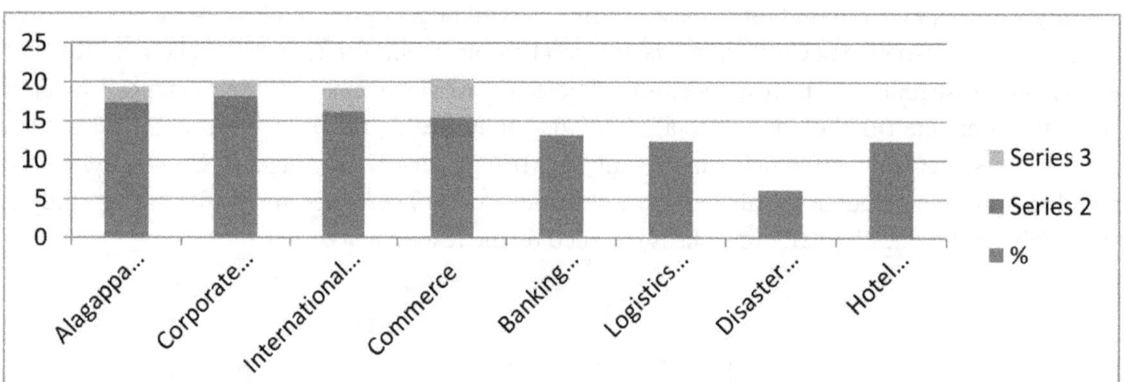

Table 2. Gender age wise respondents faculty members and research scholars in the faculty of management

Gender	Frequency (n=340)	Percentage(n=340)
Male	238	70
Female	102	30
Age	Frequency (n=340)	Percentage(n=340)
20-25	229	67.35
<25	11	0.03

Table 3. Types of social media use by the faculty member sand research

Name of Social Media	Frequency (n=340)	Percentage(n=340)
Academia.edu	158	46.47
Classmates.com	52	15.29
Facebook	324	95.29
Flickr	15	4.41
Fotolog	08	2.35
Google+	210	61.76
LinkedIn	318	93.52
LiveJournal	62	18.23
Netlog	12	3.52
Twitter	292	85.88
MySpace	42	12.35
Cyworld	39	11.47
Bebo	42	12.35
YouTube	285	83.82
Badoo	21	6.17
BlackPlanet	14	4.11

* Multiple preferences in the respondents

different type of social medias for sharing scholarly communication. The study witnesses the multiple responses about usage the different social Medias listed in the table. Majority of 324(95.29%) of them preferred face book, it follows Linked In with 318 (93.52%), Twitter with 292 (85.88%), YouTube with 285 (83.82%) and Academia with 158 (46.47%) in the respondents have preferred the Social Medias. The study incorporates on the respondents are given preferences given to the Flickr, Fotolog, Netlog, Badoo and Black Planet by less than 10% of them used (Figure 2)

The study has been found from table 4, about the source of Information acquired by the respondents to get knowledge of Information Literacy to share scholarly information at faculty of Management in Alagappa University.

The study has been witnessed the majority (64.70%)of the respondents acquired information from Internet, which is a helpful resource for all research scholars, It follows Video/Audios (56.47%), Journals/

Figure 2. Types of social media use by the faculty member sand research scholars in the faculty of management

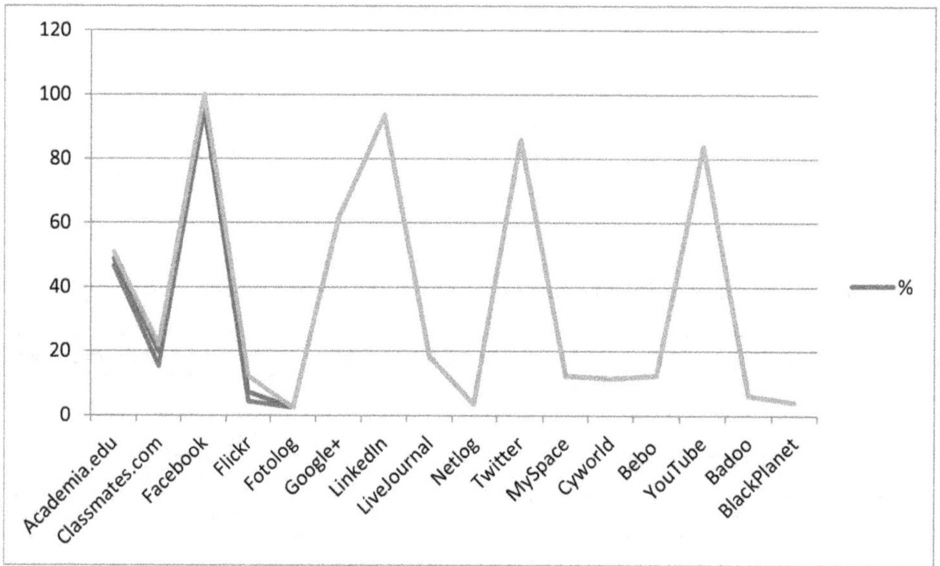

Table 4. Sources of information acquired by the respondents through social media

Sources	Frequency (n=340)	Percentage(n=340)
Internet	220	64.70
Journals/papers	187	55.0
Videos/Audios	192	56.47
Conference proceedings	125	36.76
Textbooks	174	51.17
Reference books	113	33.23
E-books	126	37.05
E-journals	142	41.76
E-magazines	102	30.0
Blogs	71	20.88

* Multiple preferences in the respondents

Table 5. Usefulness of information literacy skills by use social medias among faculty members and research scholars

Description	Very Good	Good	Moderate	Poor	Very Poor
Save time	116(34.11)	135(39.70)	48(14.11)	26(16.47)	15(0.44)
Improve quality of Research work	138(40.58)	98(28.82)	67(19.70)	19(0.55)	18(0.52)

Table 6. Research social media platforms use among faculty members and research scholars in the faculty of management

Type of Platforms	Very Highly Satisfied	Highly Satisfied	satisfied	Less satisfied	Not Satisfied
Academia	98 (28.82)	89(26.17)	72(21.17	42(12.35)	39(11.47)
Linked In	137(40.29)	76(22.35)	54(15.88)	39(11.47)	34(10.00)
Research Gate	235(69.11)	47(13.82)	27(0.79)	19(0.55)	12(0.35)
Mendeley	187(55.0)	87(2.55)	51(15.0)	19(0.55)	14(0.41)
Google Scholar	145(42.64)	127(37.5)	32(0.94)	24(7.05)	12(0.35)
Web of Science	152(4.47)	117(34.41)	29(0.85)	31(0.91)	11(0.32)
Scopus	141(41.47)	119(35.0)	37(10.88)	24(0.70)	19(0.55)

papers (55%), Textbooks (51.17%), E-journals (41.76%), E-books (37.05%), Reference books (33.23%), E-magazines (30.0%) and Blogs (20.88%).

The Information Literacy skills needed among the respondents in the study, to share use Social Medias among Teachers and Research Scholars in the faculty of Management. Table 5 has been analyzed about usefulness with information Literacy skill shall be useful for disseminating scholarly information among the respondents.

It is found the on the opinion with '' Save time'' the respondents replied by majority of 135 (39.70%) of them have noticed as Good, it follows very Good (34.11%), Moderate (14.11%), Poor (16.47%) and Very Poor (0.44%). The study has been explored about usefulness of information Literacy skill shall be useful for disseminating scholarly information among the respondents, the concerning about ''Improve quality of research work'' as the opinion was given by the respondents in the study. The majority in the respondents (40.58%) of them have noticed as Very Good, it follows Good (28.82%), Moderate (19.70%), Poor (0.55%) and very Poor (0.52%).

Table 6 analyzes the Research Social Media platforms use among Faculty members and Research Scholars in the faculty of Management in Alagappa University. The study has been witnessed about research social medias were used by the respondents, the majority of them have noticed as very Highly satisfied by 28.82%, 40.29%, 55%, 42.64%, 4.47%, and 41.47% of them used the research social media platforms Academia, LinkedIn, Research Gate, Mendeley, Google Scholar, Web of Science and Scopus respectively.

The respondents have replied their opinion on use of Research Social Media platforms, the respondents 11.47%, 10%, 0.35%, 0.41%, 0.35%, 0.32% and 0.55% of them have witnessed as ''Not Satisfied'' with Academia, LinkedIn, Research Gate, Mendeley, Google Scholar, Web of Science and Scopus respectively.

Table 5 presents the information about Purposes of use Social Medias on Information Literacy Skill acquired among Faculty members and Research Scholars in the faculty of Management in Alagappa University. The list of opinions are provided the purposes on Prepare /Share Assignment on Research, majority of 39.41% of them replied ''Highly satisfied '' and lowest 3.52% of them with ''Not satisfied''. The Second purposes on Examination preparation, the majority of 28.52% of them have noticed ''Very Highly Satisfied'' and less no of respondents 11.47% of them have preferred '' Satisfied''. Third purposes on ccontinuous Assessment preparation, Majority of 40.29% them have noticed '' Very Highly Satisfied'' and 6.17% in the respondents have witnessed ''Less satisfied''.

Table 7. Purposes of use social medias acquired the research information among faculty members and research scholars

Purposes	Very Highly Satisfied	Highly Satisfied	Satisfied	Less Satisfied	Not Satisfied
Prepare/share Assignment	116(34.11)	134(39.41)	42(12.35)	30(8.82)	12(3.52)
Examination preparation	97(28.52)	74(21.76)	39(11.47)	86(25.29)	44(1.29)
Continuous Assessment preparation	137(40.29)	91(26.76)	74(21.76)	21(6.17)	17(5.0)
Publishing papers	158(46.47)	81(23.82)	68(20.0)	31(9.11)	02(0.58)
For chatting	116(34.11)	81(23.82)	93(27.35)	39(11.47)	11(3.23)
Commenting on friends profile	118(34.70)	72(21.17)	81(23.82)	47(13.82)	12(3.52)
Share/get information on Academic and Research information	131(38.52)	53(15.58)	49(14.41)	38(11.17)	68(20.0)
Upload and share pictures and photographs	135(39.70)	68(20.0)	57(16.76)	49(14.41)	31(09.11)
Share Citations	127(37.35)	81(23.82)	51(15.0)	34(10.0)	47(13.82)
Share Bibliographical records	119(35.0)	74(21.76)	62(18.23)	29(8.52)	56(16.47)

The study has been noticed that majority in the respondents have opted their choices by 46.47%, 34.11%, 34.70%, 38.52%, 39.70%, 37.35% and 35% in the respondents have replied as Very Highly Satisfied with the purposes on Publishing papers,For chatting,Commenting on friends profile, Share/get information on Academic and Research information,Upload and share pictures and Photographs,Share Citations and Share Bibliographical records respectively.

The analysis for study has been witnessed the respondents have opined ''Very Highly Satisfied '' excluded the purpose ''Examination Preparation'' others more than 100 respondents. The opinion on '' Highly Satisfied '' all in the purposed have witnessed above 50 respondents. The opinion on '' Satisfied '' the purposes excluded ''Prepare/share Assignment'',''Examination Preparation'', Share/get information on Academic and Research information'' others have replied by above 50. The opinion on '' Less satisfied'' the purpose excluded ''Examination preparation'' others by less than 50. The opinion on '' Not Satisfied'' the purposes excluded ''Share/get information on Academic and Research information'' and ''Share Bibliographical records'' have witnessed by the respondents less than 50.

Table 8 shows the peculiar challenges accessed by the faculty members and research scholar while get the information through Social Medias in the faculty of Management in Alagappa University. The result finds that the respondents they have faced challenge with '' Time not affordable to use Social Medias'' by 134 (39.41%) of them replied ''Very Strongly Agreed'', it follows, the respondents have noticed on Less Agreed 923.82%), Strongly Agreed (18.23%), Agreed (10%) and No comment (8.52%). The second challenge on ''Not allowed to use social media in our Class Students on use of social media'', the responders have witnessed their preference as majority of 38.23% of them have stated '' Very Strongly Agreed'', followed by (20.0%), (16.76%),(13.82%), (11.17%) of them have noticed as ''Strongly Agreed'', ''No comment'', ''Agreed'' and ''Less Agreed'' respectively.

The Third challenge was faced by the respondents with ''insufficient access to the Internet and inability to subscribe to internet data '' was observed by the majority of 93(27.35%) of them stated ''Very Strongly Agreed'', it followed by 72(21.17%),68(20.0%), 58(17.05%) and 49(14.41%) of them have witnessed Strongly Agreed, Less Agreed, No comment and Agreed respectively. The fourth challenge was faced by the respondents with ''My privacy is often tampered with on social media'', the majority of

Table 8. Challenges faced while use social medias on information literacy skill acquired among faculty members and research scholars

Types of Challenges	Very Strongly Agreed	Strongly Agreed	Agreed	Less Agreed	No comment
Time not affordable to use Social Medias	134(39.41)	62(18.23)	34(10.0)	81(23.82)	29(8.52)
Not allowed to use social media in our Class	130(38.23)	68(20.0)	47(13.82)	38(11.17)	57(16.76)
Insufficient access to the internet and inability to subscribe to internet data	93(27.35)	72(21.17)	49(14.41)	68(20.0)	58(17.05)
My privacy is often tampered with on social media	81(23.82)	57(16.76)	31(09.11)	68(20.0)	103(30.29)
Social medias URLs/websites are blocked on my school's internet server	93(27.35)	68(20.0)	49(14.41)	31(09.11)	99(29.11)
Don't have smartphone for using Social Media	68(20.0)	74(21.76)	47(13.82)	38(11.17)	113(33.23)
In sufficient data storage for using Social Media	200 (58.82)	53(15.58)	49(14.41)	21(6.17)	17(5.0)

103(30.29%) of them have stated ''No Comment'', the compound of 69.70% of respondents have noticed ''Very Strongly Agreed'', ''Strongly Agreed'', ''Less Agreed'' and '' Agreed'' respectively. The fifth changes of '' Social Medias URLs/websites are blocked on my school's internet server'' was faced by the respondents group share 64.41% of them have noticed ''Very Strongly Agreed'', ''Strongly Agreed'', ''Agreed'' and ''Less Agreed'', the majority 99(29.11) of them have replied as '' No Comment''.

The Sixth challenge was ''don't have smartphone for using Social Media'', the majority 113(33.23%) of them have '' No opinion'', 66.76% of them collective responses regard to ''Very Strongly Agreed'', ''Strongly Agreed'', ''Agreed'' and ''Less Agreed'' respectively. The Sixth challenges were '' In sufficient data storage for using Social Media'' faced by the respondents in the faculty of Management in Alagappa University, the majority 58.82% of them have noticed ''Very Strongly Agreed'', the rest of 41.17% of them have reported '' Strongly Agreed'', ''Strongly Agreed'', ''Agreed ''and ''Less Agreed''

MAJOR FINDINGS

1. More than 67% in the respondents were in the age group between 20 and 25.
2. It has found Linked In with 318 (93.52%), Twitter with 292 (85.88%), YouTube with 285 (83.82%) and Academia with 158 (46.47%) in the respondents have preferred the Social Medias.
3. The majority (64.70%) of the respondents acquired information from Internet, which is a helpful resource for all research scholars.
4. The study has been analysed Information Literacy skill shall be useful for disseminating scholarly information among the respondents, the concerning about ''Improve quality of research work'' as the opinion was given by the respondents in the study.
5. About research social medias were used by the respondents, the majority of them have noticed as very Highly satisfied by 28.82%, 40.29%, 55%, 42.64%, 4.47%, and 41.47% of them used the research social media platforms Academia, LinkedIn, Research Gate, Mendeley, Google Scholar, Web of Science and Scopus respectively.
6. The purposes on Prepare /Share Assignment on Research, majority of 39.41% of them replied ''Highly satisfied '' and lowest 3.52% of them with ''Not satisfied''.

7. The respondents they have faced challenge with '' Time not affordable to use Social Medias'' by 134 (39.41%) of them replied ''Very Strongly Agreed''.

CONCLUSION

This study attempted to analyze the information literacy required to share social media to disseminate scientific information among respondents from the Faculty of Management in the Alagappa University. Today, social media allows us to share all kinds of information that we can access anytime, anywhere in the world. Social media is the primary place where information is used and shared. In particular, researchers and faculty have an urgent need for information literacy to obtain academic communication on the Internet regarding book materials and non-book papers. Information literacy is a basic need for everyone in the modern technological age, as new technical challenges emerge from the field of technology experimentation every day. Therefore, teachers and scientists needed information literacy skills. Based on the above analysis and observations, this study concludes that most students are aware of the sources of information available in the library. Respondents' most preferred shared social media platforms are teacher and faculty research social media by Facebook, WhatsApp, YouTube, Academia, Research Gate, Academia, Mendeley and Google+. The majority of respondents were 46.47%, 34.11%, 34.70%, 38.52%, 39.70%, 37.35% and 35%. Respondents said they were "very satisfied" with goals such as publishing a paper, chatting and commenting on friends' profiles, and sharing and receiving academic and scientific information. Respondents faced the challenge of "not enough internet access to subscribe to internet data". The majority (27.35%) of the 93 respondents answered that they "strongly agree".

REFERENCES

Ahmad, A. (2011). Rising of social network websites in India overview. *International Journal of Computer Science and Network Security, 11*(2), 155–158.

Baskaran, C. (2020). *Research Patterns on the Social Networks and Media: A Scientometric Portrait.* Handbook of Research on Emerging Trends and Technologies in Library and Information Science, 189-207. IGI Global. https://www.igi-global.com/chapter/research-patterns-on-the-social-networks-and-media/241564

Baskaran, C. (2021). An Exponential trend and Mapping of Research in Web 2.0, *Library Philosophy and Practice (e-journal),* 5008. https://digitalcommons.unl.edu/libphilprac/5008

Bhatt, R. K., & Kumar, A. (2014). Student opinion on the use of social networking tools by libraries. *The Electronic Library, 32*(5), 594–602. https://www.emerald.com/insight/content/doi/10.1108/EL-09-2012-0110/full/html. doi:10.1108/EL-09-2012-0110

Bicen, H., & Cavus, N. (2010). The most preferred social network sites by students. *Procedia: Social and Behavioral Sciences, 2*(2), 5864–5869. doi:10.1016/j.sbspro.2010.03.958

Boud, D., Keogh, R., & Walker, D. (1985). Writing and Reflection. In *Reflection: Turning Experience into Learning*. Nichols Publishing Company. https://craftingjustice.files.wordpress.com/2017/04/david-boud-rosemary-keogh-david-walker-reflection_-turning-experience-into-learning-routledge-1985-pp-1-165.pdf

Khot, N., & Bansode, N. (2017). Use of social networking sites by research scholars of Shivaji University, Kolhapur. *International Journal of Research in Library Science*, *3*(1), 81–90.

Madhusudhan, M. (2012). Use of social networking sites by research scholars in the University of Delhi: A study. *The International Information & Library Review*, *44*(2), 100–113. doi:10.1080/10572317.2012.10762919

Perryman, B. E. (2011). Use of social networking as an instructional aide in higher education [Doctoral dissertation, Oklahoma State University]. https://hdl.handle.net/11244/7193

Rosie, A. (2000). Online pedagogies and the promotion of "deep learning". *Information Services & Use*, *20*(2/3), 109–116. doi:10.3233/ISU-2000-202-306

Ruleman, A. B. (2012). Comparison of student and faculty technology use. *Library Hi Tech News*, *29*(3), 3. doi:10.1108/07419051211241877

Smith, K. (2019). 123 Amazing social media statistics and facts. *brandwatch*, https://www.brandwatch.com/blog/amazing-social-media-statistics-and-facts/

Chapter 13
User Attitudes on Social Media Among the Research Scholars With Social Science Backgrounds in the Universitie: A Questionnaire-Based Study

P. Pitchaipandi
Alagappa University, Karaikudi, India

ABSTRACT

The objectives of the present study were to investigate the influences of demographics and social media site preferences. The data were analyzed by using SPSS 23 statistical analysis software package. The results of the study were obtained by applying the normality test, for simple percentage analyses on the obtained data. When the chi-square analysis between the opinions of social science scholars concerning the level of use of social media and the level of use of academic research was examined, a positive and significant analysis was found between them. The sample comprised 623 social science scholars. The numbers of this study were: male 335 (53.8%) and female 288 (46.2%), 13.0%, and 12.9% of respondents preferred WhatsApp at Annamalai University. 12.0% of the respondents prefer YouTube from Alagappa University.

INTRODUCTION

Nowadays, researchers also prefer to conduct their research on social networking sites due to their growing popularity. A variety of social networking sites, including Academia.edu, Facebook, and Research Gate, among others. Enables contemporary researchers to improve their ability to acquire, share, and disseminate information. These social networking sites, according to many academics, help students keep up with the latest developments in their fields. According to Boyd & Ellison (2007), SNSs are profile-based websites that allow users to view, visit, and share their lists of social connections with other

DOI: 10.4018/978-1-6684-8805-8.ch013

members. The creation, use, and application of data and information are all aspects of research. The examination industry has developed a number of systems with the intention of encouraging specialists to share information. Academic journals and gatherings, as well as quality assurance components like arrangement boards, peer audit, distribution, and surveys, are included in this category.

In the form of a diagram, we might discuss this as follows: In this model we have four stages (ID, creation, quality attestation and dispersing) which are upheld by a collection of social affiliations and kinds of joint exertion. In this context, the term "joint effort" refers to efforts made by all parties involved in research, including scientists, administrators, funders, and the general public. The exploration network's ability to share information and learn from other people's efforts depends on each stage. Each of these phases may benefit from the presence of social tools. However, they might also challenge the methods by which research is carried out. After discussing each of these phases, we examine how online life can challenge and improve current thinking and practice. In today's world, social networking sites have emerged as the new method of communication.

REVIEW OF LITERATURE

According to Al-Mukhaini et al., (2014) discussed the social media boosts self-motivation and creates an engaging, personalized learning environment. According to Akçayr (2017), the primary motivating factor for them to use social media for educational purposes was an efficient and quick means of technology communication. Baskaran (2019) observed how researchers at a few Tamil Nadu universities used social media and other media. Out of the 273 people who participated in the study, 87 (31.9%) of them came from Annamalai University, and 16.8% and 15% of them were male and female, respectively. According to Al-Alwan (2014), a survey of 671 Jordanian students found that parental involvement has an indirect impact on academic performance and that student engagement has a direct impact on academic performance. The statements of social media for collaborative learning, the elements of students' academic performance, and the indicators of interactivity with peers and teachers were adapted from (AL-Rahmi & Othman, 2013).

Roslina Wan, Othman, and others 2016) mentioned that students' interactions with friends and family have been dominated by social media, which influences their self-concepts. Taking into account the wide use of the web-based entertainment, it is trusted that this study has given new experiences to grasp the possible worth of virtual entertainment on undergrads. Applications/Improvements: Realizing the effects of social media on student performance, further research should focus on the use of this media for academic purposes. Basri Hassan, Sharyna Shafawi, 2018) have examined that examination the parts that drive users" responsibility with online life and certifiable library use from the two clients and librarians" perspectives. The revelations showed that Data Quality, Saw Intelligence, and Saw Net Advantage have precisely shown to be significant marks of clients.

According to a study conducted by Levula, A., and Harré, M. (2016), informal organization estimates do have a significant impact on MH, with social segregation (SI) having the greatest impact (=0.198, p0.001), followed by social associations (SCs) (=0.141, p0.001), and finally social trust (ST) (=0.071, p0.001). Josserand and co. 2017) mentioned that the discoveries made during exploration are two-fold. In spite of previous investigations, the authors first discover that individuals' social capital contributes to both investigation and misuse at the specialty unit level. Littman and others 2016) inspected the assessment unveil to continuing and make and further develop Social Channel Chief (SFM), an open source application helping specialists gathering data from Twitter's Programming interface for their investigation.

Liu and co. 2018) emphasized the "211 task's dissemination of SNs used by 112 libraries in Chinese colleges. After looking at the library sites of 112 Chinese "211 Project," the distribution of SNSs used by virtual networks, portable applications, customized administration, and application stage was directed. Meenakshi and others 2017) inspected that there are ten things utilized including reaction adequacy and conduct viability to report about the ideal interest group via web-based entertainment for wellbeing efforts. Age or ethnicity is used to determine who the intended audience is. In total, 382 Kuwaiti citizens were contacted for responses, which were then analyzed for the purpose of research, according to Bin Salamh Bin (2019). 87% of respondents believe that social media is a successful tool for maintaining a company's reputation, and the majority of respondents (82%) are at least occasionally affected by news and information on social media. Due to the specific nature of the research focusing on social networks, Baskaran (2020) analyzed that the majority of the publications, 2,207 (18.48 percent), were published in the communication field. The area of computer communications and information systems comes in second, with 1,172 publications (9.81 percent). The majority of the 134 publications, or 1.96 percent, was written by researchers from the University of California systems, according to Baskaran (2021). Zhang Y was the most prolific author, contributing 16 (0.23 percent) of the publications in the Web 2.0 field.

OBJECTIVES OF THE STUDY

1. To investigate the use of various sorts of SNs among the respondents in the State Universities of Tamil Nadu.
2. To observe the utilization of Devices of SNs among the respondents and Place where use SNs and Media among the respondents.
3. To discover the Time inclination to utilize SNs by the respondents in a day and Period of involvement with use SNs and Media among the respondents.
4. To evaluate the inclination of Research Citation Indexes (RCI) and instruments for Contents and Texts through SNs.
5. To investigate the inclination of SNs for sharing examination data and Preference of research interpersonal organizations among respondents.

HYPOTHESES

H1: There is no significant difference between the respondents from Selected State Universities and shared academic Contents in the State Universities in Tamil Nadu.

H2: There is a significant difference between the respondents from Selected State Universities and they utilized SNs for Citations.

H3: There is a significant difference the respondents from Selected State Universities and they utilized SNs for tweeting the examination thoughts in the State Universities of Tamil Nadu.

METHODOLOGY

The study was planned to conduct the survey among the respondents, the purpose of the current investigation was to disseminate and gather information among the researchers conducting the investigation as Respondents from ten state universities in Tamil Nadu. This investigation has used a variety of sources, including online databases, reports, and other sources. The examination attempts to take apart the data collected from the respondents in the ten State Universities in Tamilandu. Only full-time Ph.D. students are included in the current count of residents. The data plan to collect from Alagappa University, Annamailai University, Bharathidasan University, Bharathiar University, Gandhigram Rural University, Madurai Kamaraj University, Manonmaniam Sundaranar University, Periyar University, and University of Madras are the universities. Out of ten universities in Tamil Nadu, a total of 1,000 surveys were sent to respondents, of which 623 were completed by respondents in the universities' social science departments. To depict and condense data accumulated from the respondents. In addition, the chi-square test, the one-sample t test, the one-sample KS test of normality, and the Kruskal-Wallis H test were used to examine the factors considered between the factors. The endeavor to investigate percentage instrument was also used to determine the majority of utility on specific factors.

RESULTS AND DISCUSSION

Research Areas Wise Respondents in Gender

Table 1 explains that Research territories insightful respondents reacted from the Gender. The investigation is seen that Total no of 623 respondents reacted to this examination, out of which Male 335 (53.8%) and 288 (46.2%) of the respondents Contributions of this study.

Table 1 shows the Gender by the respondents in the Research areas in Selected Universities of Tamil Nadu.

Male: Majority of the 1179 (54.8%) respondents are gender of Male participants, 10.4% are research areas of Economics; Followed by 70.6% are Education/Physical Education and 40.0% Management/ Commerce; 62.8% Sociology/ Social Work, 47.1% Library and Information Science while 51.6% History/Geography; 57.5% Philosophy/Psychology, while 62.5% Public Administration, 80.0% Anthropology/ Criminology, 59.4% of Journalism and Mass Communication; 54.7% are from research areas from Political Science and 55.0% of them respondents contributions of Others areas in respectively.

Female: Most of the 973 (45.2%) respondents are the gender of Female respondents, Followed by 39.5% of respondents from Economics; 29.4% are Education/Physical Education, 60.0% from Management/ Commerce, 37.2% from Sociology/ Social Work, 52.9% from Library and Information Science, 48.4% History/Geography, 42.5% Philosophy/Psychology, 37.5% Public Administration, 20.0% Anthropology/ Criminology, 40.6% Journalism and Mass Communication, 45.3% Political Science and 45.0% Others research areas respondents respectively.

Table 1. Research area vs. gender of the respondents

Research area	Gender		Total
	Male	Female	
Economics	135(60.5%)	88(39.5%)	223 (10.4%)
Education/Physical Education	255(70.6%)	106(29.4%)	361 (16.8%)
Management/ Commerce	210 (40.0%)	315 (60.0%)	525 (24.4%)
Sociology/ Social Work	123 (62.8%)	73 (37.2%)	196 (9.1%)
Library and Information Science	88 (47.1%)	99(52.9%)	187 (8.7%)
History/Geography	82 (51.6%)	77 (48.4%)	159 (7.4%)
Philosophy/Psychology	73 (57.5%)	54 (42.5%)	127 (5.9%)
Public Administration	5 (62.5%)	3 (37.5%)	8 (0.4%)
Anthropology/ Criminology	12 (80.0%)	3 (20.0%)	15 (0.7%)
Journalism and Mass Communication	41 (59.4%)	28 (40.6%)	69 (3.2%)
Political Science	35 (54.7%)	29 (45.3%)	64 (3.0%)
Others	120 (55.0%)	98 (45.0%)	218 (10.1%)
Total	1179 (54.8%)	973 (45.2%)	2152(100.0%)

Figure 1. Research area vs. gender of the respondents

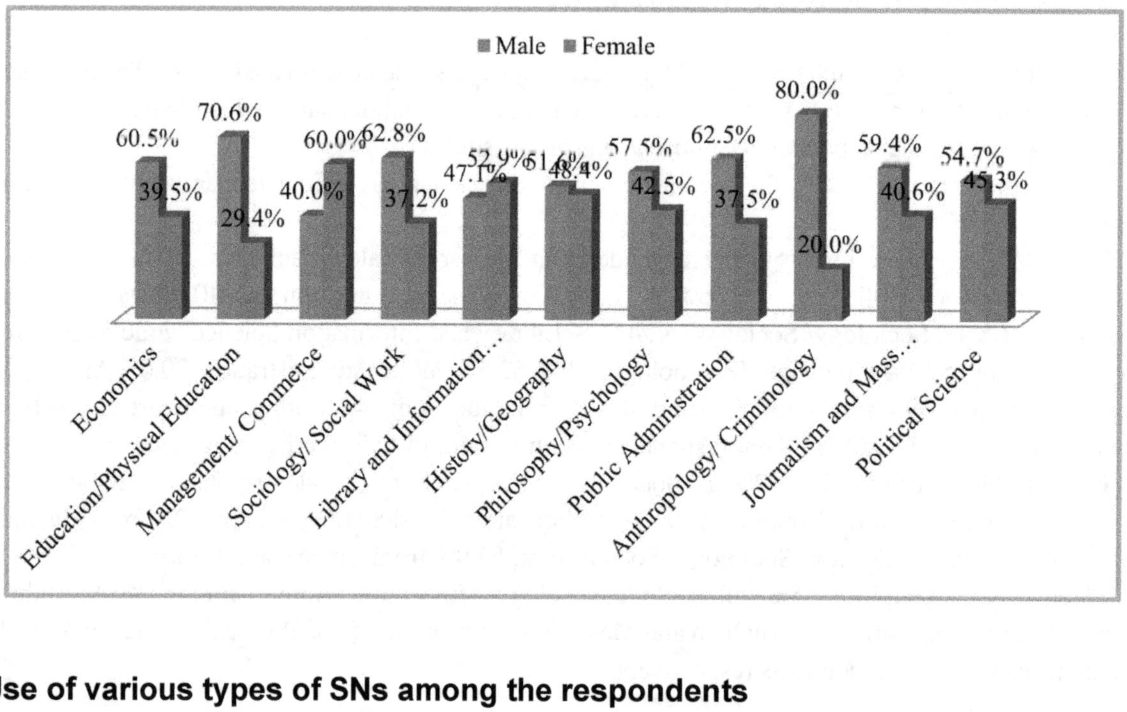

Use of various types of SNs among the respondents

The respondents knew and use of various kinds of SNs and Media for looking through academic data among the respondents in the State Universities in Tamil Nadu. Table 2 shows that there are eight kinds of informal organizations are recorded and respondents they thought about them Social Media and Networks.

Table 2. Use of social networks vs. selected universities of the respondents

		Name of the University										
		Alagappa University	Annamalai University	Bharathidasan University	Bharathiyar University	Central University of Tamil Nadu	Gandhigram Rural University	Madurai Kamaraj University	Manonmaiam Sundaranar University	Periyar University	University of Madras	Total
Face book	Count	51	61	44	53	43	43	42	43	54	39	473
	%	10.8	12.9	9.3	11.2	9.1	9.1	8.9	9.1	11.4	8.2	
	Total(%)	8.2	9.8	7.1	8.5	6.9	6.9	6.7	6.9	8.7	6.3	75.9
Twitter	Count	24	20	17	22	16	24	21	21	25	17	207
	%	11.6	9.7	8.2	10.6	7.7	11.6	10.1	10.1	12.1	8.2	
	Total(%)	3.9	3.2	2.7	3.5	2.6	3.9	3.4	3.4	4.0	2.7	33.2
You Tube	Count	58	50	41	52	50	56	46	49	60	37	499
	%	11.6	10.0	8.2	10.4	10.0	11.2	9.2	9.8	12.0	7.4	
	Total(%)	9.3	8.0	6.6	8.3	8.0	9.0	7.4	7.9	9.6	5.9	80.1
Tumblr Messenger	Count	24	20	15	27	17	24	18	29	32	17	223
	%	10.8	9.0	6.7	12.1	7.6	10.8	8.1	13.0	14.3	7.6	
	Total(%)	3.9	3.2	2.4	4.3	2.7	3.9	2.9	4.7	5.1	2.7	35.8
WhatsApp	Count	67	74	47	56	54	64	47	59	62	41	571
	%	11.7	13.0	8.2	9.8	9.5	11.2	8.2	10.3	10.9	7.2	
	Total(%)	10.8	11.9	7.5	9.0	8.7	10.3	7.5	9.5	10.0	6.6	91.7
Googleplus	Count	49	55	27	41	19	39	30	42	42	29	373
	%	13.1	14.7	7.2	11.0	5.1	10.5	8.0	11.3	11.3	7.8	
	Total(%)	7.9	8.8	4.3	6.6	3.0	6.3	4.8	6.7	6.7	4.7	59.9
Instagram	Count	22	20	16	29	30	35	18	22	22	20	234
	%	9.4	8.5	6.8	12.4	12.8	15.0	7.7	9.4	9.4	8.5	
	Total(%)	3.5	3.2	2.6	4.7	4.8%	5.6	2.9	3.5	3.5	3.2	37.6
Reddit	Count	19	7	4	9	17	10	11	11	8	8	104
	%	18.3	6.7	3.8	8.7	16.3	9.6	10.6	10.6	7.7	7.7	
	Total(%)	3.0	1.1	.6	1.4	2.7	1.6	1.8	1.8	1.3	1.3	16.7
	Count	73	87	51	62	60	65	50	63	70	42	623
	Total(%)	11.7	14.0	8.2	10.0	9,6	10.4	8.0	10.1	11.2	6.7	100.0

Face Book: The respondents announced that they were utilized SNs in the chose Universities in Tamil Nadu. A greater part 61 (12.9%) of them utilized" face book from Annamalai University and 54(11.4%) of the respondents' contributions of Periyar University, 53(11.2%) Bharathiyar University. 51(10.8%) of the respondents utilized Social Networks and Media from of the respondents from Alagappa University.

Twitter: The greater part 25(12.1%) of the respondents from Periyar University, Followed by 24 (11.6%) of the respondents structure Alagappa University and Gandhigram Rural University, they utilized '' Twitter''. It has been seen that rest of 22(10.6%) of the respondents utilized from Bharathiyar University, 21(10.1) of them utilized ''Twitter'' from Madurai Kamaraj University, Manonmaiam Sundaranar University, 17(8.2%) Bharathidasan University and University of Madras, besides about the 16(7.7%) of the respondents from Central University of Tamil Nadu.

You Tube: Majority 60 (12.0%) of them utilized from Periyar University and 58(11.6%) of the respondents from Alagappa University,56(11.2%) from Gandhigram Rural University, 50(10.0%) of the respondents from Annamalai University and Central University of Tamil Nadu. 49(9.8%) Manonmaiam Sundaranar University, 46(9.2%), Madurai Kamaraj University 41(8.2%) and Bharathidasan University and 37(7.4%) from University of Madras.

Tumblr/Messenger: most of them detailed that "Not utilize" this Social Networks, it estimates 35.8% of them higher than utilized Tumblr/Messenger. Consequently this Tumblr/Messenger is a not well known one and offers infrequently data through Social Networks.

WhatsApp: The dominant part 11.7% of the respondents answered that they were "Use" WhatsApp, since this Social Media is generally famous among all classifications of individuals, in like manner the respondents the academic data sharing insightful substance by means of this. Most of the respondents utilized 64 (11.2%) Gandhigram Rural University.

Google +: The respondents utilization of Google + Social systems larger part 373 (59.9%) of them announced "Use" in the chose State Universities of Tamil Nadu.

Instagram: Majority of respondents 234(37.6%) Utilized in research purpose, followed by 35(15.0%) of the respondents using the Instagram from Gandhigram Rural University and 30 (12.8%) respondents from Central University of Tamil Nadu.

Reddit: Most of the respondents 104 (16.7%) use of Reddit, Whereas 19(18.3%) respondents Utilized from Alagappa University and 17(16.3%) of the respondents Utilized from Central University of Tamil Nadu.

Opinion on Use of Internet Browser among the respondents

The researchers achieving academic data as far as composing research analysis, for reporting the new ideas to the exploration. It clarifies the respondents about their supposition on use Internet Browsers by the social science research scholars in the State Universities in Tamil Nadu. The investigation examined that the respondents selected their inclination us the Internet programs. The majority 594 (95.3%) of the respondents favored being used Google Chrome to look through the analyzed the data. It followed by 190 (30.5%) of the respondents picked Mozilla and 172 (27.6%) of them use Internet Explorer; out of 623 respondents in the State Universities in Tamil Nadu.

As the shows the Table 3 Opinion of Use of Internet Browser Vs. One-Sample t Test as the p value of variables of the Mozila (t=-70.691, df=622, p=001), Internet Explorer (t=-71.188, df=622, p=.001), Google Chrome (t=-231.256, df=622, p=001) and Other (t=-95.669, df=622, p=.001) a chi-square test was conducted. The results of the test and reveals that there is a significant association between gender of the respondents and their scholars at Selected Universities, as the p value is less than the significant level of 0.05 in all the cases. The null hypothesis is rejected.

Table 3. Opinion of use of internet browser vs. one-sample t test

One-Sample Test						
	Test Value = 3					
	t	df	Sig. (2-tailed)	Mean Difference	95% Confidence Interval of the Difference	
					Lower	Upper
Mozila	-70.691	622	.000	-1.305	-1.34	-1.27
Internet Explorer	-71.188	622	.000	-1.276	-1.31	-1.24
Google Chrome	-231.256	622	.000	-1.953	-1.97	-1.94
Other	-95.669	622	.000	-1.088	-1.11	-1.07

Preference of Research Citation Indexes (RCI) Among the Respondents

The exploration researchers of social science typically concentrate on distributing the papers in different journals, whereas they like to refer to the reference Index and h-Index (RCI). It is recorded that over 90% of the respondents used for research, 584 (93.7%) and 442 (70.9%) of them stated "Google Scholar" and "Scopus" as high preference. The examination is additionally explicit that 362 (58.1%) and 339 (54.4%) of the respondents recorded "CiteSeer" and "Math scinet" separately. Further, the investigation talked that 372(59.7%) and 319(51.2%) of the respondents preferred "EBSCO and Pro Quest. Among the 115(18.5%) and 124(19.9%) Get CITED and Other RCI recorded in the investigation. As can be seen in table 3, it was found out when the One Sample KS Test of Normality values concerning the level of use of the Research Citation Index Resources by the Social Science Scholars was examined that the most number of replies given to the items.

The data set on 'frequency of use of Research Citation Index Resources is normally distributed or not, one sample KS test was conducted by the researcher. Table 8 shows the descriptive statistics and the results of One sample KS Test. It is inferred that the data on frequency of use of Research Citation Index Resources is not normally distributed, as the p value is less than the significant level of 0.05. So, the null hypothesis is rejected and the alternative hypothesis is accepted. The test results convey that the researcher has to conduct non-parametric tests on these individual library resources.

The Identification of Publication Skills for SNs Among the Respondents

As can be seen in the table 6, shows the level of agreement of the respondents with regards to various competencies required to Identification of Publication Skills. It was found out when the Frequency Distribution of the opinions of the social science scholars about research ethics were analyzed that most of the respondents given to the resources. I have knowledge of available UGC recognized Social Science Journals, I have good knowledge about Impact factor, Citation, H-Index, G-Index, Cite Score etc. The Self Citation and its effects, I have proficiency in anti-plagiarism software, I Possess Knowledge on Guidelines to write review of Literature and I am Proficient in style manuals and writing skills to adopt publisher styles.

Table 6: shows the level of agreement of the respondents with regards to various competencies required to Identification of Publication Skills.

I have knowledge of available UGC recognized Social Science Journals: Out of 155 respondents who strongly Agree to get the required information and publications skills from I to have knowledge

Table 4. Use of research citation index resources: One Sample KS test of normality

		Google Scholar	CiteSeer	GetCITED	Math scinet	Scopus	Web of Science	EBSCO	Pro Quest	Other RCI
N		623	623	623	623	623	623	623	623	623
Normal Parameters	Mean	1.06	1.46	1.82	1.42	1.29	1.39	1.49	1.40	1.80
	Std. Deviation	.242	.498	.388	.494	.454	.487	.500	.491	.400
Test Statistic		.539	.364	.498	.383	.448	.399	.347	.391	.492
Asymp. Sig. (2-tailed)		.000[c]	.000[c]	.000[c]	.000[c]	.000[c]	.000[c]	.000[c]	.000[c]	.000[c]

Table 5. The identification of publication skills: frequency distribution

Publication Skills	Response									
	SA		A		N		DA		SDA	
	Count	%	Count	%	Count	%	Count	%	Count	%
I have knowledge of available UGC recognized Social Science Journals	155	24.9	143	23.0	216	34.7	48	7.7	61	9.8
I have good knowledge about Impact factor, Citation, H-Index, G-Index, Cite Score etc.	121	19.4	190	30.5	183	29.4	51	8.2	78	12.5
I am aware of Self Citation and its effects	111	17.8	157	25.2	218	35.0	62	10.0	75	12.0
I have proficiency in anti-plagiarism software	96	15.4	145	23.3	231	37.1	68	10.9	83	13.3
I Possess Knowledge on Guidelines to write review of Literature	177	28.4	147	23.6	105	16.9	23	3.7	171	27.4
I am Proficient in style manuals and writing skills to adopt to publisher styles	243	39.0	161	25.8	119	19.1	46	7.4	54	8.7

(Note: SA= Strongly Agree, A=Agree, N=Neutral, DA= Disagree, SDA= Strongly disagree)

of available UGC recognized Social Science Journals (24.9%). Out of 143 (23.0%) respondents who agreed to get the required information. 216(34.7%) of respondents who required publications skills from Neutral, 48(7.7%) of respondents who required publications skills from Disagree and 61 (9.8%) of respondents who required publications skills strongly disagree. Thus, it is clear that the respondents who get the required information from UGC-recognized Social Science Journals have a level of agreement of Strongly Agree and Agree are Neutral about the availability of UGC-recognized Social Science Journals.

I have good knowledge about Impact factor, Citation, H-Index, G-Index, Cite Score etc.: Out of 190 respondents get required publication skills from I have good knowledge about Impact factor, Citation, H-Index, G-Index, Cite Score etc. level of Agree, 30.5% Whereas 183 (29.4%) are Neutral, with Cite score. Out of 121(19.4%) respondents who get required Publication skills Strongly Agree, 78 (12.5%) are strongly disagree and 62(10.0%) are Disagree with these Cite Score.

I am aware of Self Citation and its effects: Out of 218 respondents who get required Publication skills from I am aware of Self Citation and its effects Neutral, 25.2% are Agreed and 111 (17.8%) are level of Agreed for Strongly Agree. Out of 75 respondents who get required information from Publication skills of Self Citation and its effects (12.0%) are Strongly disagree, 68 (10.9%) are Disagree with these Self Citation and its effects.

I have proficiency in anti-plagiarism software: Majority of the 231 respondents who get required proficiency of (37.1%) are Neutral and 145 (23.3%) are Agree with these skills. Out of 96 respondents who get required proficiency Publication skills for Strongly Agree (15.4%). 13.3% are strongly disagree and 10.9% are Disagree with these proficiency.

I Possess Knowledge on Guidelines to write review of Literature: 28.4% (177) of Social Science Scholars are Strongly Agree and 27.4% are Strongly disagree, 23.6% (147) are Agree, 16.9% are Neutral and 3.7% of social science scholars are Disagree with various social networks for publication skills.

I am Proficient in style manuals and writing skills to adopt to publisher styles: Out of 243 respondents who get required Proficient in style manuals and writing skills to adopt to publisher styles Strongly Agree (39.0%), 25.8% are Agree with those style manuals. Out of 119 respondents who get required publication skills (19.1%) are Neutral, 8.7% are Strongly disagree and 7.4% are Disagree with these style manuals.

6. CONCLUSION

The study discussed above the respondents observes the useful and impact of sharing Social Networks (SNs) media in the selected eight state Universities of Tamil Nadu. The study could remain the analysis would embark thoroughly justified based on the survey undertaken in the selected Universities. The analysis seen that Total no of 623 respondents responded to this assessment, out of which 73 (11.7%) of them responded from Alagappa University. It followed by Annamalai University 87 (14.0%), 51(8.2%) of the respondents from Bharathidasan University, Bharathiyar University 62(10.0%), Central University of Tamil Nadu 60(9.6%), Gandhigram Rural University 65 (10.4%), Madurai Kamaraj University 50(8.0%), Manonmaiam Sundaranar University 63(10.1%), Periyar University 70(11.2%) and University of Madras 42(6.7%). The respondents reported that they were used SNs in the picked Universities in Tamil Nadu. The majority of the respondents most preferences of WhatsApp (91.7%) from highly contributions of Annamalai University and Alagapa University, among the overall 473 (75.9%) contributions of 92.9% of the use of Social Media from University of Madras, 86.3% of them used" face book from Bharathidasan University while 85.5% Bharathiyar University and 84.0% Madurai Kamaraj University, 77.1% Periyar University.

The larger share 499 (80.1%) of the respondents structure they are used ''You Tube''. The significant offer of the respondents favored being utilized 95.3% of the respondents favored being used Google Chrome to look through the analyzed the data to glance through the corresponding to the study. the larger share 584 (93.7%) and 442 (70.9%) of the respondents favored "Google Scholar" and "Scopus" as high preference of research distinguishing proof apparatuses utilized individually. Among them Male scholars shared research data from their 244 (72.8%) group sites. 216(34.7%) of respondents who required publications skills from Neutral, 218 respondents who get required Publication skills from I am aware of Self Citation and its effects Neutral, 231 respondents who get required proficiency of (37.1%) are Neutral. The study concentrates quantitative information from the respondents in the Selected State Universities in Tamil Nadu. The examination centers the respondents achieve the Scholarly Information through Social Networks significance of Social Science in the chose ten Universities in Tamil Nadu. The investigation made it clear and finds that the analysts need to have cell phone for Research applications establishment in wording make specialists bunch inside and off the limits of the Universities. The examination could find that WhatsApp, YouTube and Facebook were utilized dominatingly contrast with other Social Networks in Selected State Universities in Tamil Nadu. Consequently specialists typically share their exploration data and are watching current data from YouTube too. The University must give free Wi-Fi availability to utilize the SNs. The investigation find that greater part of the respondents were utilizing SNs in their home contrast with other spot, it find that the analyst may get all the more extra time separated from working hours at the Institutions.

Most of the respondents highly preferred on Ref. works, End Note and Mendeley Reference the executives Software, so as to the reference the board exceptionally needful to deal with the assets where incorporate the Citations utilizing different Citation style needful and surveys in their exploration. By and by, the specialists are more enthusiasm to check their Citations on their exploration work and h-Index on the motivation behind Google Scholar is essential one it tends to be enlisted effectively by each analyst. The investigation broke down that more respondents' exceptionally favored Google researcher for examinations their h-Index, Citations and research effect of the Publications of sociology Scholars.

REFERENCES

Akçayır, G. (2017). Why do faculty members use or not use social networking sites for education? *Computers in Human Behavior*, *71*, 378–385. doi:10.1016/j.chb.2017.02.028

Al-Alwan, A. F. (2014). Modeling the relations among parental involvement, school engagement and academic performance of high school students. *International Education Studies*, *7*(4), 47–56. doi:10.5539/ies.v7n4p47

Al-Mukhaini, E. M., Al-Qayoudhi, W. S., & Al-Badi, A. H. (2014). Adoption of social networking in education: A study of the use of social networks by higher education students in Oman. *Journal of International Education Research*, *10*(2), 143–154. doi:10.19030/jier.v10i2.8516

Al-Rahmi, W. M., & Othman, M. S. (2013). Evaluating student's satisfaction of using social media through collaborative learning in higher education. *International Journal of Advances in Engineering and Technology*, *6*(4), 1541–1551.

Baskaran, C. (2019). Scholarly information share through Social Networks (SNs) and medias among social science scholars in selected State Universities in Tamilnadu. *International Journal of Library and Information Studies*, *9*(3), 83–92.

Baskaran, C. (2020). Research Patterns on the Social Networks and Media: A Scientometric Portrait. Handbook of Research on Emerging Trends and Technologies in Library and Information Science, pp.189-207. http://doi.org/ doi:10.4018/978-1-5225-9825-1.ch014

Baskaran, C. (2021). An Exponential trend and Mapping of Research in Web 2.0. Library Philosophy and Practice (e-journal). 5008. https://digitalcommons.unl.edu/libphilprac/5008

Bin Salamh, F. M. (2019). Protecting organization reputations during a crisis: Emerging social media in risk and crisis communication. *Indian Journal of Science and Technology*, *12*(18), 1–7. doi:10.17485/ijst/2019/v12i18/144603

Boyd, D. M., & Ellison, B. N. (2007). Social networking sites: Definition, history and scholarship. *Journal of Computer-Mediated Communication*, *13*(1), 210–230. doi:10.1111/j.1083-6101.2007.00393.x

Josserand, E., Schmitt, A., & Borzillo, S. (2017). Balancing present needs and future options: How employees leverage social networks with clients. *The Journal of Business Strategy*, *38*(1), 14–21. doi:10.1108/JBS-01-2016-0003

Levula, A., & Harré, M. (2016). Social Networks and Mental Health: An egocentric perspective. *Mental Health Review (Brighton)*, *21*(3), 161–173. doi:10.1108/MHRJ-10-2015-0029

Littman, J., Chudnov, D., Kerchner, D., Peterson, C., Tan, Y., Trent, R., Vij, R., & Wrubel, L. (2016). API-based social media collecting as a form of web archiving. *International Journal on Digital Libraries*, *19*(1), 21–38. doi:10.100700799-016-0201-7

Liu, L., Zhang, L., & Ye, P. (2018). Research on the application of SNS in University Libraries. *The Electronic Library*, *36*(2), 369–386. doi:10.1108/EL-05-2016-0120

Meenakshi, S. P., Sivakumar, P., & Ilango, P. (2017). Effective health promotion interventions for targeted audience using social networking sites: A systematic review. *Indian Journal of Science and Technology*, *10*(36), 1–09. doi:10.17485/ijst/2017/v10i36/112402

Roslina Wan Othman, W., Fariha Mohd Apandi, Z., & Haslinda Ngah, N. (2016). The uses of social media on student's communication and self concepts among TATIUC students. *Indian Journal of Science and Technology*, *9*(17). doi:10.17485/ijst/2016/v9i17/88730

Chapter 14
Preference and Opinion on Social Media for Disseminating Scholarly Information Among Research Scholars in the Universities in Tamil Nadu:
A Study

P. Pitchaipandi
Alagappa University, Karaikudi, India

ABSTRACT

The chapter explores how social media and networks disseminate and share scholarly information among the research scholars in the selected state universities of Tamil Nadu in India. The researcher has adopted a questionnaire method in order to collect data from respondents in the Universities, entertaining only the full-time research scholars. The result of the study demonstrates a majority 87 (14.0%) of them responded from Annamalai University. 47(14.0%) of them were male and 40 (13.9%) of them female. Of a total of 623 respondents, 44 (10.8%) of them responded from Annamalai University in the age group 26-35 years. The majority 594 (95.3%) of the respondents preferred to use "Google Chrome" to search various types of research information; the majority 571(91.7%) of the respondents use of WhatsApp; 499(80.1%) of the respondents used YouTube..

INTRODUCTION

The term social media research encompasses any form of research that uses data from social media sources. Research in this environment can be divided into two types: the use of social media as a research tool (e.g., surveys on social media platforms) and research on the activities and content of social media itself. Social media research differs from other forms of online research, such as Internet-based surveys

DOI: 10.4018/978-1-6684-8805-8.ch014

or website reviews, in the social nature of the data extracted from platforms specifically set up for this purpose. As with any new form of research, several methodological issues must be considered to ensure that rigour is maintained. These practical issues are discussed in this section and in the subsequent chapters of this guide. As with other methods, research using social media should be considered a tool in a researcher's toolbox. While some research questions may only use social media research methods, many will benefit from the use of social media research in addition to other methods (SMRG, 2016). Social media users: Social media users are not representative of the population (Ruths and Jurgen, 2014). Therefore, bias exists and it may be difficult to extrapolate the results to the general population. However, this characteristic can be helpful when research focuses on a group that is particularly active on a social media platform, and when there is an opportunity to obtain data on groups that are more likely to be unresponsive to other research methods. Platforms can host numerous automated 'bots' and professionally managed accounts that impersonate human users. Therefore, large studies should attempt to filter out the results from such anomalous sources during the analysis.

LITERATURE REVIEW

Bianchi and Phillips (2005) discussed mobile phone use as banned or illegal under certain circumstances and in some jurisdictions. Nevertheless, some people still use their mobile phones despite recognized safety concerns, legislation, and informal bans. Drawing on potential predictors from addiction literature, this study sought to predict usage and, specifically, problematic mobile phone use from extraversion, self-esteem, neuroticism, gender, and age. To measure problem use, a Mobile Phone Problem Use Scale was devised and validated as a reliable self-report instrument against the Addiction Potential Scale and overall mobile phone usage levels. Problem use was a function of age, extraversion, and low self-esteem but not neuroticism. Humphreys (2007) reported on the social and behavioral norms of dodgeball use. A comparison between social networking sites and Dodgeball highlights some of the communicative differences between mobile technology and the Internet. The findings of the study suggest that dodgeball use can influence the way informants experience public space and social relations therein. Folorunso, Vincent, Adekoya, and Ogunde. (2010) argued that students at the University of Agriculture, Abeokuta, Nigeria. We tested the attributes of Dissemination of Innovations empirically using social networking sites (SNS) as the target innovation. It was used to test the influence on attitude and trust regarding SNS and to determine how attitude would impact the intention to use it.

Shafique, Anwar & Bushra, M. (2010) found that trend of social media usage among students of the Islamia University of Bahawalpur (IUB). Use social media to upload, share, and tag images of keywords. The networks are academic services for social media websites, such as Cite Like, Technorati, Connotea, Blogger, Twitter, and Wikis, which are used by students for communication, research work, and online learning educational purposes of the knowledge society. Mohamed & Sumitha (2011) found the use of social networking sites majority of the students were aware of social networking sites and use these sites for responsive communication. Orkut was the most popular and used social networking site than Facebook and MySpace. However, many students used these sites for academic purposes. Awareness programs should be provided to educate students about the applications, benefits, and reassurances associated with networking sites. Stern and Willits (2011) examined the usage of Web 2.0, technologies in developing education learning management systems and new media such as Word press, Buddy press, and Twitter into a tri-fold approach that extends beyond the physical classroom walls to build a community of learning technologies to the shared learning experience between faculty and students modern course technologies such as grade books and stimulating the re-imagined learning technology stage that integrates the best tools of the LMS with the further shared, democratizing social media.

Mashups, AJAX, Bittner & Müller (2011) examined the properties of research information systems and research networking tools. Research information systems (RISs) are software tools that provide access to research-related information. Such systems contain information about researchers, research institutions, research projects, and research outputs. The requirements of research information systems are driven by various economic and social factors. Sutko & Silva (2011) focused the classification of eponymous and anonymous location-aware interfaces by investigating how these applications may require us to rethink our understanding of urban sociability, particularly how we coordinate and communicate in public spaces. We argue that common assumptions made about location-aware mobile media, namely their ability to increase one's spatial awareness and encourage one to meet more people in public spaces, might be fallacious due to pre-existing practices of sociability in the city. Amy Gonzales & Jeffrey T. Hancock (2011) revealed that in contrast to previous work on OSA, becoming self-aware by viewing one's own Facebook profile enhances self-esteem rather than diminishes it. Participants who updated their profiles and viewed their own during the experiment also reported greater self-esteem, which lends additional support to the Hyper personal Model. These findings suggest that selective self-presentation in digital media, which leads to intensified relationship formation, also influences impressions of the self.

Parveen (2011) explored the social networking site (SNS) Facebook to work as an effective tool for imparting information or knowledge and proving awareness among Library and Information Science (LIS) Professionals of University Libraries of the U.P.. The research focused on the use of Facebook by Library and Information Science Professionals. The uniqueness of this social networking site is the sharing of information among users, ranging from the highly personal to academic interests of the participants. Singh and Gill (2011) analyzed web applications and Internet Technology usage for educational instructional technology. Social networking sites are preferred by the research scholars' data and services in a form that allows remixing by others, creating social network effects through" architecture of participation", This research is only confined to the M.Phil and Ph.D. scholars study that several tools of Web 2.0 like Blogs, Wikis, RSS (Really Simple Syndication), Instant Messaging, Social Networking, Podcasting, Tagging, Social Bookmarking Services. Lee, DH (2013) attempts to focus on the newly emerging mobile communication practices that have resulted from smartphones allowing people to access their SNSs anytime and anywhere. To understand how the mobility and immediate accessibility of smartphones have affected the communication practices of SNS, this study will pay particular attention to the use of Twitter via smartphones. It explored this through 49 Korean Twitter users' everyday communication and social interaction practices via their smartphones. Based on qualitative interview data, this study discusses how mobile social media forms a pseudo-aural space for volatile but self-expediential social networks, and how it shapes a new sociality. This suggests that sociality strays from existing social practices and has new cultural significance. Xiao Wang, (2014) examines why sports spectators use social media (e.g., Facebook and Twitter) while attending sporting events. Based on a cross-sectional survey of 466 participants who generally attended games with friends or family, structural equation modeling analysis revealed that social media use through a mobile device during sporting events served four functions: information-based and game-pleasure-based utilitarian functions, an ego-defensive function, and a value-expressive function. The first three functions, descriptive norms and self-efficacy, predicted attitudes toward social media use. Baskaran (2019) analysed the majority (83.5%) of respondents using "Whaatsapp" out of respondents. The respondents reported (77.7%) and Smartphone/Mobiles (73.6%).

Baskaran (2019) analyses the majority (83.5%) of respondents using "Whaatsapp" out of respondents. The respondents reported (77.7%) and (73.6%) using devices tools "Laptop" and Smartphone/Mobiles are respectively.

Radhakrishnan and Baskaran (2019) have examined the authorship productivity, the applicability of Lotka's law, Price's square root law, and Pareto's 80/20 rule. The Chi-square value was 225.65, which is higher than the table value in the analysis of Lotka's law. Baskaran (2020) explored 11, 941 records on social networks and media retrieved from the Web of Science database during the study period. The predominant records were 2,576 (21.57%) published in 2018, followed by 2,281 (19.10%) published in 2017. Radhakrishnan and Baskaran (2020) conducted research on most of the publications shared by social media Twitter. Moreover, readers prefer to read articles through Mendeley's Reference Manager. The study reveals that there is a moderate correlation between the citation and altmetric scores. Pitchaipandi (2020) explained the Department wise participants using social media access places whereas time spent, social media tools, different types of using Internet Browsers and Social media devices for using the period and purpose of using social media.

OBJECTIVES OF THE STUDY

1. To analyze the profile of the samples and age group-wise respondents of the study among research scholars in the State Universities of Tamil Nadu.
2. To explore the research area, Respondents in the Selected State Universities of Tamil Nadu.
3. To observe the Prefer the different types of SNs among the respondents
4. To study the use of research citation indices (RCI) by respondents in the State Universities of Tamil Nadu.

METHODOLOGY

The respondents to the study in the category of research scholars include only the full-time scholars in the selected state universities of Tamil Nadu, India. The researcher has decided to collect data from ten selected universities, viz., Alagappa University, Bharathidasan University, Bharathiar University, University of Madras, Periyar University, Annamalai University, Madurai Kamaraj University, Manonmaniam Sundaranar, Central University of Tamil Nadu, and Gandhigram Rural University, all of which and the respondents determined only from the departments were social science researchers in the Department of Economics, Education/Sports Education, Management/Commerce, Sociology/Social Work, Library and Information Science, History/Geography, Philosophy/Psychology, Public Administration, Anthropology/ Criminology, Journalism and Mass Communication, Political Science, and others. The researcher collected data from the first 623 doctoral scholars she met in her department on the day of data collection. The data were collected using a questionnaire.

The researcher selected the random samples. The collected data were simplified using single, double, or triple tables. The tabulated data were analyzed using the simple percentage method used in MS Excel tools and SPSS 23 version analyses. A distinction is made between location-based mobile social networks and non-location-based mobile social networks to show that social media mobility is indeed much broader than just location. Second, several important theoretical issues for future research are identified, including micro-, meso-, and macro-level theories. Finally, methodological challenges and opportunities are reflected, culminating in a call for multidisciplinary research programs to fully understand the role of mobile social media in today's world (Lee Humphreys, 2010). It is discussed that addressing this challenge requires attention to the mobile nature of these devices and a focus on well-defined, widespread opportunities. The second part of this article discusses ethical, philosophical, and theoretical issues related to the methodological possibilities of collecting large amounts of behavioral data with software-based mobile devices (Boase, 2013).

ANALYSIS AND INTERPRETATIONS

Research is a predominant factor in the universities, and scholars have obtained Scholarly information from various forms of documents. To find the study, scholars search for relevant information in social networks and media (SNs), as shown in Fig. 1. Table 1 shows that the respondents are accredited by state universities with grade "A" from "NAAC" A total of 623 respondents were included in this study. The study analyzed 87 (14.0%) respondents from Annamalai University, of whom 47 (14.0%) were male and 40 (8.2%) were female. The study found that more than 14% of the respondents from Alagappa University (11.7%) out of the 47 (14.0%) and 26 (9.0%) male and female respondents, respectively; 65 (10.4%) of the respondents from Gandhigram Rural University, 31 (9.3%) male and 34 (11.8%) female respondents, Periyar University had 11.2 percent of respondents, 33 (9.9%) male and 37 (12.8%) female. Of the (10.1%) respondents, 29 (8.7%) were male and 34 (11.8%) were female. Of the (10.0%) total respondents, 28 (8.4%) were male and 34 (11.8%) were female. Of the 60 (10.0%) Central University of Tamil Nadu respondents, 37 (11.0%) were male and 23 (6.25%) were female. (8.2%) of the respondents at Bharathidasan University were male, 29 (8.7%) were male, and 22 (7.6%) were female. At Madurai Kamaraj University (8.0%), 32 (9.6%) male and 18 (6.25%) female students responded. The study find out of the ten universities in Tamil Nadu, less than 10% of the respondents answered Madras University (6.7%). Table 3 shows the seven types of social media listed and the respondents' comments on the dissemination of the research task.

The researchers obtained scientific information about the use of social media, among the social science researchers in the selected state universities in Tamil Nadu, Table 2 shows that out of the 623 respondents, 87 (14.0%) reported from Annamalai University whose age category is below 25. This is followed by Alagappa University with 11.7%, Periyar University with 11.2%, Gandhigram Rural University with 10.4%, Manonmaniam Sundaranar University with 10.1%, and Bharathiar University with 10.0%. The Central University of Tamil Nadu (9.6%), Bharathidasan University (8.2%), and University of Madras (6.7%) of the respondents were included in this study. The Universities are together share more than 60% of the respondents in this study. The figure 2 shows that these universities are important centers of higher education in Tamil Nadu. The remaining respondents are from other universities in the region.

1. Alagappa University

The majority of 54 (13.2%) of the respondents were in the age group of 26–35 years and were from Alagappa University.

Table 1. Selected universities of the respondents vs. gender

Selected Universities	Frequency	Gender		Total
		Male	Female	
Alagappa University	Count	47	26	73
	%	14.0	9.0	11.7
Annamalai University	Count	47	40	87
	%	14.0	13.9	14.0
Bharathidasan University	Count	29	22	51
	%	8.7	7.6	8.2
Bharathiyar University	Count	28	34	62
	%	8.4	11.8	10.0
Central University of Tamil Nadu	Count	37	23	60
	%	11.0	6.25	9.6
Gandhigram Rural University	Count	31	34	65
	%	9.3	11.8	10.4
Madurai Kamaraj University	Count	32	18	50
	%	9.6	6.25	8.0
Manonmaiam Sundaranar University	Count	29	34	63
	%	8.7	11.8	10.1
Periyar University	Count	33	37	70
	%	9.9	12.8	11.2
University of Madras	Count	22	20	42
	%	6.6	6.9	6.7
Total	Count	335	288	623
	%	53.8	46.2	100

Figure 1. Selected universities of the respondents vs. gender

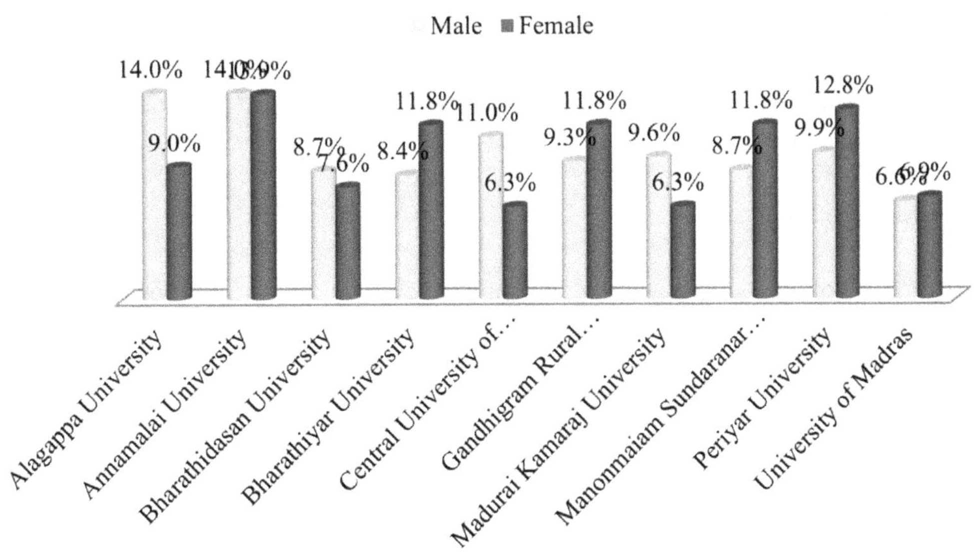

Table 2. Age group wise respondents of the study

Selected Universities	Frequency	Age group				Total
		Below 25 years	26-35 years	36-45 years	Above 45 years	
Alagappa University	Count	15	54	4	0	73
	%	11.5	13.2	5.5	0	11.7
Annamalai University	Count	25	44	16	2	87
	%	19.1	10.8	21.9	20	14.0
Bharathidasan University	Count	6	39	6	0	51
	%	4.6	9.5	8.2	0	8.2
Bharathiyar University	Count	14	44	4	0	62
	%	10.7	10.8	5.5	0	10.0
Central University of Tamil Nadu	Count	12	40	7	1	60
	%	9.2	9.8	9.6	10	9.6
Gandhigram Rural University	Count	18	40	5	2	65
	%	13.7	9.8	6.8	20	10.4
Madurai Kamaraj University	Count	8	33	8	1	50
	%	6.1	8.1	11.0	10	8.0
Manonmaiam Sundaranar University	Count	13	38	11	1	63
	%	9.9	9.3	15.1	10	10.1
Periyar University	Count	13	47	9	1	70
	%	9.9	11.5	12.3	10	11.2
University of Madras	Count	7	30	3	2	42
	%	5.3	7.3	4.1	20	6.7
Total	Count	131	409	73	10	623
	%	21.0	65.7	11.7	1.6	100

2. Annamalai University

The majority of respondents (19.1%), followed by (10.8%), (21.9%), and (20%) of the age group under 25, 26–35, 36–45, and over 45

3. Bharathidasan University

The majority (9.5%) of respondents were in the 26–35 age group. This was followed by six (8.2%) of the respondents who were between 36 and 45 years old.

4. Bharathiar University

The majority (10.8%) of respondents were in the 26–35 age group. (5.5%) and (0%) of the respondents were between 36 and 45 and over 45 years old, respectively.

5. Central University of Tamil Nadu

The majority (9.8%) of respondents belonged to the age group 26–35 years, 12 (9.2%) and 7 (9.6%) of respondents were between under 25 years and 36–45 years, respectively.

6. Gandhigram Rural University

The majority of 40 (9.8%) respondents are between 26 and 35 years old. This is followed by 18 (13.7%) and 5 (6.8%) of the respondents between under 25 years old and 36–45 years old, respectively.

7. Madurai Kamaraj University

The majority (8.1%), 11.0%), and 6.1%) of the respondents reported being in the age groups of 26–35, 36–45, and under 25, respectively.

8. Manonmaniam Sundaranar University

The majority (9.3%) of the respondents were in the age group of 26–35 years, followed by (9.9%) and (15.1%) of the respondents in the age group of 36–45 and below 25 years.

9. Periyar University

The majority of 11.5%, 9.9%, and 12.3% of the respondents indicated that the age group of the respondents is between 26 and 35 years and less than 25 years between 36 and 45 years, respectively.

10. University of Madras

The study found that (6.0%) of them, 6%, and 4% of the respondents were reported in the age groups between 26-35, 36 and 45, and over 45 years, respectively.

Figure 2. Age group wise respondents of the study

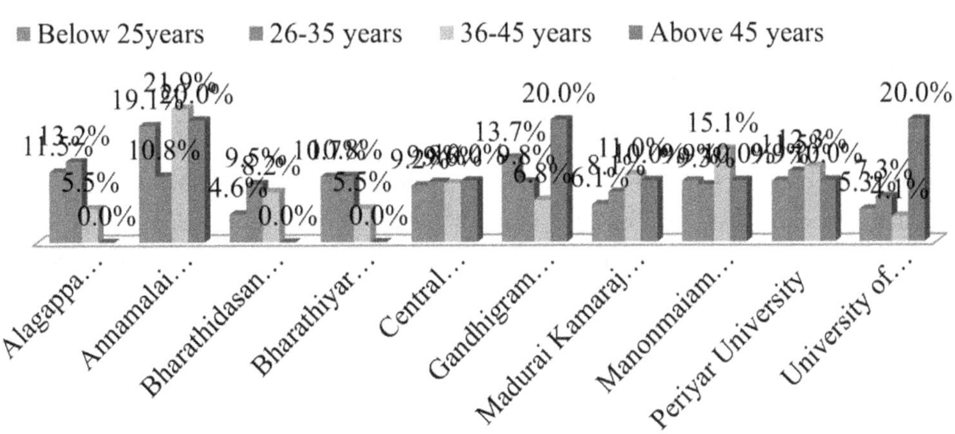

Table 3. Research areas of the respondents vs. selected universities

Research areas	Selected Universities										Total
	Alagappa University	Annamalai University	Bharathidasan University	Bharathiyar University	Central University of Tamil Nadu	Gandhigram Rural University	Madurai Kamaraj University	Manonmaiam Sundaranar University	Periyar University	University of Madras	
Economics	7 (9.6%)	4 (4.6%)	9 (17.6%)	11 (17.7%)	1 (1.7%)	4 (6.2%)	11 (22%)	1 (1.6%)	10 (14.3%)	9 (21.4%)	67 (10.8%)
Education/ Physical Education	23 (31.5%)	10 (11.5%)	1 (2.0%)	14 (22.6%)	9 (15.0%)	16 (24.6%)	4 (8%)	10 (15.9%)	6 (8.6%)	6 (14.3%)	99 (15.9%)
Management/ Commerce	23 (31.5%)	20 (23.0%)	11 (21.6%)	12 (19.4%)	11 (18.3%)	10 (15.4%)	7 (14%)	23 (36.5%)	23 (32.9%)	7 (16.7%)	147 (23.6%)
Sociology/ Social Work	2 (2.7%)	1 (1.1%)	17 (33.3%)	9 (14.5%)	3 (5.0%)	3 (5.0%)	10 (20%)	3 (4.8%)	3 (4.3%)	5 (11.9%)	56 (9.0%)
Library and Information Science	7 (9.6%)	9 (10.3%)	4 (7.8%)	2 (3.2%)	11 (18.3%)	2 (3.1%)	6 (12%)	3 (4.8%)	9 (12.9%)	1 (2.4%)	54 (8.7%)
History/ Geography	4 (5.5%)	17 (19.5%)	8 (15.7%)	2 (3.2%)	5 (8.3%)	0 (0.0%)	0 (0%)	11 (17.5%)	2 (2.9%)	0 (0.0%)	49 (7.9%)
Philosophy/ Psychology	0 (0.0%)	12 (13.8%)	0 (0.0%)	9 (14.5%)	1 (1.7%)	2 (3.1%)	4 (8%)	19 (1.6%)	9 (12.9%)	1 (2.4%)	39 (6.3%)
Public Administration	0 (0.0%)	1 (1.1%)	0 (0.0%)	0 (0.0%)	0 (0.0%)	0 (0.0%)	0 (0%)	0 (0.0%)	0 (0.0%)	1 (2.4%)	2 (0.3%)
Anthropology/ Criminology	0 (0.0%)	0 (0.0%)	0 (0.0%)	0 (0.0%)	0 (0.0%)	0 (0.0%)	0 (0%)	2 (3.2%)	0 (0.0%)	3 (7.1%)	5 (0.8%)
Journalism and Mass Communication	0 (0.0%)	0 (0.0%)	0 (0.0%)	0 (0.0%)	4 (6.7%)	0 (0.0%)	2 (4%)	9 (14.3%)	7 (7.1%)	2 (4.8%)	24 (3.9%)
Political Science	0 (0.0%)	12 (13.8%)	0 (0.0%)	0 (0.0%)	1 (1.7%)	1 (1.5%)	3 (6%)	0 (0.0%)	0 (0.0%)	0 (0.0%)	17 (2.7%)
Others	7 (9.6%)	1 (1.1%)	1 (2.0%)	3 (4.8%)	14 (23.3%)	27 (41.5%)	3 (6%)	0 (0.0%)	1 (1.4%)	7 (16.7%)	64 (10.3%)
Total	73 (11.7%)	87 (14.0%)	51 (8.2%)	62 (10.0%)	60 (9.6%)	65 (10.4%)	50 (8.0%)	63 (10.1%)	70 (11.2%)	42 (6.7%)	623 (100%)

1. **Alagappa University:** Table 3 analysed the majority 23 (31.5%) of the respondents responded from dept. of Education/Physical Education and Management/ Commerce to the study. It followed 7(9.6%) of them respondents from Library and Information Science, Economics and Others research areas. Dept. of History/ Geography 4(5.5%), 2(2.7%) Sociology/ Social work.
2. **Annamalai University:** The study shows the respondents reported from Annamalai University withnessed that respondents most of them 20(23.0%) from dept. of Management/Commerce, 17(19.5%) of the respondents History/Geography and 12(13.8%) Philosophy/Psychology and Political Science While 10(11.5%) Education/Physical education, 9(10.3%) of the respondents Library and Information Science, 4(4.6%) of the respondents from Economics and (1.1%) Sociology/ Social Work, Public Administration and Other departments contribution of this study.
3. **Bharathidasan University:** The study found that most of the respondents witnessed from Dept. of Sociology/ Social Work 17(33.3%). Among the Management/Commerce 11(21.6%), 9(17.6%) of the respondents contributions of the departments in Economics, 8(15.7%) of the respondents are History/Geography, 4(7.8%) Library and Information Science and 1(2.0%) of the respondents are participated from Education/Physical Education and Other departments contributions of this study.
4. **Bharathiar University:** It has been reported that majority of the 14(22.6%) respondents have reported from dept. of Education/Physical Education. Followed by 12(19.4%) of the respondents

from Management/Commerce.11(17.7%) of the respondents contributions of them department of Economics, 9(14.5%) of the respondents from Sociology/ Social Work and Philosophy/Psychology. 3(4.8%) of the respondents from participants are Others, furthermore 2(3.2%) of the respondents from Library and Information Science and History/Geography.

5. **Central University of Tamil Nadu:** Data presents on the research areas wise respondents of the study, which explains University of Madras majority of the respondents contributions of them department of 14(23.3%) of the respondents contributions of departments in Others, 11(18.3%) of them from Management/Commerce and Library and Information Science, 9(15.0%) of the respondents in department of Education/Physical Education, 5(8.3%) of the respondents from History/Geography, 4(6.7%) of them from Journalism. The study has witnessed Mass Communication, 3(5.0%) of the respondents from Sociology/ Social Work, further, 1(1.7%) of the respondents from Economics, Philosophy/ Psychology and Political Science.

6.Gandhigram Rural University: The study most of the respondents contribution of departments of others in 27(41.5%), among the 16(24.6%) of the respondents are Education/Physical Education, Management/ Commerce 10(15.4%), 4(6.2%) of the respondents contributions of Economics, 3(5.0%) Sociology/ Social Work and 2(3.1%) of them have responded from Library and Information Science and Philosophy/Psychology besides 1(1.5%) of the respondents of Political Science.

7. **Madurai Kamaraj University**: The study found that above 11(22%) of the respondents participated from dept. of Economics. Followed by 10(20%) Sociology/ social work with 7(14%) of the respondents contributions of Management/ Commerce. 6(12%) of the respondents are Library and Information Science, 4(8%) of the respondents from Education/Physical Education and Philosophy/Psychology, 3(6%) of the respondents participants from departments of Political Science and Other departments, 2 (4%) of the respondents of Journalism and Mass Communication.

8. **Manonmaniam Sundaranar University**: The study observed that above one percent of the respondents and whole 63 (10.1%) of them out of 623 witnessed the respondents from departments of Management/ Commerce 23(36.5%). Followed by 19(1.6%) of the respondents contributions of them department of Philosophy/Psychology, 10(15.9%) of the respondents of Education/Physical Education. 9(14.3%) of Journalism and Mass Communication, 3(4.8%) of the respondents of Sociology/ Social Work and Library and Information Science, 2(3.2%) of the respondents participants of the departments of Anthropology/ Criminology along with 1(1.6%) of the respondents from the area of Economics of them was reported to the study. The respondents had aware on usage with different types of Social Medias towards search scholarly information by the respondents in the State Universities in Tamil Nadu.

9. **Periyar University**: Above one Percent of the Respondents reports from the Research areas 23(32.9%) in the Dept. of Management/ Commerce with 10(14.3%) of the respondents contributions of research areas in Economics with 9(12.9%) Library and Information Science, Philosophy/ Psychology among the 7(7.1%) of respondents contribution of Journalism and Mass Communication.6 (8.6%) Education/Physical Education, 3(4.3%) Sociology/ Social Work 1(1.4%) of the respondents have responded from ''other research'' areas in contribution of them respondents respectively.

10. **University of Madras**: The most of the respondents contribution of research areas in Economics 9(21.4%), Followed by 7(16.7%) of the respondents participants of the departments of Management/ Commerce and Other departments. 6(14.3%) of the respondents from Education/Physical Education, and 5(11.9%) of the respondents from Sociology/ Social Work, 3(7.1%) of the respondents from Anthropology/ Criminology, 2(4.8%) of the respondents from Journalism and Mass Communication and 1(2.4%) of the respondents from research areas in Library and Information Science, Philosophy/ Psychology and Public Administration respondents respectively.

1. **Face Book:** Table 4 presents the respondents opted opinion about Facebook used by Majority 61 (12.9%) of them equally preferred from Annamalai University. Whereas 54 (11.4%) of them using SNs in Periyar University and 53 (11.2%) Bharathiar University, then it followed by 51 (10.8%) of the respondents used Facebook for accessing research information from Alagappa University.
2. **Twitter:** The majority 25 (12.1%) of the respondents they has accessed twitter form Periyar University followed by 24 (11.6%) Alagappa University and Gandhigram Rural University. On the other has been noticed that rest of 22 (10.6%) of them used ''Twitter'' by the respondents from Bharathiar University, 21(10.1%) Madurai Kamaraj University and Manonmaniam Sundaranar University,
3. **You Tube:** The preference of You Tube by the respondents of Selected State Universities of Tamil Nadu. It has witnessed that majority 60 (12.0%) of them preferred YouTube by the respondents in Periyar University and 56 (11.2%) of them used in Gandhigram Rural University, 52 (10.4%) and 50 (10.0%) of them used by the respondents in Annamalai University and Central University of Tamil Nadu respectively.
4. **Tumblr/Messenger**: The respondents prefer to access Tumble/Messenger social media for collecting information to research in the selected Universities. Out of 623 respondents 223 (35.8%) of them preferred Tumble/Messenger social media by the respondents. It has been noticed that majority of 32 (14.3%) the respondents preferred from Periyar University whereas 29(13.0%) Manonmaniam Sundaranar University.
5. **Whatsapp:** Majority 91.7% of the respondents replied that they had preferred Whatsapp, on causes on it is a most popular social media among irrespective of people, likewise the respondents they had disseminated the scholarly information for sharing the scholarly contents via this. It has been witnessed that most of 74 (13.0%) of the respondents preferred Whattsapp in Annamalai University, it followed by 67(11.7%) of the respondents preferred in Alagappa University. 64 (11.2%) of the respondents from Gandhigram Rural University,62 (10.9%) of them Periyar University respectively.
6. **Google +:** Google + Social Media preferred with 373 (59.9%) respondents out 623 whole for study was reported among State Universities of Tamil Nadu. The study found it 55(14.7%) of the respondents mostly preferred Google + in Annamalai University. It has been notice that Google+. where the respondents preference given to the Social medias to sharing Scholarly Information.
7. **Instagram:** The is to notice that out of 623 respondents 234 (37.6%) of them preferred to access Instagram for sharing research related task. It has been witnessed that Majority 35 (15.0%) of the respondents from Gandhigram Rural University preferred Instagram social media, it followed by 30 (12.8%) Central University of Tamil nadu of them preferred to access for enhancing scholarly works.

Table 4. Prefer the different types of SNs among the respondents

Name of the University		Types of SNS								Total
		Facebook	Twitter	You Tube	Tumblr Messenger	WhatsApp	Googleplus	Instagram	Reddit	
Alagappa University	Count	51	24	58	24	67	49	22	19	73
	%	10.8%	11.6%	11.6%	10.8%	11.7%	13.1%	9.4%	18.3%	
	Total (%)	8.2%	3.9%	9.3%	3.9%	10.8%	7.9%	3.5%	3.0%	11.7%
Annamalai University	Count	61	20	50	20	74	55	20	7	87
	%	12.9%	9.7%	10.0%	9.0%	13.0%	14.7%	8.5%	6.7%	
	Total (%)	9.8%	3.2%	8.0%	3.2%	11.9%	8.8%	3.2%	1.1%	14.0%
Bharathidasan University	Count	44	17	41	15	47	27	16	4	51
	%	9.3%	8.2%	8.2%	6.7%	8.2%	7.2%	6.8%	3.8%	
	Total (%)	7.1%	2.7%	6.6%	2.4%	7.5%	4.3%	2.6%	.6%	8.2%
Bharathiyar University	Count	53	22	52	27	56	41	29	9	62
	%	11.2%	10.6%	10.4%	12.1%	9.8%	11.0%	12.4%	8.7%	
	Total (%)	8.5%	3.5%	8.3%	4.3%	9.0%	6.6%	4.7%	1.4%	10.0%
Central University of Tamil Nadu	Count	43	16	50	17	54	19	30	17	60
	%	9.1%	7.7%	10.0%	7.6%	9.5%	5.1%	12.8%	16.3%	
	Total (%)	6.9%	2.6%	8.0%	2.7%	8.7%	3.0%	4.8%	2.7%	9.6%
Gandhigram Rural University	Count	43	24	56	24	64	39	35	10	65
	%	9.1%	11.6%	11.2%	10.8%	11.2%	10.5%	15.0%	9.6%	
	Total (%)	6.9%	3.9%	9.0%	3.9%	10.3%	6.3%	5.6%	1.6%	10.4%
Madurai Kamaraj University	Count	42	21	46	18	47	30	18	11	50
	%	8.9%	10.1%	9.2%	8.1%	8.2%	8.0%	7.7%	10.6%	
	Total (%)	6.7%	3.4%	7.4%	2.9%	7.5%	4.8%	2.9%	1.8%	8.0%
Manonmaiam Sundaranar University	Count	43	21	49	29	59	42	22	11	63
	%	9.1%	10.1%	9.8%	13.0%	10.3%	11.3%	9.4%	10.6%	
	Total (%)	6.9%	3.4%	7.9%	4.7%	9.5%	6.7%	3.5%	1.8%	10.1%
Periyar University	Count	54	25	60	32	62	42	22	8	70
	%	11.4%	12.1%	12.0%	14.3%	10.9%	11.3%	9.4%	7.7%	
	Total (%)	8.7%	4.0%	9.6%	5.1%	10.0%	6.7%	3.5%	1.3%	11.2%
University of Madras	Count	39	17	37	17	41	29	20	8	42
	%	8.2%	8.2%	7.4%	7.6%	7.2%	7.8%	8.5%	7.7%	
	Total (%)	6.3%	2.7%	5.9%	2.7%	6.6%	4.7%	3.2%	1.3%	6.7%
Total	Count	473	207	499	223	571	373	234	104	623
	% of Total	75.9%	33.2%	80.1%	35.8%	91.7%	59.9%	37.6%	16.7%	100.0%

8. **Reddit:** The study has been listed seven Social Medias preferred by the respondents in the State Universities in Taminadu. It had analyzed based on data noted that "Others" types is given for getting result on the study, 104(16.7%) of the respondents preferred to access other non- listed Social Medias out of 623 respondents in the selected Universities of Tamil Nadu.

Table 5. Use of research citation indexes (RCI) vs. selected universities of the respondents

Name of the University		Research citation indexes									Total
		Google Scholar	CiteSeer	GetCITED	Math scinet	Scopus	Web of Science	EBSCO	Pro Quest	Other RCI	
Alagappa University	Count	70	39	1	54	42	40	47	54	11	358
	%	12.0%	11.5%	.9%	14.9%	9.5%	10.5%	14.7%	14.5%	8.9%	
	Total (%)	2.3%	1.3%	.0%	1.8%	1.4%	1.3%	1.5%	1.8%	.4%	11.8%
Annamalai University	Count	74	42	2	53	54	42	7	69	14	357
	%	12.7%	12.4%	1.7%	14.6%	12.2%	11.0%	2.2%	18.5%	11.3%	
	Total (%)	2.4%	1.4%	.1%	1.7%	1.8%	1.4%	.2%	2.3%	.5%	11.7%
Bharathidasan University	Count	46	23	1	23	33	23	34	49	3	235
	%	7.9%	6.8%	.9%	6.4%	7.5%	6.0%	10.7%	13.2%	2.4%	
	Total (%)	1.5%	.8%	.0%	.8%	1.1%	.8%	1.1%	1.6%	.1%	7.7%
Bharathiyar University	Count	62	62	4	62	62	62	62	62	24	462
	%	10.6%	18.3%	3.5%	17.1%	14.0%	16.2%	19.4%	16.7%	19.4%	
	Total (%)	2.0%	2.0%	.1%	2.0%	2.0%	2.0%	2.0%	2.0%	.8%	15.2%
Central University of Tamil Nadu	Count	56	28	7	31	42	37	30	39	16	286
	%	9.6%	8.3%	6.1%	8.6%	9.5%	9.7%	9.4%	10.5%	12.9%	
	Total (%)	1.8%	.9%	.2%	1.0%	1.4%	1.2%	1.0%	1.3%	.5%	9.4%
Gandhigram Rural University	Count	63	26	7	25	43	36	28	17	13	258
	%	10.8%	7.7%	6.1%	6.9%	9.7%	9.4%	8.8%	4.6%	10.5%	
	Total (%)	2.1%	.9%	.2%	.8%	1.4%	1.2%	.9%	.6%	.4%	8.5%
Madurai Kamaraj University	Count	42	21	3	26	20	17	21	10	12	172
	%	7.2%	6.2%	2.6%	7.2%	4.5%	4.5%	6.6%	2.7%	9.7%	
	Total (%)	1.4%	.7%	.1%	9%	.7%	.6%	.7%	.3%	.4%	5.7%
Manonmaiam Sundaranar University	Count	61	35	35	35	54	54	35	30	11	350
	%	10.4%	10.3%	30.4%	9.7%	12.2%	14.1%	11.0%	8.1%	8.9%	
	Total (%)	2.0%	1.2%	1.2%	1.2%	1.8%	1.8%	1.2%	1.0%	.4%	11.5%
Periyar University	Count	69	38	36	36	61	43	40	28	11	362
	%	11.8%	11.2%	31.3%	9.9%	13.8%	11.3%	12.5%	7.5%	8.9%	
	Total (%)	2.3%	1.3%	1.2%	1.2%	2.0%	1.4%	1.3%	.9%	.4%	11.9%
University of Madras	Count	41	25	19	17	31	28	15	14	9	199
	%	7.0%	7.4%	16.5%	4.7%	7.0%	7.3%	4.7%	3.8%	7.3%	
	Total (%)	1.3%	.8%	.6%	.6%	1.0%	.9%	.5%	.5%	.3%	6.5%
	Count	584	339	115	362	442	382	319	372	124	3039
	Total (%)	19.2%	11.2%	3.8%	11.9%	14.5%	12.6%	10.5%	12.2%	4.1%	100.0%

Research scholars with a social science background usually concentrate on publishing their research papers in non-indexed journals. The study finds to discuss the respondents' preferences on the research citations shared on social media by the respondents in the State Universities of Tamil Nadu. Table 5 discusses that a very large number of 584(19.2%) of the respondent's preference given to "Google Scholar'', the next preference of the respondents with 442 (14.5%) "Scopus.'' The study has reported that more than 20%, 382(12.6%), 372(12.2%) 362(11.9%), and 339(11.2%) of the respondent's preference their choice of "Web of science" "Pro Quest" "Math Scinet" and "Cite Seer" respectively. The study discussed that less than 10% of whom 124(4.1%) and 115(3.8%) respondents replied their preference of non-familiar citation Indexes are "Others" and "Get CITED" respectively. Further, the respondents have also familiar With "Google Scholar" and "Cite Seer" research citation Indexes which combined together 70(12.0%) 54(14.9%), i.e. 47(14.7%),42(9.5%), and 40(10.5%) of them witnessed that preference of "Google Scholar", "Math Scinet", and Pro Quest, i.e. EBSCO, Scopus, and Web of Science respectively to disseminate through Social Medias on their research contribution. 74(12.7%), 69(18.5%),54(12.2%),53(14.6%), and 42(12.4%) of them respondents preference of Google Scholar, Pro Quest, Scopus, Math scinet and CiteSeer respectively.

The study can be witnessed that respondents had preferred the Research Citation Indexes which has been showing as the majority of the respondents preferred Google Scholar and Math scinet, and Pro-Quest in Alagappa University and Annamalai University. Of the respondents from Bharathidasan University witnessed Pro Quest 49(13.2%) and Google Scholar 46(7.9%) of them, Bharathiar University 62(18.3%) of them preferred Google Scholar, Cite Seer, Math Scinet, Scopus, Web Science, EBSCO, and Pro Quest, Similarly, the respondents from Central University of Tamil Nadu respondents most reference of 56(9.6%) and 42(9.5%) of the Google Scholar and Scopus. Gandhigram Rural University of the respondents' have their preference 63(10.8%) and 43(9.7%) of the Google Scholar and Scopus. Madurai Kamaraj University of the respondents preferred 42(7.2%) Google Scholar. Manonmaiam Sundaranar University of the respondents preferred 61(10.4%) and 54(14.1%) Google Scholar and Scopus, Web of Science. Periyar University and the University of Madras had been noticed they have mostly preferred by 69(11.8%) and 61(13.8%) of them with 41(7.0%) and 31(7.0%) of them have preferred Google Scholar and Scopus respectively.

CONCLUSION

The study was analyzed based on quantitative data from the respondents in the Selected State Universities in Tamil Nadu. The study focuses on the respondents' attain Scholarly Information through Social Networks relevance of Social Science in the selected ten Universities in Tamil Nadu. The result finds researchers need to have smartphones to explore and create research groups within and of Universities. The study could be witnessed that WhatsApp, YouTube, and Facebook were used predominantly compare to other Social Networks in Selected five State Universities in Tamil Nadu. Therefore researchers usually share their research information and are watching current information from YouTube also. The University has to be providing free Wi-Fi connectivity for the use of the SNS. The study find that the majority of the respondents were using SNS in their homes compared to other places, it find that the researcher may get more free time apart from working hours at the Institutions where include the Citations using various Citation styles needful and reviews in their research. Presently, researchers are more interested to count their Citations in their research work, and h-Index on the purpose of Google Scholar is the primary one it can be registered easily by every researcher. The study analyzed that more respondents highly preferred Google Scholar for analyses of their h-Index, Citations, and research impact of the Publications of Social Science Scholars. The present context of research to an individual depends on the research information and communication disseminated by the Social Media on which twitted and Mendeley on the research publications score can be counted in order to achieve Altmetrics score will distinct to a researcher for their contribution.

The chapter has been explored the scholars who consume scholarly information via Social Media on the report of the study. The self-report data correlate only moderately with the server log data, indicating low criterion validity. The categorical self-report measure asking respondents to estimate how often they use their mobile phones fared better than the continuous self-report measure asking them to estimate their mobile phone activity yesterday. Multivariate exploratory analysis further suggests that it may be difficult to identify under- and over-reporting using demographic variables alone (Boase & Ling, 2013). This conceptualizes the new spatial logic created by the social use of location-aware mobile technologies, analyzing how mobile communities are formed by the mapping of social networks in urban spaces. It explores two main areas with the goal of understanding how locative mobile social networks (LMSNs) challenge the traditional logic of networks. First, it conceptualizes LMSNs by comparing them to (1) traditional transportation and communication networks, and (2) mobile social networks (MSNs). Second, the paper discusses the potential social implications of LMSNs, such as privacy, surveillance, and social exclusion) Adriana de Souza e Silva & Jordan Frith, 2010). The various forms of attachment across a working population in Africa, It is aimed at establishing the forms of attachment to the mobile phone and their possible socio-demographic variations. Data was collected from 374 used-clothing traders in Accra, using a mixed-methods approach. Employing the conversion framework from domestication theory, a thematic analysis showed that attachment to mobile phone use is culturally specific. Principal component analysis revealed five types of attachment with significant variations across age, gender, and level of education (Asante, 2018).

REFERENCES

Amy, L. (2011). Mirror, Mirror on my Facebook Wall: Effects of Exposure to Facebook on Self-Esteem. *Cyberpsychology, Behavior, and Social Networking*, *14*(1-2), 79–83. Advance online publication. doi:10.1089/cyber.2009.0411 PMID:21329447

Asante, R.K.B. (2018). Exploration of the forms of mobile phone attachment among traders in Ghana. *Mobile media & Communication*, *7* (1), 24-40. doi:10.1177/2050157918764015

Baskaran, C. (2019). Scholarly information share through Social Networks (SNs) and Medias among social science scholars in selected State Universities in Tamil Nadu. *International Journal of Library and Information Studies*, *9*(3), 83–92.

Baskaran, C. (2020). Research Patterns on the Social Networks and Media: A Scientometric Portrait. Handbook of Research on Emerging Trends and Technologies in Library and Information Science, p.19. doi:10.4018/978-1-5225-9825-1.ch014

Baskaran, C. (2020). Research enrichment of the faculty members in Alagappa University, India: the metrics based on WoS (Web of Science) and SCOPUS. *Library Philosophy and Practice (e-journal)*. 3726. https://digitalcommons.unl.edu/libphilprac/3726

Bianchi, A., & Phillips, J.G. (2005). Psychological predictors of problem mobile phone use. *cyberpsychology & Behavior and Social Networking*, *8*(1), 39-51. doi:10.1089/cpb.2005.8.39

Bittner, S., & Müller, A. (2011). Social networking tools and research information systems: Do they compete. *Webology, 8*(1), 1-8. https://www.webology.org/2011/v8n1/a82.html

Boase, J. (2013). Implications of software-based mobile media for social research. *Mobile Media & Communication*, *1*(1), 57–62. doi:10.1177/2050157912459500

Boase, J., & Ling, R. (2013). Measuring Mobile Phone Use: Self-Report Versus Log Data. *Journal of Computer-Mediated Communication*, *18*(4), 508–519. doi:10.1111/jcc4.12021

de Souza e Silva, A., & Frith, J. (2010). Locative Mobile Social Networks: Mapping Communication and Location in Urban Spaces. *Mobilities*, *5*(4), 485–505. doi:10.1080/17450101.2010.510332

Folorunso, O., Vincent, R. O., Adekoya, A. F., & Ogunde, A. O. (2010). Diffusion of innovation in social networking sites among university students. *International journal of computer science and security*, *4*(3), 361-372.

Humphreys, L. (2007). Mobile Social Networks and Social Practice: A Case Study of Dodgeball. *Journal of Computer-Mediated Communication*, *13*(1), 341–360. doi:10.1111/j.1083-6101.2007.00399.x

Humphreys, L. (2010). Mobile social media: Future challenges and opportunities. *Mobile Media & Communication*, *1*(1), 20–25. doi:10.1177/2050157912459499

Lee, D. H. (2013). Smartphones, mobile social space, and new sociality in Korea. *Mobile Media & Communication*, *1*(3), 269–284. doi:10.1177/2050157913486790

Mohamed, H. K., & Sumitha, E. (2011). Perception and Use of Social Networking Sites by the Students of Calicut University. *DESIDOC Journal of Library and Information Technology*, *31*(4).

Parveen, N. (2011). Use of social networking site (Facebook) in making awareness among the library and information science professionals of university libraries of UP: A case study. *International Journal of Digital Library Services*, *1*(1), 9–17.

Pitchaipandi, P. (2020). Impact and Usage of Social Media among the Post Graduate Students of Arts in Alagappa University, Karaikudi, India. Measuring and Implementing Altmetrics in Library and Information Science Research. IGI Global. doi:10.4018/978-1-7998-1309-5.ch010

Radhakrishnan, S., & Baskaran, C. (2019). Authorship Productivity and Applicability of Lotka's Law in Phytochemistry Literature. *Library Philosophy and Practice (e-journal)*. 3618. https://digitalcommons.unl.edu/libphilprac/3618

Radhakrishnan, S., & Baskaran, C. (2020). Phytochemistry Literature: An Altmetrics Analysis (2020). *Library Philosophy and Practice (e-journal)*. 4048. https://digitalcommons.unl.edu/libphilprac/4048

Ruths, D., & Jurgen, P. (2014). Social media for large studies of behavior. *Science*, *346*(6213), 1063–1064. doi:10.1126cience.346.6213.1063 PMID:25430759

Shafique, F., Anwar, M., & Bushra, M. (2010). Exploitation of Social media among University students: A case study. *Webology*, *7*(2), 34–47.

Singh, K. P., & Gill, M. S. (2011). *Using social media for social research: An introduction*. Use of social networking sites by the research scholars: Social Media Research Group. https://assets.publishing.service.gov.uk/government/uploads/system/uploads

Stern, D. M., & Willits, M. D. (2011). *Social media killed the LMS: Re-imagining the traditional learning management system in the age of blogs and online social networks*. In Educating Educators with Social Media Emerald Group Publishing Limited. doi:10.1108/S2044-9968(2011)0000001020/full/pdf

Sutko, D. M., & Silva, A. D. S. (2011). Location-aware mobile media and urban sociability. *New Media & Society*, *13*(15), 807–823. doi:10.1177/1461444810385202

Wang, X. (2014). Using attitude functions, self-efficacy, and norms to predict attitudes and intentions to use mobile devices to access social media during sporting event attendance. *Mobile Media & Communication*, *3*(1), 75–90. doi:10.1177/2050157914548932

Chapter 15
Exploring Opportunities for Utilizing Open Educational Resources (OERs) Towards Promoting Inclusive Education Settings in Tamil Nadu

K. Sathish Kumar
https://orcid.org/0000-0002-6036-7395
Alagappa University, Karaikudi, India

R. Jeyshankar
https://orcid.org/0000-0002-0186-0657
Pondicherry University, Puducherry, India

M. Muniyasamy
https://orcid.org/0000-0001-7481-9957
Alagappa University, Karaikudi, India

P. Natarajan
https://orcid.org/0009-0008-1452-6157
SRM Arts and Science College, Chennai, India

ABSTRACT

This chapter explores the potential of using open educational resources (OERs) to promote inclusive education settings in Tamil Nādu. The research method adopted for this study is a descriptive survey among with 275 research scholars. The chapter aims to understand the awareness, availability, and utilization of OERs among research scholars in Tamil Nādu, and to identify the opportunities for using OERs to promote inclusive education settings in the region. The survey mode under questionnaire includes questions such as the awareness of OERs, the sources from which research scholars access OERs, and their utilization in teaching and learning. The findings of the survey will help to gain an understanding of the

DOI: 10.4018/978-1-6684-8805-8.ch015

usage of OERs and the potential of using them to promote inclusion in Tamil Nādu. The results of this survey will also be beneficial in understanding the current status of OER usage and the opportunities for utilizing them to improve the quality of education in the region.

INTRODUCTION

Inclusive education refers inclusion of all the students irrespective of their sex, race, color, poverty, disability and they have been given equal opportunity in education and to be considered as being an integral part of the learning community (Bali et al, 2020; Hockings et al, 2012; Jingyu et al, 2020). Education is the fundamental right of a child and it is very important for individual development as well as national development. In the modern era, education is considered as the key to success. Education holds a vital role in the development of a nation and it is the right of every individual to receive quality education Ajegbomogun et al, (2017). In the present digital era, Open Educational Resources (OERs) are playing a crucial role in enhancing the quality of education and providing universal access to educational materials (Baskaran and Ramesh, 2019; Prasad and Baskaran, 2019). OERs are defined as any type of educational materials that are freely accessible, openly licensed, and able to be used, reused, shared, remixed and distributed by others Roeder et al, (2017). OERs are making a major impact on the way educational materials are produced, shared, and used Sathish Kumar, (2020). OERs offer a great opportunity for teachers to provide quality education to all students regardless of their economic status and geographical location. The concept of Open Educational Resources (OERs) is gaining traction in the modern educational landscape and is quickly becoming a popular tool for enhancing the educational experience of students in all levels of education, from primary school to post-secondary (Walsh, 2020). OERs are digital resources – such as multimedia, textbooks, lesson plans and other materials – that are freely available for teachers, students, and other educators to use and adapt to their own needs OERs are often used to reduce the cost of materials for students and to provide access to educational materials to those who may not otherwise have access (Binu P.C & Baskaran, 2020). The use of OERs has become increasingly important in the field of inclusive education. Inclusive education is defined as "the process of strengthening the capacity of the educational system to respond to the diverse needs of all learners" (Das, 2011; Roeder et al, 2017). This process involves a shift away from the traditional model of education, which often leaves some students behind, to a more holistic approach that seeks to meet the needs of all learners. In this context, OERs can be seen as an effective tool for promoting inclusive education (Lo et al, 2022). OERs can be used to create a more equitable learning environment by providing access to educational materials to all students, regardless of their socio-economic status, educational background, or other factors (Almaiah et al, 2020). Additionally, OERs can provide teachers with the flexibility to adapt the materials to fit the needs of their students, which can help make learning more engaging and productive for all learners (Kim et al, 2019; Zhang et al, 2020). This paper will explore the potential opportunities of utilizing OERs to promote inclusive education settings in the state of Tamilnadu, India. In this context, the paper will look at the current state of inclusive education in Tamilnadu, the challenges that are faced in promoting inclusive education, and the potential of OERs to address these challenges (Kumar and Mahendraprabu, 2021; Sathish Kumar et al, 2021). It will also examine how OERs can be effectively utilized to promote inclusive education in Tamilnadu, and the potential impacts that this could have on students and the wider educational system. Finally, the paper will discuss the possible implications of using OERs to promote inclusive education in the state Miyazoe and Anderson, (2013). Tamilnadu is a state in the South of India with a population of more than 72 million. It is one of the most populous states in the country and is home to a diverse range of cultural, religious and linguistic groups. In terms of education, Tamil Nadu has a high literacy rate and has seen an increase in enrolment in schools and colleges in recent years. However, there are still significant disparities in access to education and educational resources between different socio-economic groups, leading to disparities in educational outcomes. In this context, the use of OERs to promote inclusive education in Tamilnadu has the potential to address some of these disparities by providing access to educational resources to all students, regardless of their socio-economic status, educational background, or other factors. Additionally, OERs can provide teachers with the flexibility to adapt the materials to fit the needs of their students, which can help make learning more engaging and productive for all learners (Perryman et al, 2014 ;Baskaran, 2019). analyzed that use of social network and Medias by the Research scholars in Selected Universities of Tamilnadu. To highest 87 (31.9%) of respondents Annamalai University, (16.8%) and (15%) of them Male and Female respectively out of 273 total respondents responded to the study

In recent times, Open Educational Resources (OERs) have emerged as a powerful and viable tool to promote inclusive education settings in Tamilnadu. OERs are publicly available educational resources that can be used, modified, and shared freely (Karunanayaka et al, 2016; King, 2017). OERs have a number of advantages over traditional educational resources, such as being freely available, easily accessible, and providing a more diverse range of educational opportunities. In this context, this paper seeks to explore the possibilities and opportunities for utilizing OERs to promote inclusive education settings in Tamilnadu. Inclusive education is a concept that focuses on providing education to all persons, regardless of their abilities or disabilities (Altınay et al, 2016 ;Zhang et al, 2020). This approach has been gaining increasing importance in recent times, as it seeks to reduce educational disparities between different groups of people. This is especially true in the case of Tamilnadu, where there is a significant gap between the educational opportunities provided to persons with disabilities and those provided to persons without disabilities. In this context, it is important to explore ways in which OERs can be used to bridge this gap and promote inclusive education settings in Tamilnadu (Kumar et al, 2021; Mahendraprabhu,et al, 2021; Sathish Kumar, et al 2021) Baskaran, (2020) analyzed that majority of the publications, 2,207 (18.48%), were published from communication area, on account of the specific nature of the research concentrating on Social networks. Computer communications information systems area scores second with 1,172 (9.81%) of the publications. This paper will begin by presenting an overview of OERs and discussing their potential benefits. It will then examine the current state of inclusive education in Tamilnadu and discuss the challenges faced by persons with disabilities when it comes to accessing educational opportunities. Following this, the paper will evaluate the potential of OERs to promote inclusive education in the state and discuss ways in which OERs can be effectively utilized to bridge the existing gap between educational opportunities for persons with and without disabilities. Finally, the paper will conclude with a discussion of the implications of utilizing OERs to promote inclusive education in Tamil Nadu. Overall, this paper seeks to explore the opportunities and possibilities for utilizing OERs to promote inclusive education settings in Tamilnadu. By examining the potential of OERs to bridge the gap between educational opportunities for persons with and without disabilities, this paper will provide a valuable contribution to the understanding of the role of OERs in promoting inclusive education in Tamil Nadu. Baskaran, (2021). explored the majority 134 (1.96%) of the publications was contributed by the researchers from the University of California systems. Zhang Y was the top author was contributed 16(0.23%) of the publications in the field of Web 2.0.

SIGNIFICANCE OF THE STUDY

Education is the backbone of a country to preserve our culture and transmit it to the upcoming generation. The socio-economic, political, cultural, etc. all the development of a country is possible only through the quality education. To provide quality education link education with free is very important. To overcome those problems, it is very important to have some innovations in the education systems, which are most reliable and effective for the education system. The education system requires psychological, philosophical, technological advancement. For the development of a country, it is very important to educate all the people; therefore, nowadays the concept of inclusive education is very much popular. But it is very difficult to provide education with free resources for all the students in a common classroom. Therefore, it is very important to conduct such a study related to the challenges of using open access for inclusive education. But there are very few studies have been done related to this topic. Therefore, this study is carried out by the researcher to improve the quality of education through open access including all the students in a common classroom.

REVIEW OF LITERATURE

In the literature, online and distance learning are mentioned in different contexts. In the realization of online teaching and classes, the affective and cognitive dimension of learning as a consequence of the perception of the online presence of others and relationships with them was investigated Pugoy et al, (2016). This is especially important for working with children with disabilities because interactions with teachers and peers have a significant impact on the stated dimensions of their learning, but also on social development and inclusion (Conole, 2012; Pearson et al, 2019). 10 years ago digital technologies in education focused on creating, sharing, and accessing instructional content in digital forms, including online courses, digital libraries, games, and apps (Di Bitonto et al, 2015; Doğantan, 2020; Kim, 2020). Digitizing educational content, bringing devices to school, and one-off stand-alone learning apps were basic steps in the drive toward bringing technology into classrooms (He, 2020; Major et al, 2018). All these phenomena have their special and significant impact on the use of modern technologies with students with disabilities and should be further explored in the field of special education and rehabilitation (Alvi and Gupta, 2020; Hultgren, 2014). An important issue of online learning of students with disabilities is the absence or very reduced socialization. Teachers, parents, and students have noted ways in which online education can uniquely support the development of social competence in students through their ability to potentiate the strengths of individual students rather than exacerbate their weaknesses (Abdallah, 2018; de Souza Rodrigues et al, 2020).

The focus for education in general has a stage towards an inclusive movement that recognizes differences in students. However, there are problems in preparing learning plans such as low professionalism of teachers, weak teacher competence, and the boredom of teachers in preparing learning plans (Martin, 2018; Whyte et al, 2014). Baskaran (2019) analysed the majority (83.5%) of respondents using "Whaatsapp" out of respondents. The respondents reported (77.7%) and Smartphone/Mobiles (73.6%). Baskaran (2019) analyses the majority (83.5%) of respondents using "Whaatsapp" out of respondents. The respondents reported (77.7%) and (73.6%) using devices tools "Laptop" and Smartphone/Mobiles are respectively. Radhakrishnan and Baskaran (2019) have examined the authorship productivity, the applicability of Lotka's law, Price's square root law, and Pareto's 80/20 rule. The Chi-square value was

225.65, which is higher than the table value in the analysis of Lotka's law. Baskaran (2020) explored 11, 941 records on social networks and media retrieved from the Web of Science database during the study period. The predominant records were 2,576 (21.57%) published in 2018, followed by 2,281 (19.10%) published in 2017. Radhakrishnan and Baskaran (2020) conducted research on most of the publications shared by social media Twitter. Moreover, readers prefer to read articles through Mendeley's Reference Manager. The study reveals that there is a moderate correlation between the citation and altmetric scores. According to international experience, inclusive education requires a system approach to the coordinated policies and practices, covering a number of thematic areas such as teacher education, early intervention, development of curriculum and evaluation practices in addition to using ICTs (Brown and Croft, 2020; Hodgkinson-Williams and Paskevicius, 2012).

The activities of international organizations such as UNESCO IITE and the European Agency for Special Needs and Inclusive Education can be crucial in providing this kind of education (Conole, 2012; Pearson et al, 2019). Education in the online environment, in the context of the educational needs of students with disabilities, includes educational measures, procedures, methods, and techniques applied over the Internet, through educational means of modern technology (computer, mobile phone, tablet, etc.) that use this connection, and whose purpose is to improve the knowledge, skills, abilities, and involvement of students Abdallah, (2018). We emphasize that the use of television and radio broadcasts, as well as telephone conversations and correspondence methods (classic mail), belongs to the use of distance learning techniques, which are not necessarily e-learning techniques, so these techniques or only some of them are not always part of education in an online environment (Buhl et al, 2018; Gadd and Weedon, 2017; Kuliya and Usman, 2020; Verkuyl et al, 2020 ; Xia, 2013).

METHODOLOGY

The present study aims to analyze the challenges of using OER for inclusive education in Tamilnadu. So, the researcher has used a descriptive survey research method. It was used for investigation and collection of the data to study the relationship between OER and inclusive education in Tamilnadu. In the present study, a sample of 275 research scholars from the higher educational institutions with the help of a simple random sampling technique is used for the selection of the sample. In the present study, a self-made questionnaire has used for collecting data. To analyze the data, statistical tools such as SPSS (Statistical Package for Social Sciences) version 23.0 were used. The responses to each survey question were coded as numerical values and organized in a spreadsheet in order to easily identify and analyze patterns. The data was then analyzed in order to identify key trends and insights.

ANALYSIS AND INTERPRETATION OF DATA

Demographic Description

The study population includes a research scholars presently engaged in the research work at the Tamilnadu. Table 1 contains the Universities wise data related to the number of research scholars enrolled, questionnaire distributed among them and collected by fully filled up samples in the selected educational institutions in Tamilnadu.

Table 1. Demographic description of research scholars

SI. No	Name of the State Institutions	Questionnaire Distributed	Questionnaire Received
1	Alagappa University, Karaikudi	47	42
2	National Institute for Empowerment of Persons with Multiple Disabilities, Chennai	42	29
3	The Spastics Society of Tamilnadu, Chennai	32	28
4	Tamil Nadu Open University, Chennai	35	25
5	Mother Teresa Womens University, Kodaikanal	32	21
6	Tamil Nadu Teachers Education University, Chennai	26	18
7	Avinashilingam Institute for Home Science and Higher Education for Women, Coimbatore	28	24
8	NKT National College of Education for Women, Chennai	30	29
9	National Institute for the Empowerment of Persons with Visual Disabilities Divyangjan Regional Centre, Chennai	41	31
10	Kalasalingam Academy of Research and Education, Virudhunagar	32	28
Total		**345**	**275**

It is clear from the above Table 1 that 345 questionnaires were distributed among the research scholars of the university and 275 filled questionnaires were received for this present study. This distribution of questionnaire was taken place on the basis of stratified random sampling among the college wise and department wise population. Data shows very high response rate from the respondents' side for fill up questionnaire. The demographic description of research scholars in gender wise are given below in Table 2 and Figure 1.

Table 2. Demographic description of research scholars in gender wise

Gender	Male	Female
Samples	135 (49.09%)	140 (50.90%)

Figure 1. Demographic description of research scholars in gender wise

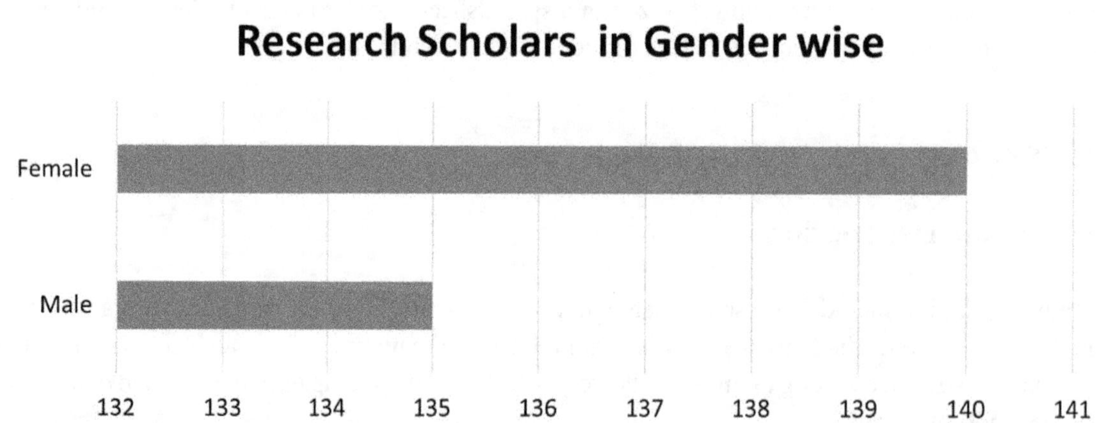

Table 3 analyzes the statistical techniques adopted to measure the **Statement. 1** shows the responses to the statement 'Now a day's education is linked with the online education. It has been seen that the student's responses are 45%, 34%, 6%. 15% and 0% accordingly it means strongly agree, agree, undecided, strongly disagree, and disagree for the above-mentioned statement. So, it can be concluded that maximum students strongly agree with the statement that nowadays education is linked with the online.

Statement.2 shows the responses of the statement 'ICT makes the teaching learning process more effective'. In this statement 39% students are strongly agree with this statement, 54% students are agree, 4%, 2% and 1% students are responses undecided, strongly disagree and disagree accordingly. So, we can conclude that ICT makes teaching learning process is more effective.

Statement.3 shows the responses of the statement 'It is very important to include all the students in one classroom for the development of the country'. In this statement 22% students are strongly agree with this statement, 66% students are agree, 4% students are undecided, strongly disagree and disagree. From the responses we can conclude that inclusive classroom is very important for the development of the country.

Statement.4 shows the response of the statement 'It is very difficult to provide education for the students with special need with the help of ICT in a common classroom along with the normal children'. In this statement 29% students are strongly agree with this statement, 40% students are agree, 7% students are undecided, 19% are strongly disagree and 5% are disagree. So we can conclude from the diagram that it is very difficult to provide education for the students with special need with the help of ICT in a common classroom along with the normal children.

Statement.5 shows the responses of the statement 'In Tamilnadu lack of technology, internet access, and qualified trainers etc. are the biggest challenges in providing ICT-based educational services to the masses. In this statement 61% students are strongly agree with this statement, 34% students are agree, 4% students are undecided, 1% are strongly disagree and no one is disagree with the statement. So we can conclude that In Tamilnadu lack of technology, internet access, and qualified trainers etc. are the biggest challenges in providing ICT-based educational services to the masses.

Table 3. Opportunities for utilizing open educational resources (OER) in universities

Sl. No	Statement	S. A	A	N	D	S. D	Mean	Standard Deviation
1.	Now a day's education is linked with online learning.	45%	34%	6%	15%	0%	4.42	0.88
2.	Online learning makes the teaching-learning process more effective.	39%	54%	4%	2%	1%	4.26	0.86
3	It is very important to include all the students in one classroom for the development of the country.	22%	66%	4%	4%	4%	4.32	0.74
4.	It is very difficult to provide education for the students with special needs with the help of online in a common classroom along with normal children.	29%	40%	7%	19%	5%	4.23	0.88
5.	Department of Special Education and Rehabilitation Science's lack of technology, internet access, and qualified trainers, etc. are the biggest challenges in providing online-based educational services to the masses.	61%	34%	4%	1%	0%	4.45	0.75
6.	In Department of Special Education and Rehabilitation Science institution has provided education online in an inclusive setup.	0	5%	6%	38%	51%	4.37	1.06

Statement.6 shows the responses of the statement 'In Tamilnadu all the institution has provided education Trough ICT in an inclusive set up'. In this statement no students are strongly agree with this statement, 5% students are agree, 6% students are undecided, 38% are strongly disagree and 51% disagree with the statement. Therefore, we can conclude that in Tamilnadu educational institution are not establish in an inclusive set up along with the ICT.

Table 4 examination statistical measure made up on the **Statement. 1** shows the responses to the statement librarians should play advisory roles to faculties, by trying to influence integration of OER in their courses. In this statement 39% students are strongly agree with this statement, 54% students are agree, 4%, 2% and 1% students are responses undecided, disagree and strongly disagree accordingly. So, we can conclude from that in librarians should play advisory roles to faculties, by trying to influence integration of OER in their courses.

Statement. 2 shows the responses to the statement Librarians can offer advice to institutions, academic Staff and students. In this statement 22% students are strongly agree with this statement, 66% students are agree, 4%, 4% and 4% students are responses undecided, disagree and strongly disagree with the statement. Therefore, we can conclude that Librarians can offer advice to institutions, academic Staff and students.

Statement. 3 shows the responses to the statement Academic libraries can incorporate OER in their collections, selecting and suggesting relevant OER for their institutional curriculum and their users. In this statement 29% students are strongly agree with this statement, 40% students are agree, 7%, 19% and 5% students are responses undecided, disagree and strongly disagree. From the responses we can conclude that Academic libraries can incorporate OER in their collections, selecting and suggesting relevant OER for their institutional curriculum and their users.

Statement. 4 shows the responses to the statement Academic libraries should imbibe the outreach skills, subject based guides, Seminars and workshops. In this statement 61% students are strongly agree with this statement, 34% students are agree, 4%, 1% and 0% students are responses undecided, disagree and strongly disagree with the statement. So, we can conclude that Academic libraries should imbibe the outreach skills, subject based guides, Seminars and workshops.

Statement. 5 shows the responses to the statement Including OER as part of the library's strategic plan is a way of promoting OER. In this statement 0% students are strongly agree with this statement, 5% students are agree, 6%, 38% and 51% students are responses undecided, disagree and strongly disagree

Table 4. The ways of promoting open educational resources (OER) in universities

Sl. No	Item	SA	A	N	D	SD	Mean	Standard Deviation
1.	Librarians should play advisory roles to faculties, by trying to influence integration of OER in their courses	39%	54%	4%	2%	1%	4.22	0.84
2.	Librarians can offer advice to institutions, academic Staff and students	22%	66%	4%	4%	4%	4.24	0.72
3.	Academic libraries can incorporate OER in their collections, selecting and suggesting relevant OER for their institutional curriculum and their users.	29%	40%	7%	19%	5%	4.34	0.86
4.	Academic libraries should imbibe the outreach skills, subject based guides, Seminars and workshops.	61%	34%	4%	1%	0%	4.54	0.79
5.	Including OER as part of the library's strategic plan is a way of promoting OER.	0	5%	6%	38%	51%	4.73	1.02

with the statement. So, we can conclude that Including OER as part of the library's strategic plan is a way of promoting OER.

FINDINGS OF THE STUDY

The researcher has used self-made questionnaire along with taken interview from some students who are pursuing their higher education. After the analysis of data and interview the major findings of the study are as follows.

It is very important to include all the students in one classroom for the development of the country.

1. It is very difficult to provide education for the students with special needs with the help of OER in a common classroom along with the normal children. Therefore, the school should build up the necessary physical infrastructure for special children in respect of their special needs.
2. In Tamilnadu lack of technology, internet access, and qualified trainers, etc. are the biggest challenges in providing OER-based educational services to them asses.
3. In the Tamilnadu there are many poor families are living. Students who belong from poor family may not have their personal computer at home and with no access to the internet facilities.
4. In the Tamilnadu, there are multilingual students in a class. So, teacher faces various challenges in dealing with students while using ICT.
5. One of the primary challenges in the use of ICT in education is the lack of knowledge and skills among teachers.
6. Developing countries like India the proficiency of English language is not too high expect the urban areas and major world class educational software produced in English language. So, there is a barrier to maximize to profit of ICT among the students of rural area.

SUGGESTIONS

After the deep analysis, the investigator has emerged out the following suggestions-

1) Train and skill teachers should appoint in all educational institutions. Because without the proper training no teacher is efficient in effectively using OER in the teaching-learning situation.
2) Proper awareness should develop among all the members of the society irrespective of their caste, creed, color, weaker section, or any other disabilities.
3) Government should provide sufficient infrastructures such as computers, projectors, and internet facilities to all the educational institutions.
4) Attitude of teachers, parents, policymakers, and peers towards special needs children should change. All of the members of society should ensure that the special needs children can't feel inferior in educational institutions.
5) Proper curriculum should be implemented in school that all the students can learn together irrespective of any barriers. The curriculum should focus on the aim of universalization of the education system.

6) School should develop all the facilities for the special needs' children along with the proper resource room in the school and school infrastructure should develop according to the needs of special needs children.

CONCLUSION

Open educational resources and education are closely related to each other. OER makes the whole teaching-learning process effective manner. From the above discussion, we can conclude that Tamilnadu has lots of challenges of using open-access materials for inclusive education. The problem is raised due to different reasons, such as lack of technology, internet access, and qualified trainers, multilingual and multicultural classroom, the gap between policies and implementation, etc. But for the development of the educational sector for a country, Using OER in the classroom and the inclusion of all the students both are the demand of the time. Therefore, all the stakeholders of education such as teachers, parents, policymakers, the administration should work collaboratively to remove the barriers of an inclusive classroom. So, on the side of policymakers, they should adopt some long-term policy to remove such kind of barriers.

REFERENCES

Abdallah, A. K. (2018). Parents Perception of E-Learning in Abu Dhabi Schools in United Arab Emirates. *IJASOS- International E-Journal of Advances in Social Sciences, 4*(10), 30–41. doi:10.18769/ijasos.415513

Ajegbomogun, F. O., Okunlaya, R. O. A., & Alawiye, M. K. (2017). Analytical study of E-learning resources in national open University of Nigeria. *Education and Information Technologies, 22*(5), 2403–2415. doi:10.100710639-016-9548-z

Almaiah, M. A., Al-Khasawneh, A., & Althunibat, A. (2020). Exploring the critical challenges and factors influencing the E-learning system usage during COVID-19 pandemic. *Education and Information Technologies, 25*(6), 5261–5280. doi:10.100710639-020-10219-y PMID:32837229

Altınay, Z., Ossiannilsson, E., Kalaç, M. O., Başarı, G., Aktepebaşı, A., & Altınay, F. (2016). Establishing a framework on OER practices for ICT competence of disabled citizens. *The Turkish Online Journal of Educational Technology, 15*(3), 68–72.

Alvi, M., & Gupta, M. (2020). Learning in times of lockdown: How Covid-19 is affecting education and food security in India. *Food Security, 12*(4), 793–796. doi:10.100712571-020-01065-4 PMID:32837637

Bali, M., Cronin, C., & Jhangiani, R. S. (2020). Framing open educational practices from a social justice perspective. *Journal of Interactive Media in Education, 2020*(1), 1–12. doi:10.5334/jime.565

Baskaran, C. (2019). Scholarly information share through Social Networks (SNs) and Medias among social science scholars in selected State Universities in Tamil Nadu. *International Journal of Library and Information Studies, 9*(3), 83–92.

Baskaran, C. (2020) Research Patterns on the Social Networks and Media: A Scientometric Portrait. Handbook of Research on Emerging Trends and Technologies in Library and Information Science, p.19. IGI Global. doi:10.4018/978-1-5225-9825-1.ch014

Baskaran, C. (2020). Research enrichment of the faculty members in Alagappa University, India: the metrics based on WoS (Web of Science) and SCOPUS. *Library Philosophy and Practice (e-journal).* 3726. https://digitalcommons.unl.edu/libphilprac/3726

Brown, M., & Croft, B. (2020). Social annotation and an inclusive praxis for open pedagogy in the college classroom. *Journal of Interactive Media in Education, 2020*(1), 1–8. doi:10.5334/jime.561

Buhl, M., Andreasen, L. B., & Pushpanadham, K. (2018). Upscaling the number of learners, fragmenting the role of teachers: How do massive open online courses (MOOCs) form new conditions for learning design? *International Review of Education, 64*(2), 179–195. doi:10.100711159-018-9714-1

Conole, G. (2012). Fostering social inclusion through open educational resources (OER). *Distance Education, 33*(2), 131–134. doi:10.1080/01587919.2012.700563

Das, A. (2011). Emergence of open educational resources (OER) in India and its impact on lifelong learning. *Library Hi Tech News, 28*(5), 10–15. doi:10.1108/07419051111163848

de Souza Rodrigues, M. A., Chimenti, P., & Nogueira, A. R. R. (2020). An exploration of eLearning adoption in the educational ecosystem. *Education and Information Technologies, 26*(1), 585–615. doi:10.100710639-020-10276-3

Di Bitonto, P., Pesare, E., Rossano, V., & Roselli, T. (2015). Smart learning environments using social network, gamification and recommender system approaches in e-health contexts. *Smart Innovation. Systems and Technologies, 41*, 491–500. doi:10.1007/978-3-319-19875-0_44

Doğantan, D. E. (2020). An interactive instruction model design with role play technique in distance education: A case study in open education system. *Journal of Hospitality, Leisure, Sport and Tourism Education, 27*, 100268. doi:10.1016/j.jhlste.2020.100268

Gadd, E., & Weedon, R. (2017). Copyright ownership of e-learning and teaching materials: Policy approaches taken by UK universities. *Education and Information Technologies, 22*(6), 3231–3250. doi:10.100710639-017-9583-4

He, J. (2020). Research and practice of flipped classroom teaching mode based on guidance case. *Education and Information Technologies, 25*(4), 2337–2352. doi:10.100710639-020-10137-z

Hockings, C., Brett, P., & Terentjevs, M. (2012). Making a difference-inclusive learning and teaching in higher education through open educational resources. *Distance Education, 33*(2), 237–252. doi:10.1080/01587919.2012.692066

Hodgkinson-Williams, C., & Paskevicius, M. (2012). The role of postgraduate students in co-authoring open educational resources to promote social inclusion: A case study at the University of Cape Town. *Distance Education, 33*(2), 253–269. doi:10.1080/01587919.2012.692052

Hultgren, A. K. (2014). Internationalization of Higher Education in East Asia. *Journal of Interactive Media in Education, 2014*(2), 1–2. doi:10.5334/jime.aa

Jingyu, W., Yuping, B., Yihzong, W., Zhihui, L., Xiangzheng, D., Islam, M., & Managi, S. (2020). Measuring inclusive wealth of China: Advances in sustainable use of resources. *Journal of Environmental Management, 264*, 110328. doi:10.1016/j.jenvman.2020.110328 PMID:32224293

Kim, D., Lee, I. H., & Park, J. H. (2019). Latent class analysis of non-formal learners' self-directed learning patterns in open educational resource repositories. *British Journal of Educational Technology, 50*(6), 3420–3436. doi:10.1111/bjet.12746

Kim, S. (2020). How a company's gamification strategy influences corporate learning: A study based on gamified MSLP (Mobile social learning platform). *Telematics and Informatics, 57*, 101505. doi:10.1016/j.tele.2020.101505

King, T. W. (2017). Postgraduate students as OER capacitators. *Open Praxis, 9*(2), 223. doi:10.5944/openpraxis.9.2.566

Kuliya, M., & Usman, S. (2020). Perceptions of E-learning among undergraduates and academic staff of higher educational institutions in north-eastern Nigeria. *Education and Information Technologies, 26*(2), 1787–1811. doi:10.100710639-020-10325-x

Kumar, K. S., & Mahendraprabu, M. (2021).. . *Perception of Oer and Oep, 1*(1), 1–26. doi:10.5281/zenodo.4705868#.YH8vr1I2Qz4.mendeley

Kumar, K. S., Prabu, M. M., Kalaiyarasan, G., Ramnath, R., Kumar, N. S., Mookkiah, M., Kumar, M. A., & B, R. (2021). *Open Educational Practices For Research Effectiveness To Research Scholar, 3*(4),113-121. Zenodo. doi:10.5281/ZENODO.4681771

Lo, C.-K., Tlili, A., & Huang, X. (2022). education sciences The Use of Open Educational Resources during the COVID-19 Pandemic : A Qualitative Study of Primary School Mathematics Teachers in Hong Kong. *Education Sciences, 12*(744), 1–14. doi:10.3390/educsci12110744

Mahendraprabhu, K., Sathish Kumar, M., & Mani, P. S. K. (2021). Open Educational Resources and their Educational Practices in Higher Education Open Educational Resources and their Educational Practices in Higher Education. *Mukt Shabd Journal, 10*(2), 527-540-. doi:10.5281/zenodo.4682079

Major, L., Warwick, P., Rasmussen, I., Ludvigsen, S., & Cook, V. (2018). Classroom dialogue and digital technologies: A scoping review. *Education and Information Technologies, 23*(5), 1995–2028. doi:10.100710639-018-9701-y

Martin, B. (2018). Faculty technology beliefs and practices in teacher preparation through a TPaCK lens. *Education and Information Technologies, 23*(5), 1775–1788. doi:10.100710639-017-9680-4

Miyazoe, T., & Anderson, T. (2013). Interaction Equivalency in an OER, MOOCS and Informal Learning Era. *Journal of Interactive Media in Education, 2013*(2), 1–15. doi:10.5334/2013-09

P.C. B., & Baskaran, C. (2020). The Faculty Members And Research Scholars Attain Academic And Research Thoughts By Accessing Electronic Information In The Universities Of Kerala, India. *Library Philosophy and Practice (e-Journal), 3739*, 1–27. https://digitalcommons.unl.edu/libphilprac/3739

Pearson, V., Lister, K., McPherson, E., Gallen, A. M., Davies, G., Colwell, C., Bradshaw, K., Braithwaite, N., & Collins, T. (2019). Embedding and sustaining inclusive practice to support disabled students in online and blended learning. *Journal of Interactive Media in Education, 2019*(1), 1–10. doi:10.5334/jime.500

Perryman, L.-A., Hemmings-Buckler, A., & Seal, T. (2014). Learning from TESS-India's Approach to OER Localisation Across Multiple Indian States. *Journal of Interactive Media in Education, 2014*(2), 1–11. doi:10.5334/jime.af

Pugoy, D.L., R. A., D.L. Habito, C., & B. Figueroa, R. (2016). Hybrid online/offline mobile solutions for accessing open educational resources in areas with poor internet connectivity. *Asian Association of Open Universities Journal, 11*(2), 182–196. doi:10.1108/AAOUJ-09-2016-0030

Roeder, I., Severengiz, M., Stark, R., & Seliger, G. (2017). Open Educational Resources as a Driver for Manufacturing-related Education for Learning of Sustainable Development. *Procedia Manufacturing, 8*, 81–88. doi:10.1016/j.promfg.2017.02.010

Sathish Kumar, K. M. M. (2020). Awareness of swayam programme among the Postgraduate students. In LAP LAMBERT Academic Publishing, 1 (1). doi:10.1108/AAOUJ-07-2016-0022

Sathish Kumar, K., Mahendraprabu, M., Kalaiyarasan, G., Ramnath, R., Sasi Kumar, N., & Mani, M. (2021). Social Media as an Open Educational Practice Tools and Challenges. *Library Philosophy and Practice, 2021*(1), 1–24. doi:10.5281/zenodo.4705256

Sathish Kumar, K., Mahendraprabu, M., Kalaiyarasan, G., Ramnath, R., Sasi Kumar, N., & Mookkiaha, M. (2021). A vision of teaching learning practices in mathematics education through open educational resources. *International Journal of Education and Teaching, 1*(2), 1–30. doi:10.51483/IJEDT.1.2.2021.30-36

Verkuyl, M., Lapum, J. L., St-Amant, O., Bregstein, J., & Hughes, M. (2020). Healthcare students' use of an e-textbook open educational resource on vital sign measurement: A qualitative study. *Open Learning, 38*(1), 22–37. doi:10.1080/02680513.2020.1835623

Walsh, A. J. (2020). Removing barriers: Textbook affordability and OER at Sinclair community college. *RSR. Reference Services Review, 48*(3), 385–396. doi:10.1108/RSR-03-2020-0011

Whyte, S., Schmid, E. C., van Hazebrouck Thompson, S., & Oberhofer, M. (2014). Open educational resources for CALL teacher education: The iTILT interactive whiteboard project. *Computer Assisted Language Learning, 27*(2), 122–148. doi:10.1080/09588221.2013.818558

Xia, J. (2013). Let us take a Yale open course: A Chinese view of open educational resources provided by institutions in the West. *Journal of Computer Assisted Learning, 29*(2), 122–137. doi:10.1111/j.1365-2729.2012.00477.x

Zhang, X., Tlili, A., Nascimbeni, F., Burgos, D., Huang, R., Chang, T. W., Jemni, M., & Khribi, M. K. (2020). Accessibility within open educational resources and practices for disabled learners: a systematic literature review. In Smart Learning Environments 7(1),1-19. Springer. doi:10.118640561-019-0113-2

Chapter 16
Social Networking Skills in Information Literacy Towards Disseminating Scholarly Communication Among Research Scholars at Alagappa University:
A Case Study

Sivagami Murugappan
Alagappa University, Karaikudi, India

R. Jeyshankar
https://orcid.org/0000-0002-0186-0657
Pondicherry University, Puducherry, India

ABSTRACT

The study aims to identify the fundamental principles and utilization of social networking skills (SNS) in information literacy (IL) by research scholars of Alagappa University. It intends to compare the earlier educational level, dissemination of information level after employing SNS in research. It also frameworks the challenges recognized by the scholars in accessing SNS and profound changes in information environments and outlines of generation and necessity of scientific information in new modes of assessment, which predominantly focuses on methods and standards of scientific research work. Data was collected by distributing questionnaires through Google form and print form among research scholars of various departments and received 71 responses. The study aims to identify the fundamental principles and utilization of social networking skills (SNS) in information literacy (IL) by research scholars of Alagappa University. It intends to compare the earlier educational level, dissemination of information level and after employing SNS in research.

DOI: 10.4018/978-1-6684-8805-8.ch016

INTRODUCTION

Social media and social networking platforms are extensively used in scholarly communication to share and disseminate information (Manca and Ranieri, 2017) and are considered to be essential for providing the right information to the right users at the right time in a customizable, personalized, and collaborative way to the users, which is vital for the information professional. Researchers can associate with other researchers in one's field, and stay up-to-date on recent signs of progress in one's area of proficiency. To apprise themselves, they use many platforms such as Twitter, LinkedIn, and Research Gate to promote their work, engage with other researchers, and receive feedback on their findings (Work et al, 2015). These social platforms are highly utilized in the academic field for accessing rapid responses and interacting with each other, aiding to publicize new activities such as publications of their works, conferences, meetings, webinars, and exchanging research questions and answers.

Besides, many journals and conferences now use social media to promote their events and publications, which can help researchers stay informed about upcoming opportunities to present their work and engage with their peers. They can be used in information literacy to help individuals to locate, evaluate, and effectively use information. Social media platforms are used to follow experts and organizations in a particular field, which help to determine and discover new resources. Social networking platforms, such as LinkedIn, can be used to connect with professionals to seek out experts for research advice or collaboration. Too many are raised in any online communities and discussion forums are available on social media platforms, which can be used to find and engage with like-minded individuals and to ask for feedback on research or other information-related projects.

Overall, social media and social networking can be powerful tools for individuals and scholars to connect, collaborate, find, evaluate, disseminate, and use information effectively in their research. Hence, this study aims to identify the significance of social networking skills to researchers to progress in academic and personal and extensive usage of academic media in retrieving information literacy. Barriers are raised to accessing these social media for scholarly communication and its impact on knowledge-sharing activities.

LITERATURE REVIEW

Al-Aufi and Fulton (2014) resolute that 71% of academics are using social media for informal scholarly communication, proposing that the impact of social media is important and more research is essential to comprehend its influence on scholars and aims to learn more about this phenomenon. That effect might be robust in countries where wireless connectivity is universal, such as Kuwait, which has 99% coverage and more than 90% of the population uses social media. Gu and Widén-Wulff, (2011) determined that the LSE blog is a reputable source that is used as a substitute for a peer-reviewed journal article as open-access content. It is different from other academic social media such as Research Gate and Academia.edu, which control the accessibility of the full text of scholarly works on their sites. Despite these platforms, social media are entirely different, which causes a shift in scholarly communication; hence they have various tools and services that support research and it also improves the organizational research profile (Ali et al., 2017). It supports knowledge propagation through open access and other characteristics such as search competence, personalization, and logical tools like data about a number of sights, downloads, reads, and citations (Williams and Woodacre, 2016).

Academic institutions always utilize social media to keep them updated and to know about the announcements of conferences, and webinars held nationally and internationally. It also aids researchers in collaborating with the same institution or communicating with other institutions globally (Al-Aufi and Fulton, 2014). According to Nadez (2012), most researchers use Facebook, Twitter, and Instagram to distribute, disseminate and associate with other professionals to share their acquired knowledge and to track other researchers' activities. Social media like Research Gate, Academia.edu, and Mendeley have fascinated the consideration of academics. According to Jeng et al. (2015) and Thelwall and Kousha (2014), these websites support scholars to build their specialized networks and expose their research work, as well as enable their research activities smoothly. The profiles that academics post on these websites offer them access to opportunities, such as providing and locating publications, generating online research groups, dealing with published documents and citations, enquiring and responding to research questions, and collaborating on research and publications Increased. Elsayed (2015) investigated academic social networks and analyzed that 93% of researchers are highly influenced and willing to publish their works, share their consequences and communicate with each other. But to the contrary, the study by Donlean (2016) found that one-third of researchers and professionals are not using academic tools, instead, they use self-development and enlargement of their social networks. It also found that there are some barriers to using social media, like lack of interest, lack of skills, lack of time, etc. For this, Donlean suggested three ways, such as sharing of knowledge, practical training and institutional dissertations and publishing career opportunities on social media will increase the usage of social media among academics. Manca and Ranieri (2017) surveyed and found that Italian academic professionals very rarely use social media for scholarly communication. Social media are highly used for self–development, to extend their collaboration with other authors, and to increase their personal prominence. Mansour (2015) resolute a study in Kuwait to analyze the usage of social networking sites with the help of library professionals and found that 63.3% are utilizing social media for efficient dissemination and sharing of their knowledge, communication, and information and to create formal and informal collaboration. He also determined some barriers such as lack of awareness, reliability, taking more time to access, and inadequate training in their use. Baskaran (2020) analyzed that majority of the publications, 2,207 (18.48%), were published from communication area, on account of the specific nature of the research concentrating on Social networks. Computer communications information systems area scores second with 1,172 (9.81%) of the publications.

The review of pertinent studies discussed the use of social media by researchers contributed to the progress of the conceptual framework that directed this study. The framework accentuates the use of wide-ranging social media, such as Twitter and Facebook, and academic social media such as Research Gate and academia.edu. The literature review has also highlighted the part of social media in scholarly communication activities, such as sharing research papers, reporting scholarly events, structuring research networks, inspiring collaborations, pursuing research use, etc. These discussions were allowed in mind while designing and collecting data to conduct the study.

PROBLEM OF STATEMENT

This study investigated the use of social media in progressing social networking skills for scholarly communication among research scholars. The integrated review of the literature recommends the usage of social media in academics and barriers to access but no study was conducted about the impact of social media on information literacy, social networking skills, and knowledge-sharing activities. Therefore, this study is expected to enhance our understanding of the role of social media in scholarly communication in the context of Alagappa University.

RESEARCH QUESTIONS

RQ1. What is the level of the use of social media (e.g. Twitter, Facebook, etc.) among scholars at improving social networking skills?
RQ2. What is the range of the use of academic social media (e.g. Research Gate, Academia.edu, etc.) among scholars for enlightening information literacy skills?
RQ3. What problems are raised with accessing these social media among research scholars for scholarly communication?
RQ4. What is the impact created by social media on progressing social networking skills and knowledge-sharing activities?

Hypotheses

- There is no significant association between the level of satisfaction in using social media for sharing knowledge and residing place.
- There is no significant difference between the use of social media tools and improving social networking skills among scholars.
- There is no significant difference between academic social media and enlightening information literacy skills among scholars.
- There is no significant difference in accessing these social media among gender for scholarly communication.
- There is no significant difference between the impacts of such knowledge-sharing activities with the source utilized to share their knowledge.

METHODOLOGY

The study was designed to scrutinize social networking skills in information literacy toward disseminating scholarly communication among research scholars at Alagappa University. It represents the impact of social media and how they are deployed in educational activities. The category of tools employed by researchers to access social media, the source of sharing their knowledge, and their impact on knowledge sharing are evaluated in this study. An online survey method was adopted by distributing Google forms through e-mail and WhatsApp and also a printed questionnaire was distributed to research scholars with a trivial description. A total of 71 respondents were earmarked as the sampling structure for this study. Data derived from the questionnaire responses received either in the mode of social media or on paper were analyzed descriptively using various analyses in the SPSS statistical package.

DATA ANALYSIS AND INTERPRETATION

Data interpretation is the process of swotting collected data of social networking skills in disseminating scholarly communication and arriving at pertinent conclusions using various analytical methods which initiates with the demographic details.

The table - 1 interprets that most of the respondents are from rural areas and respondents from semi-urban is very low due to lack of facilities or poverty or English proficiency or family instability or some other physical problems faced by them. The consequences of this study depicted that the gender distribution is almost even; but the diminutive difference signifies the trend of the educational programs, in which female students are more curious to secure good positions in society than the males. From the total respondents, most of the respondents are from the arts discipline, the cause may be the researcher also belongs to the art department, and arts students are responding better than other disciplines. The analysis evidences that students fit the age group of 26-30 relatively had a healthier performance in academics than other age groups, due to their level of understanding about research and its significance to society.

Null Hypothesis: There is no significant association between level of satisfaction in utilizing social media for sharing knowledge and residing place.

Table: 2 illustrate the estimated significance value, which is more than 0.05. Hence, the null hypothesis is accepted that there is no significant difference between residing place and the level of satisfaction in utilizing social media for sharing knowledge. It may be due to now-a-days everyone using smart phones and accessing the internet in a quick way, especially respondents who are mostly from rural areas; they may have the right knowledge in accessing and sharing information in a proper way. However, it could

Table 1. Demographic information of respondents

Descriptive Statistics			
Demographic profile	**Options**	**Frequency**	**Percent**
Location	Rural	33	46.47
	Urban	25	35.21
	Semi-Urban	13	18.30
	Total	71	100
Discipline	Science	12	16.90
	Arts	30	42.25
	Management	29	40.85
	Total	71	100
Gender	Male	34	47.89
	Female	37	52.11
	Total	71	100
Age Group	20-25	14	19.71
	26-30	33	46.48
	31-35	15	21.13
	Above 35	9	12.68
	Total	71	100

Figure 1. Frequency of distribution of demographic information

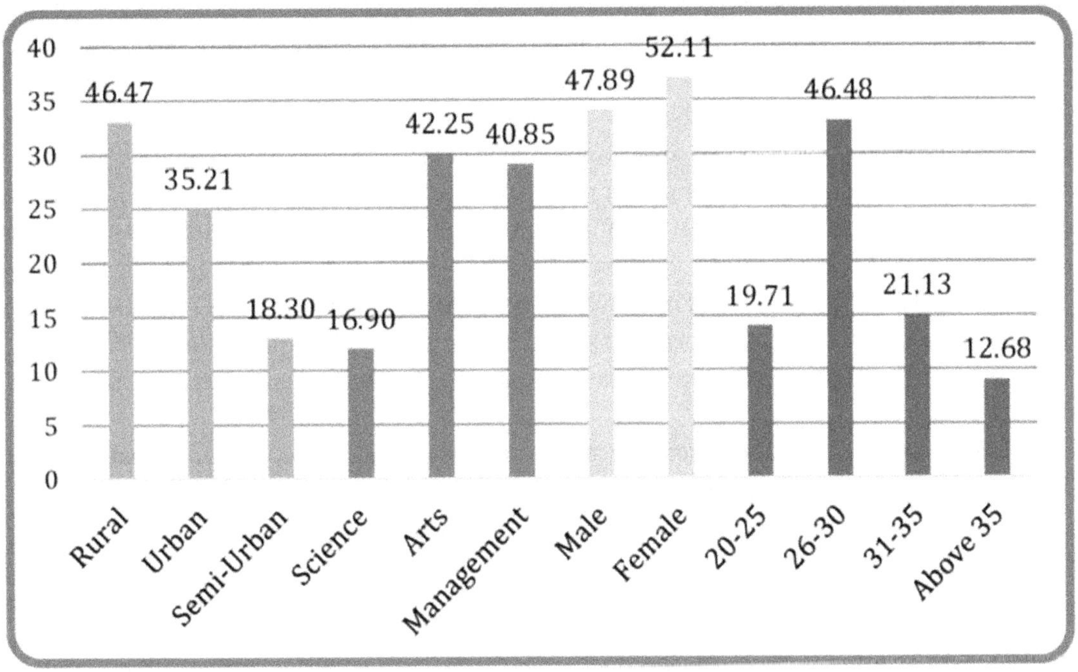

Table 2. Chi-square test for significant relationship between level of satisfaction in using social media and residing place

Correspondence Table				
Level of satisfaction in using Social Media for sharing knowledge:				
Variables	Lower Extent	Medium	To large extent	
Rural	3	12	18	33
Urban	1	9	15	25
Semi-Urban	1	6	6	13
Total	5	27	39	71
Chi-Square Tests				
	Value	df	Asymptotic Significance (2-sided)	
Pearson Chi-Square	1.110	4	.893	
Likelihood Ratio	1.148	4	.887	
Linear-by-Linear Association	.011	1	.916	
No. of Valid Cases	71			

be that other factors, such as age, income, or personal preferences, play a more significant role in determining satisfaction levels with social media and residing place.

Null Hypothesis: There is no significant difference between the use of social media tools and improving social networking skills among scholars

Table 3 depicts the value of 0.18 for R-squared means that 18% of the variance in the dependent variable is explained by the independent variables in the model. This means that 82% of the variance in the dependent variable is still unexplained by the model. A low ANOVA value of 0.070 indicates that there is not a significant difference between the means of the information literacy skills for the different social media platforms. The social media what's App alone has significant relationship with improving social networking skills of researcher since through what's App only many academic news, notification about webinars, workshops, national and international conferences are disseminated. Even many of the researchers use this social media tool for collecting data through online questionnaire with the assistance of Google forms. Therefore, it can be concluded that there is no significant impact created in social networking skills by using social media. Moreover, social media usage may not provide adequate opportunities for the development of certain skills. It's also possible that individual differences, such as prior experiences and personality traits, may play an effective role in social media usage develops social networking skills.

Null Hypothesis: There is no significant correlation between academic social media and enlightening information literacy skills among scholars.

Table 3. Regression test for significant impact on social media and social networking skills

Model Summary				
Model	**R**	**R Square**	**Adjusted R Square**	**Std. Error of the Estimate**
1	.426[a]	.181	.091	.999
a. Predictors: (Constant), Face Book, Twitter, You Tube, Messenger, Whats App, Instagram and others				

ANOVA[a]						
Model		Sum of Squares	df	Mean Square	F	Sig.
1	Regression	13.935	7	1.991	1.995	.070[b]
	Residual	62.854	63	.998		
	Total	76.789	70			
a. Dependent Variable: Social Networking Skills						
a. Predictors: (Constant), Face Book, Twitter, You Tube, Messenger, Whats App, Instagram and others						

Coefficients[a]						
Model		Unstandardized Coefficients		Standardized Coefficients	t	Sig.
		B	Std. Error	Beta		
1	(Constant)	5.115	1.237		4.136	.000
	FaceBook	-.001	.153	-.001	-.006	.996
	Twitter	.002	.165	.002	.015	.988
	You Tube	.284	.206	.188	1.379	.173
	Messenger	.100	.130	.122	.766	.446
	Whats App	-.950	.308	-.373	-3.081	.003
	Instagram	-.049	.148	-.047	-.331	.742
	Others	-.141	.142	-.137	-.993	.325
a. Dependent Variable: Social Networking Skills						

The Table 4 represents about the academic information skills and academic social media are related to each other and researchers are aware of utilization of these tools to improve their research activities.

Communication Skills: Many academic social media platforms provide opportunities for users to engage in discussions, collaborate on projects, and present their work to others, which can help improve communication skills.

Critical Thinking: Researchers can engage in discussions, read and analyze research papers, and receive feedback on their work, which can all encourage critical evaluation of information and arguments.

Time Management: Using academic social media platforms effectively can require good time management skills, as they can be time-consuming to use and require regular engagement and also used for staying organized and tracking one's own research and professional development.

Problem Solving Skills: Researchers can collaborate on projects, receive feedback on their work, and engage in discussions with others in their field, which can all encourage creative and innovative thinking. It's important to note that these platforms are not a substitute for developing these skills through direct experience and intentional practice. However, they can complement and support the development of these skills.

Null Hypothesis: There is no significant difference between accessing these social media among gender for progressing scholarly communication.

Table 5 illustrates that there is no significant gender-based differences in access to technology and the internet cannot limit individuals' ability to access social media does not persist any digital divide. Women and girls are much knowledgeable in access to the technology and infrastructure needed to ac-

Table 4. Correlation test for a significant correlation between online meeting apps and departments

Correlations						
		Communication Skills	**Critical Thinking**	**Time Management**	**Problem Solving Skill**	**Academic**
Communication Skills	Pearson Correlation	1	.513**	.673**	.224	.084
	Sig. (2-tailed)		.000	.000	.061	.484
	N	71	71	71	71	71
Critical Thinking	Pearson Correlation	.513**	1	.710**	.301*	-.076
	Sig. (2-tailed)	.000		.000	.011	.531
	N	71	71	71	71	71
Time Management	Pearson Correlation	.673**	.710**	1	.328**	.058
	Sig. (2-tailed)	.000	.000		.005	.631
	N	71	71	71	71	71
Problem Solving Skill	Pearson Correlation	.224	.301*	.328**	1	.030
	Sig. (2-tailed)	.061	.011	.005		.805
	N	71	71	71	71	71
Academic	Pearson Correlation	.084	-.076	.058	.030	1
	Sig. (2-tailed)	.484	.531	.631	.805	
	N	71	71	71	71	71

Table 5. independent samples t-test for the significant difference in barriers of accessing social media among gender

Independent Sample Test							
Variables		Levene's Test for Equality of Variances		t-test for Equality of Means			Null Hypothesis
		F	Sig.	t	df	Sig. (2-tailed)	
Lack of Internet Service	Equal variances assumed	3.696	0.059	-1.347	69	0.182	Accepted
	Equal variances not assumed			-1.332	61.922	0.188	
Lack of Knowledge	Equal variances assumed	0.006	0.937	-1.595	69	0.115	Accepted
	Equal variances not assumed			-1.591	67.669	0.116	
Lack of Authenticity	Equal variances assumed	2.374	0.128	2.082	69	0.041	Rejected
	Equal variances not assumed			2.116	63.39	0.038	
Lack of Peer Support	Equal variances assumed	3.385	0.07	2.606	69	0.011	Rejected
	Equal variances not assumed			2.663	58.128	0.01	
Lack of Institution Support	Equal variances assumed	15.762	0	3.379	69	0.001	Rejected
	Equal variances not assumed			3.48	49.669	0.001	

cess the internet and social media. Moreover, that internet services are provided to everyone in same way, there in gender difference is possible.

Lack of peer support: social media can be a powerful tool for connecting with others in one's field, but without a supportive network of peers, it may be more challenging for individuals to engage in scholarly communication and make meaningful connections.

Lack of institutional support: Institutions may not provide support or resources which can limit individuals' ability to use these platforms effectively.

Lack of authenticity: On social media, it can be challenging to verify the authenticity of information and individuals. This can lead to concerns about the reliability of information shared on these platforms, making it difficult for individuals to trust and use these tools for scholarly communication.

These barriers can impact individuals of all genders and limit their ability to engage in scholarly communication on social media. Addressing these barriers will be important for ensuring equal access to these tools and fostering a more inclusive scholarly community.

Null Hypothesis: There is no significant relationship between impacts of knowledge sharing activities with the source utilized among research scholars.

Table 6 show the estimated R square value is 0.392, indicates that the model has a forecasting accuracy of 39.2%. Further ANOVA significance value is found to be less than 0.05. Hence the model is fit. Also, the coefficient significance values are found to be less than 0.05 for friends, and webinars / workshop/conference/seminar. Therefore, it can be concluded that there is a significant impact created by sources of sharing knowledge on scholarly communication. There is no significant impact created by forums, online tutorials, and regular classes because scholarly communication is based on evidence-based research and is expected to meet rigorous standards for accuracy and reliability, which may not always, be the case in forums and tutorials. Research scholars are not attending any regular classes, so there is no significant impact created by these sources in progressing their scholarly communication.

Table 6. Regression test for significant impact on source of sharing their knowledge and impact in scholarly communication

Model Summary				
Model	**R**	**R Square**	**Adjusted R Square**	**Std. Error of the Estimate**
1	.626	.392	.345	.471
a. Predictors: (Constant), Webinars/ Workshop/Conference/Seminar, Forums, Online Tutorials Regular classes				

ANOVA						
Model		**Sum of Squares**	**df**	**Mean Square**	**F**	**Sig.**
1	Regression	9.287	5	1.857	8.366	.000
	Residual	14.432	65	.222		
	Total	23.718	70			
a. Dependent Variable: Impact of Knowledge Sharing						
b. Predictors: (Constant), Webinars/ Workshop/Conference/Seminar, Forums, Online Tutorials Regular classes						

Coefficients						
Model		**Unstandardized Coefficients**		**Standardized Coefficients**	**t**	**Sig.**
		B	**Std. Error**	**Beta**		
1	(Constant)	1.465	.493		2.972	.004
	Friends	.617	.132	.534	4.681	.000
	Forums	-.143	.098	-.186	-1.464	.148
	Online Tutorials	-.059	.090	-.092	-.653	.516
	Regular classes	-.097	.089	-.156	-1.086	.282
	Webinars/ Workshop/Conference/Seminar	.239	.100	.346	2.394	.020
a. Dependent Variable: Impact of such knowledge sharing activities						

Null Hypothesis: There is no significant correlation between online meeting apps used for sharing knowledge and departments among scholars

Table 7 estimated significance value is more than 0.05 except for Live storm and Click meeting. Hence the null hypothesis is accepted, meaning there is no significant relationship between Online Meeting Apps and departments. Only Live storm and Click meeting are having significant relationship with departments. The correlation value is calculated to be Zoom meeting (22%) Go to Webinar (22%), Live Strom (26%), Click Meeting (30%), Outlook (36%) and Go to Meeting (22%) these meetings are mostly concentrated by arts and managements scholars and our respondents are mostly from these disciplines, hence they secured a positive correlation. the specific departments within an organization that use online meeting platforms will depend on the size, structure, and needs of the organization. Google meet with (-0.016) Pearson correlation value shows a negative correlation between concerned departments. Some users may find that the user interface or features of Google Meet or Zoom are not as user-friendly as they would like, or they may prefer a different platform that better meets their needs.

Table 7. Correlation test for a significant correlation online meeting apps and departments

Online Meeting Apps		Zoom	Go To Webinar	Live storm	Click Meeting	Google Meet	Outlook	Go to Meeting	Department
Zoom	Pearson Correlation	1	.677**	0.066	0.096	.491**	-0.039	0.182	0.222
	Sig. (2-tailed)		0.000	0.585	0.427	0.000	0.749	0.128	0.062
	N	71	71	71	71	71	71	71	71
Go To Webinar	Pearson Correlation	.677**	1	.351**	.428**	.419**	0.129	.401**	0.221
	Sig. (2-tailed)	0.000		0.003	0.000	0.000	0.282	0.001	0.063
	N	71	71	71	71	71	71	71	71
Live storm	Pearson Correlation	0.066	.351**	1	.836**	0.028	.734**	.575**	.265*
	Sig. (2-tailed)	0.585	0.003		0.000	0.814	0.000	0.000	**0.026**
	N	71	71	71	71	71	71	71	71
Click Meeting	Pearson Correlation	0.096	.428**	.836**	1	0.193	.612**	.472**	.309**
	Sig. (2-tailed)	0.427	0.000	0.000		0.108	0.000	0.000	**0.009**
	N	71	71	71	71	71	71	71	71
Google Meet	Pearson Correlation	.491**	.419**	0.028	0.193	1	0.155	0.155	-0.016
	Sig. (2-tailed)	0.000	0.000	0.814	0.108		0.195	0.196	0.892
	N	71	71	71	71	71	71	71	71
Outlook	Pearson Correlation	-0.039	0.129	.734**	.612**	0.155	1	.664**	0.108
	Sig. (2-tailed)	0.749	0.282	0.000	0.000	0.195		0.000	0.368
	N	71	71	71	71	71	71	71	71
Go to Meeting	Pearson Correlation	0.182	.401**	.575**	.472**	0.155	.664**	1	0.226
	Sig. (2-tailed)	0.128	0.001	0.000	0.000	0.196	0.000		0.059
	N	71	71	71	71	71	71	71	71
Department	Pearson Correlation	0.222	0.221	.265*	.309**	-0.016	0.108	0.226	1
	Sig. (2-tailed)	0.062	0.063	0.026	0.009	0.892	0.368	0.059	
	N	71	71	71	71	71	71	71	71

CONCLUSION

In the present study, the data were collected through a questionnaire based on the social networking skills in information literacy skills and its dimensions, to identify the information literacy skills among research scholars of Alagappa University and its affiliated Colleges. The result of the study threw light on information literacy skills among research scholars and found that they have an average level of literacy skills. The study also brings the findings on information literacy skills by using social media and to identify the use of academic social media to improve their research skills, information and technology skills, competency skills, and communication skills. It also determines the academic skills like publishing skills, gaining current knowledge skills, scholarly evaluations and update information skills are to be improved through social media and various training programs. More training programs like soft skill programs, orientation programs, and technical training programs can be conducted to progress the skills of research scholars in the future which aid them to produce more qualitative research in every field. The study recognizes the areas which were needed to be concentrated upon while manipulating the future IL programs to be steered by the library and diverse institutes for the user community. This study assistances to organize various information literacy programs in various institutions, promoting and to progress information literacy skills among research scholars, and to improve the quality of research which aims to achieve advancement in knowledge, shaping public policy, improving the health and well-being of society, supporting economic development and promote social change.

REFERENCES

Al-Aufi, A., & Fulton, C. (2014). Use of social networking tools for informal scholarly communication in humanities and social sciences disciplines. *Procedia: Social and Behavioral Sciences*, *147*, 436–445. doi:10.1016/j.sbspro.2014.07.135

Ali, M., Wolski, M., & Richardson, J. (2017). Strategies for using ResearchGate to improve institutional research outcomes. *Library Review*, *66*(8/9), 726–739. doi:10.1108/LR-07-2017-0060

Baskaran, C. (2020). Research Patterns on the Social Networks and Media: A Scientometric Portrait. Handbook of Research on Emerging Trends and Technologies in Library and Information Science, pp.189-207. IGI Global. doi:10.4018/978-1-5225-9825-1.ch014

Donelan, H. (2016). Social media for professional development and networking opportunities in academia. *Journal of Further and Higher Education*, *40*(5), 706–729. doi:10.1080/0309877X.2015.1014321

Elsayed, A. (2015). The use of academic social networks among Arab researchers: A survey. *Social Science Computer Review*, *34*(3), 1–14. doi:10.1177/0894439315589146

Gu, F., & Widén-Wulff, G. (2011). Scholarly communication and possible changes in the context of social media: A Finnish case study. *The Electronic Library*, *29*(6), 762–776. doi:10.1108/02640471111187999

Jeng, W., He, D., & Jiang, J. (2015). User participation in an academic social networking service: A survey of open group users on Mendeley. *Journal of the Association for Information Science and Technology*, *66*(5), 890–904. doi:10.1002/asi.23225

Manca, S., & Ranieri, M. (2017). Exploring digital scholarship. a study on use of social media for scholarly communication among Italian academics. In A. Esposito (Ed.), *Research 2.0 and the Impact of Digital Technologies on Scholarly Inquiry* (pp. 116–141). IGI Global. doi:10.4018/978-1-5225-0830-4.ch007

Mansou, E. (2015). The use of social networking sites (SNSs) by the faculty members of the school of library and information science, PAAET, Kuwait. *The Electronic Library*, *33*(3), 524–546. doi:10.1108/EL-06-2013-0110

Nadez, G. (2012). Use of social networks for academic purposes: A case study. *The Electronic Library*, *31*(6), 781–791. doi:10.1108/EL-03-2012-0031

Thelwall, M., & Kousha, K. (2014). Academia.edu: Social network or academic network? *Journal of the Association for Information Science and Technology*, *65*(4), 721–723. doi:10.1002/asi.23038

Williams, A., & Woodacre, M. (2016). The possibilities and perils of academic social networking sites. *Online Information Review*, *40*(2), 282–294. doi:10.1108/OIR-10-2015-0327

Work, S., Haustein, S., Bowman, T. D., & Larivière, V. (2015). Social Media in Scholarly Communication. A Review of the Literature and Empirical Analysis of Twitter Use by SSHRC Doctoral Award Recipients, Study commissioned by the Social Sciences and Humanities Research Council, Montreal, Canada.

Chapter 17
Utilization and Impact of ICT in Information Literacy Expertise Among the Postgraduate Students of Alagappa University

Santhiya Gurusamy
Alagappa University, Karaikudi, India

P. Muthumari
Alagappa University, India

ABSTRACT

This study examines the present usage of ICT and its influence in information literacy skills among postgraduate students of Alagappa University Central Library users. The questionnaire survey was distributed and simple random sampling technique was used in the study. A random questionnaire was distributed to 150 students and the study adopted an exploratory research design. The adoption and use of ICT has enhanced subject knowledge, new educational principles, and technologies. Quantitative research data were entered into an MS Excel file where statistics were uploaded and analyzed using SPSS software; a comparison between postgraduate students' awareness and use of ICT knowledge. Performance difference between gender and age were two groups involved in descriptive statistics, simple percentage, and ANOVA test.

INTRODUCTION

Information Communication Technology (ICT) is one of the buzzwords of today the world of education. It has transformed society into an information society and our way of life.

It has integrated into every area of our lives. This is Describes exciting and innovative ways to provide global access to lifelong learners Information, opt-out, and support. Computers enable people to work creatively. It can be used Discussing, questioning, supporting a partner, debating, and sharing data,

DOI: 10.4018/978-1-6684-8805-8.ch017

analyzing, searching, collecting Organizing and exploring online information and the real world. ICTs play an important role in disseminating knowledge Advancement of our educational community. The library is and can be an important source of information ICT infrastructure should provide the necessary facilities to the users in such a way that they can access them Required information. The Internet is the gateway to access libraries and knowledge centers Electronic modern age offers completely different knowledge Institutions, companies, organizations, research centers and people everywhere in the world.

Literacy is traditionally thought of as the ability to read and write. Meaning of the word Expanded to include the ability to use language and access information from a variety of sources Resources in many formats such as images, printed text and CD ROMs, databases, and websites. Students must work effectively with information at all levels Bloom's Taxonomy (Remembering, Understanding, Applying, Analyzing, Evaluating, Creating, etc.,). Information literacy includes traditional skills such as reading, researching, and more the letter; but new ways of reading and writing introduced new skills. An information literate person should know why information is needed when information is a Think critically about how to find information about needs, and how to use all of these resources about the information available.

Information Literacy

The definition and definitions of information literacy can be summarized as follows.

1. To use of ICTs to retrieve and disseminate information
2. To skills to find and use information in information sources.
3. To recognizing, identifying, evaluating information needs, and using the information to acquire or expand knowledge.

The third option is the most comprehensive and useful one, as it covers both the use of ICTs and information (Boekhorst, 2003).Information literacy is important for today's students as it enhances critical thinking Methods and Thinking Skills – Making inquiries and searching for answers, Finding data, influencing hypotheses, evaluating evidence, and solving problems Motivated and effective students, successful givers, specific individuals, and, specific individuals and Skilled subjects.

LITERATURE REVIEW

The work of Boekhorst (2003) has examined Information literacy delivered over the years can be summed up in three concepts: The ICT concept refers to the ability to retrieve and use ICT Dissemination of information. The information resource concept refers to the capacity Find and use information independently or with the help of intermediaries, and finally, the information process of information literacy refers to the concept process. Recognizing the need for information, retrieving, evaluating, using, and disseminating it Information to acquire or expand knowledge. This article of Biradar (2006) has been investigates on students' use of the Internet access by the students and Faculties in Kwembu University. The results indicated that 42.1% of the students use the Internet Twice a week and 31.25% teachers use it every day. Majority of the students, as well as faculties, Use the Internet for educational purposes. Their favorite place to use the internet is next to the library Business locations. Majority of the respondents are satisfied with internet resources Services. Also, preliminary results of Thanuskodi (2012) study suggest that ICT Faculty of Annamalai University, Tamil Nadu and India. The study found that Most of the faculty members preferred internet (82.85%) and computers (75%). 75.71% Faculty members used ICT to generate new information, 83.92% for research and 80% for research. Develop skills and knowledge.

The publication of Jebamalar (2013) has been studied the use of ICT-based resources and services among users. Arts and Science Colleges affiliated to Manonmaniam Sundaranar University and 179 Research Of scholars, 31.84% of them faced slow access speed while accessing ICT based resources and Services. The work of Rohatgi, Scherer, and Hatlevik (2016) their study revealed the role of ICT self-sufficiency in students' ICT use and their new knowledge of computer and information literacy (CIL). Their findings point to a potential mechanism linking ICT use, self-efficacy, and CIL, and implications for future research and educational practice are discussed. Baskaran (2020) looked at the data and found that the majority of the publications, 2,207 (18.48%), were in the communication field due to the particular nature of the research that focused on social networks. The area of computer communications and information systems comes in second, with 1,172 publications (9.81%).

OBJECTIVES OF THE STUDY

The following objectives have been formulated for the purpose of the present study:

- To analyze the socio-demographics of the respondents in terms of gender, age, and residence wise in Alagappa University, Tamil Nadu, India.
- To compare the level of awareness Mode of access Networks Vs. ICT tools
- To determine how well information is managed and shared among postgraduates at Alagappa University and Accessing of ICT information needed.
- To investigate the frequency of using ICT-based resources and services and to find information literacy tools used among postgraduate students in Alagappa University.
- Recommending improvement measures based on the findings and solutions of the study on how ICT use supports information literacy among postgraduate students at Alagappa University.

LIMITATIONS OF THE STUDY

- The findings of the study are valid for postgraduate students' ICT skills in Alagappa University it may or may not be valid for other universities.
- A questionnaire is designed only for Post Graduates of Alagappa University.
- The present work is only a comprehensive study of ICT in information literacy skills of postgraduate students Alagappa University. The sample of this study consisted online questionnaire method, Information Literacy Assessment.

METHODOLOGY

Data was collected from Students of various departments of Alagappa University. In the present study, a questionnaire was carefully designed to collect the required data only from the postgraduate student. Totally 302 respondents received questionnaires; purpose sampling method questionnaires were distributed.

The collected data were tabulated using Excel and analyzed thereby statistical tools such as mean and simple percentages. A closed-ended questionnaire is different Sections on demographic information of the first category; there was a second type Determine library use; the third category is about cataloging skills and web skills. The questionnaire focuses on key questions such as types of libraries accessed, Frequency of visits to libraries, the purpose of library visit, use of internet for access Information, where and how users access the Internet and search for information online. When Applying web problems faced by users, knowledge of parts of the book, Awareness of the method followed for shelving books, knowledge of the item in general

The catalogue includes catalogue search capabilities, information retrieval, and feedback on the usefulness of library orientation/user education programs etc.

The language of the question paper was simple, easy, and clear according to the level Students. The questionnaire was reviewed by experts in the field and modifications were made Based on feedback and instructions. Key findings of the survey were analyzed and Summarized in the following section.

ANALYSIS AND DISCUSSIONS

Male: 143 (51%) majority of the respondents contributed to the study. 69 (48.3%) respondents using ICT tools on smart phone/mobiles, 38 (61.3%) using desktops/laptops, 18 (56.3%) i-pad and 29 (44.6%) respondents using tablet computer.

Age group: Majority of the respondents (34.4%) was in the age group of 20-25 years, 51(35.7%) using ICT tools on smart phone/mobiles, 5 (8.1%) using desktops/laptops, 17 (53.1%) i-pad and 31 (47.7%) respondents using tablet computer.

Table 1. Socio demography of the respondents vs. use of ICT tools

Socio demography		Types of tools								Total	
		Smart Phone/Mobiles		Desktop/ Laptop		I-pad		Tablet Computer			
		Count	%	Count	%	Count	%	Count	%	Count	%
Gender	Male	69	48.3	38	61.3	18	56.3	29	44.6	154	51.0
	Female	74	51.7	24	38.7	14	43.8	36	55.4	148	49.0
Total		143	47.4	62	20.5	32	10.6	65	21.5	302	100
Age Group	20-25	51	35.7	5	8.1	17	53.1	31	47.7	104	34.4
	26-30	37	25.9	9	14.5	7	21.9	16	24.6	69	22.8
	31-35	24	16.8	13	21.0	4	12.5	5	7.7	46	15.2
	Above 35	31	21.7	35	56.5	4	12.5	13	20	83	27.5
Total		143	47.4	62	20.5	32	10.6	65	21.5	302	100
Residing area	Rural	86	60.1	25	40.3	18	56.3	48	73.8	177	58.6
	Urban	29	20.3	10	16.1	5	15.6	7	10.8	51	16.9
	Semi-Urban	28	19.6	27	43.5	9	28.1	10	15.4	74	24.5
Total		143	47.4	62	20.5	32	10.6	65	21.5	302	100

Residing areas: Slightly most of the respondents are hailed from a rural background (58.6%) while (16.9%) of the respondents are from areas of Urban and (24.5%) of the respondents belong to the Semi-Urban areas (Figure 1). Majority of the respondents at rural area background 86(60.1%) using ICT tools on smart phone/mobiles, 25 (40.3%) using desktops/laptops, 18 (56.3%) i-pad and 48 (73.8%) respondents using tablet computer, due to level of understanding of research and its importance to society.

Table 3 discloses the use of mode of access network among Alagappa University by the post-students. Overall Analysis: Out of 302 respondents, 134 (44.4%) respondents use Wi-Fi network access and another set of 51 respondents (16.9%) use Broadband network access. 73 (24.2%) of the respondents using data pack while 44 (14.6) respondents using other network options.

Wi-Fi: 134 (44.4%) majority of the respondents using Wi-Fi network access. 68 (47.6%) respondents using Wi-Fi to ICT tools on smart phone/mobiles, Respectively 20 (32.3%) using desktops/laptops, 17

Figure 1. Socio demography of the respondents vs. use of ICT tools

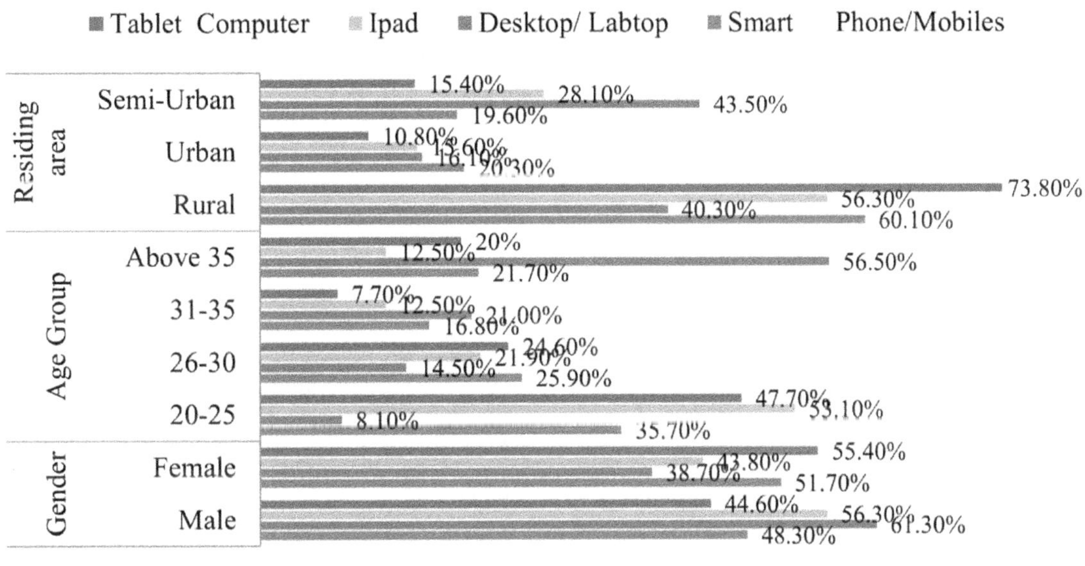

Table 2. Mode of access networks vs. ICT tools

Mode of access Network	Types of tools								Total	
	Smart Phone/Mobiles		Desktop/ Laptop		I-pad		Tablet Computer			
	Count	%	Count	%	Count	%	Count	%	Count	%
Wi-Fi	68	47.6	20	32.3	17	53.1	29	44.6	134	44.4
Broadband	29	20.3	10	16.1	5	15.6	7	10.8	51	16.9
Data Pack	27	18.9	27	43.5	9	28.1	10	15.4	73	24.2
Others	19	13.3	5	8.1	1	3.1	19	29.2	44	14.6
Total	143	47.4	62	20.5	32	10.6	65	21.5	302	100
Chi- Square Test		$\chi 2$		df					Sig.	
		33.394		9					.000	

(53.1%) i-pad and 29 (44.6%) respondents access Wi-Fi for using tablet computer (figure 2). Maybe Alagappa University has very rich Wi-Fi availability.

Chi-Square Analysis: It has been analyzed that there is an Mode of access Networks Vs. ICT tools, a chi-square test was conducted.

Null-Hypothesis: Table 2 shows the result of the test and reveals that there is a significant Mode of access Networks (X^2=33.394, df=9, and p=.001). Wi-Fi, Broadband, Data pack, other network access as the p-value is less than the significant level of 0.05. Thus, the null hypothesis is rejected and the alternative hypothesis is accepted.

Table 3 shows access to ICT information needed Vs. Area of residence of the respondents in Alagappa University, followed by post-graduate students. Overall Analysis: Out of 302 respondents, 124 (41.1%) respondents use online education from conferences/seminars/workshop and 68 respondents (22.5%) use online tutorials on social media. 56 (18.5%) of the respondents discuss with friends, while 49 (16.2%)

Figure 2. Mode of access networks vs. ICT tools

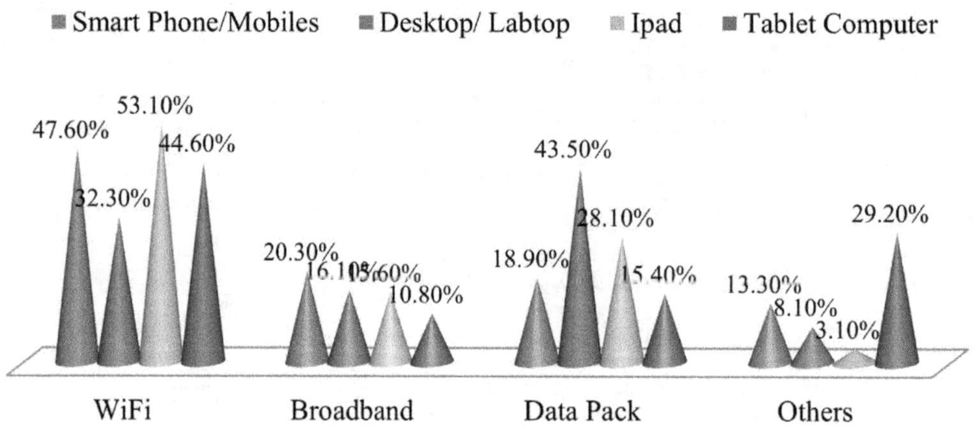

Table 3. Accessing of ICT information needed vs. residing area of the respondents

	Residing area						Total	
	Rural		Urban		Semi-Urban			
	Count	%	Count	%	Count	%	Count	%
Friends	41	23.2	10	19.6	5	6.8	56	18.5
Forums	0	0	4	7.8	1	1.4	5	1.7
Online Tutorial's	50	28.2	14	27.5	4	5.4	68	22.5
Regular Classes	30	16.9	12	23.5	7	9.5	49	16.2
Conference/Seminar/Workshop	56	31.6	11	21.6	57	77.0	124	41.1
Total	177	58.6	51	16.9	74	24.5	302	100
Chi- Square Test	**χ2**			**df**			**Sig.**	
	70.817			**8**			**.000**	

of regular classes and (1.7%) 5 of the respondents use the resources of forums in online education. It shows figure 3.

Conferences/Seminars/Workshop: More than (31.6%) 56 respondents from Rural Area, whereas (21.6%) 11 respondents from Urban area and 57 (77%) from Semi-Urban area students use online education.

Chi-Square Analysis: It has been analyzed that there is an Accessing of ICT information needed Vs. Residing area of the Respondents, a chi-square test was conducted. Table 3 shows the result of the test and reveals that there is a significant information need (X^2=70.817, df=8, and, p=.001). Friends, Forums, Online Tutorial's, Regular classes and Conference/Seminar/Workshop as the p-value is less than the significant level of 0.05. Thus, the null hypothesis is rejected and the alternative hypothesis is accepted.

Table 4 has indicated to Knowledge Source Vs. Use of ICT tools of the respondents in Alagappa University, followed by post-graduate students. Overall Analysis: Out of 302 respondents, 124 (41.1%) respondents use online education from conferences/seminars/workshop and 68 respondents (22.5%) use

Figure 3. Accessing of ICT information needed vs. residing area of the respondents

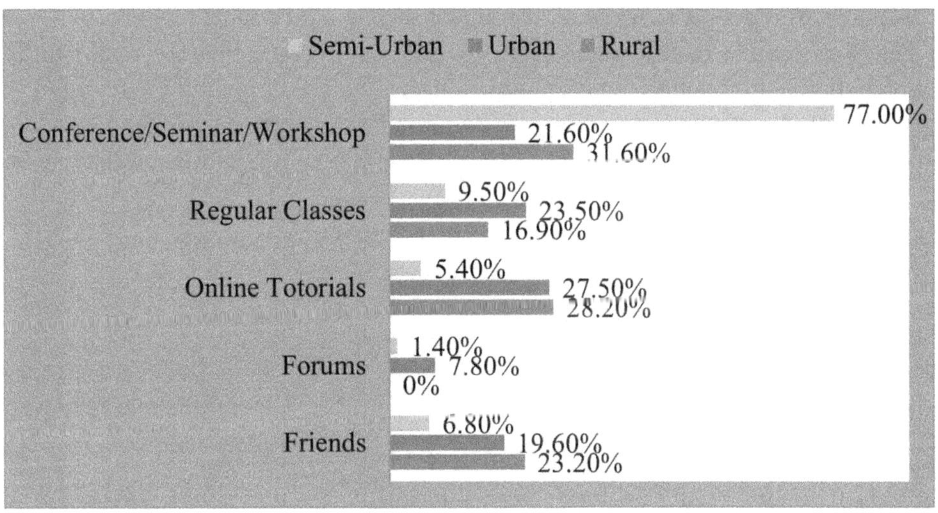

Table 4. Knowledge source vs. use of ICT tools

	Types of ICT tools								Total	
	Smart Phone/Mobiles		Desktop/ Laptop		I-pad		Tablet Computer			
	Count	%	Count	%	Count	%	Count	%	Count	%
Friends	28	19.6	10	16.1	7	21.9	11	16.9	56	18.5
Forums	4	2.8	0	0	0	0	1	1.5	5	1.7
Online tutorial's	36	2.5	17	27.4	8	25	7	10.8	68	22.5
Regular Classes	34	23.8	0	0	4	12.5	11	16.9	49	16.2
Conference/Seminar/ Workshop	41	28.7	35	56.5	13	40.6	35	53.8	124	41.1
Total	143	47.4	62	20.5	32	10.6	65	21.5	302	100
Chi-Square Tests		$\chi 2$				df			Sig.	
		35.321				**12**			**.000**	

online tutorials on social media. 56 (18.5%) of the respondents discuss with friends, while 49 (16.2%) of regular classes and (1.7%) 5 of the respondents use the resources of forums in online education. It shows figure 4.

Conferences/Seminars/Workshop: More than (28.7%) 41 respondents using Smart phone/Mobiles, whereas (56.5%) 35 respondents using Desktop/Laptop, 13 (40.6%) using I-pad, 35 (53.8%) respondents using Tablet/Computer.

Chi-Square Analysis: It has been analyzed that there is an Knowledge Source Vs. Use of ICT tools, a chi-square test was conducted. Table 4 shows the result of the test and reveals that there is a significant Knowledge Source (X^2=35.321, df=12, and p=.001). Friends, Forums, Online Tutorial's, Regular classes and Conference/Seminar/Workshop as the p-value is less than the significant level of 0.05. Thus, the null hypothesis is rejected and the alternative hypothesis is accepted.

ICT Information Literacy-Wise Analysis

Video Learning: Slightly more than half of the respondents 180 (59.6%) strongly agree and one-fourth of the respondents agree that they can use Video learning educational methods.

Webinars: The high number of respondents 112 (37.1%) say Neutral that they can gain information from Webinars.

Online classes and Quiz: The majority of the respondents 120 (39.7%) were neutral about using online classes and quizzes to get information. Majority of Similar feedback was obtained for both online classes and quizzes.

Judgment Writing Competition: High respondents 117 (38.7%) choose a neutral preferred mode of education.

Figure 4. Knowledge source vs. use of ICT tools

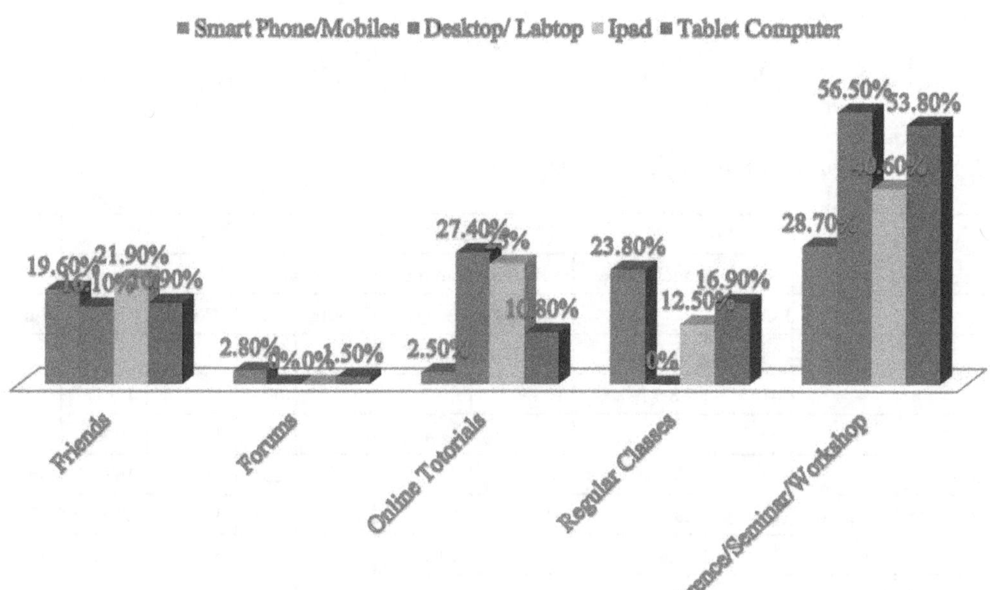

Table 5. Use of ICT in information literacy: frequency distribution

	Respondents											
	SA		A		N		DA		SDA		Total	
	Count	%	Count	%	Count	%	Count	%	Count	%	Count	%
Video Learning	180	59.6	60	19.9	46	15.2	16	5.3	0	0%	302	100.0
Webinars	82	27.2	97	32.1	112	37.1	11	3.6	0	0%	302	100.0
Online Classes	92	30.5	80	26.5	120	39.7	10	3.3	0	0%	302	100.0
Quiz	96	31.8	81	26.8	120	39.7	5	1.7	0	0%	302	100.0
Judgment Writing Competition	75	24.8	70	23.2	117	38.7	36	11.9	4	1.3	302	100.0
Certificate Courses	106	35.1	105	34.8	70	23.2	21	7.0	0	0%	302	100.0
Conference calls	66	21.9	69	22.8	145	48.0	11	3.6	11	3.6	302	100.0
Blog Articles	55	18.2	66	21.9	129	42.7	36	11.9	16	5.3	302	100.0
Electronic Simulation	58	19.2	101	33.4	128	42.4	10	3.3	5	1.7	302	100.0
Online Discussions	79	26.2	87	28.8	111	36.8	20	6.6	5	1.7	302	100.0

*(Note: SA= Strongly Agree, A= Agree, N= Neutral, DA= Disagree, SDA= Strongly disagree)

Certificate Courses: The respondents of the certification course is 105 (34.8%) agree and 106 (35.1%) strongly agree; There is very little difference between the two concepts. The concept of strongly agree is more in this.

Conference calls, Blog Articles, Electronic Simulation & Online Discussions: In all these the majority of the respondents choose neutral. Respectively, respondents are 145 (48%) in Conference calls, 129 (42.7%) in Blog articles, 128 (42.4%) in Electronic Simulation, and 111 (36.8%) in Online Discussions.

Only a few respondents disagree and strongly disagree with all competencies required to identify the need for information. High respondents chose neutral.

Null Hypothesis: There is no significant difference between gender of the respondents and their level of agreement with all the 10 ICT in Information Literacy in Utilization in central library resources and services.

Name of the Test: One-Way- ANOVA Test

Purpose: to analyze whether there is a significant difference between male and female respondents and their level of agreement with the ICT in Information Literacy they confront with Utilization of ICT resources.

Result: (Table 6):

1. There is no significant difference between Male and Female respondents and their level of agreement with the ICT in Information Literacy they confront with Utilization of ICT resources of Online Discussions p-value is (.029), as the p value is less than the significant level of 0.05. The null hypothesis is rejected and the alternative hypothesis is accepted.

Figure 5. Use of ICT in information literacy: frequency distribution

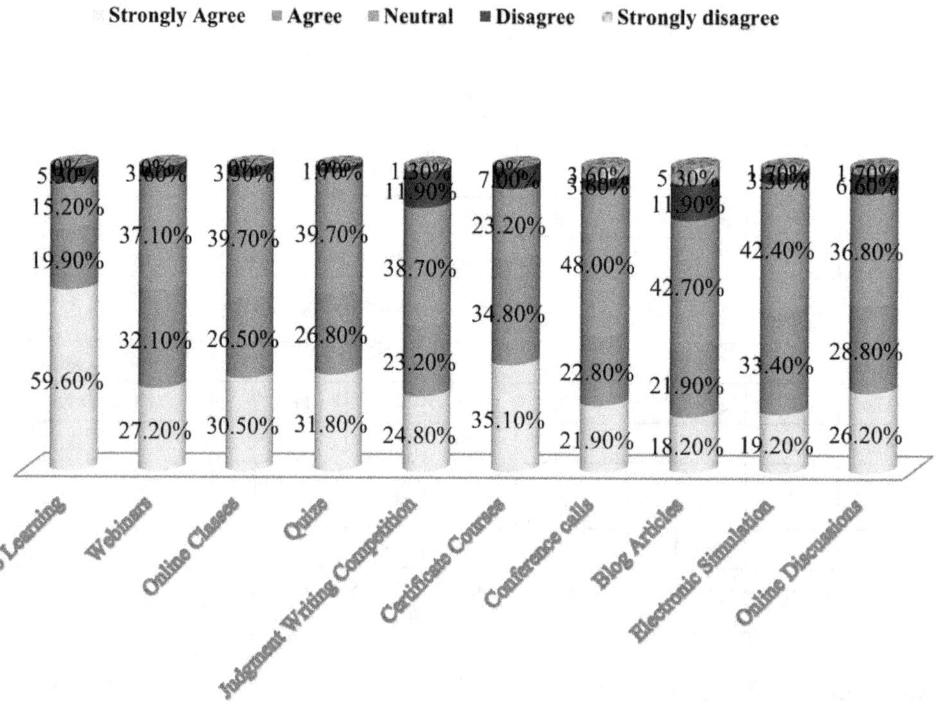

2. There is a significance difference between Male and Female respondents and their level of agreement with the ICT in Information Literacy they confront with Utilization of ICT resources of Video Learning (p-.212), Webinars (p-.468), Online Classes (p-.564), Quize (p-.663), Judgment Writing Competition (p-.061), Certificate Courses (p-.219), Conference calls (p-.052), Blog Articles (p-.516), Electronic Simulation (p-.059), as the p value is the less than 0.05%. The null hypothesis is accepted.

MAJOR FINDINGS OF THE STUDY

- The majority of the male 143 (51%) respondents contributed to the study.
- Majority of the respondents (34.4%) was in the age group of 20-25 years
- Majority of the respondents at rural area background 86(60.1%) using ICT tools on smart phone/mobiles, 25 (40.3%) using desktops/laptops, 18 (56.3%) i-pad and 48 (73.8%) respondents using tablet computer, due to level of understanding of research and its importance to society.
- 134 (44.4%) majority of the respondents using Wi-Fi network access. 68 (47.6%) respondents using Wi-Fi to ICT tools on smart phone/mobiles, Respectively 20 (32.3%) using desktops/laptops, 17 (53.1%) i-pad and 29 (44.6%) respondents access Wi-Fi for using tablet computer

Table 6. Utilization of ICT in information literacy vs. gender of the respondents: One- Way- ANOVA test

		Sum of Squares	df	Mean Square	F	Sig.
Video Learning	Between Groups	1.329	1	1.329	1.568	.212
	Within Groups	254.221	300	.847		
	Total	255.550	301			
Webinars	Between Groups	.403	1	.403	.529	.468
	Within Groups	228.643	300	.762		
	Total	229.046	301			
Online Classes	Between Groups	.271	1	.271	.333	.564
	Within Groups	244.100	300	.814		
	Total	244.371	301			
Quiz	Between Groups	.148	1	.148	.191	.663
	Within Groups	232.025	300	.773		
	Total	232.172	301			
Judgment Writing Competition	Between Groups	3.717	1	3.717	3.532	.061
	Within Groups	315.714	300	1.052		
	Total	319.430	301			
Certificate Courses	Between Groups	1.309	1	1.309	1.519	.219
	Within Groups	258.572	300	.862		
	Total	259.881	301			
Conference calls	Between Groups	3.682	1	3.682	3.797	.052
	Within Groups	290.861	300	.970		
	Total	294.543	301			
Blog Articles	Between Groups	.489	1	.489	.423	.516
	Within Groups	346.889	300	1.156		
	Total	347.377	301			
Electronic Simulation	Between Groups	2.769	1	2.769	3.585	.059
	Within Groups	231.724	300	.772		
	Total	234.493	301			
Online Discussions	Between Groups	4.602	1	4.602	4.838	.029
	Within Groups	285.335	300	.951		
	Total	289.937	301			

- Most of the (31.6%) 56 respondents from Rural Area, whereas (21.6%) 11 respondents from Urban area and 57 (77%) from Semi-Urban area students preferred Conference/Seminars/Workshops related online education.
- There is no significant difference between gender of the respondents and their level of agreement with all the 10 ICT in Information Literacy in Utilization in central library resources and services.

SUGGESTIONS

- The present study reveals that there is an extended arrangement for Alagappa University students should use the resources. The main objective is to assess the level of students' ICT and information skills in optimal use of ICT facilities, resources, and services.
- Most of the users are aware of the services but not much of the ICT tools and procedures and they are confusing. Therefore, they must be well trained to find and use them.
- It has been observed from the analysis that information literacy programs are vital in university departments.

CONCLUSION

This research provided a Current Utilization and Impact of ICT in Information Literacy among Post Graduate students of Alagappa University: Data was collected through a well-structured e-questionnaire and observational system. Various tools were used to find the data a solution similar to the analysis of variance for the ANOVA test. The study suggested that there is a need to improve student's access to ICT in one direction training programs for rural students improved and improved utilization of ICT resources their educational level.

The strategies used in meeting their information literacy should be expressed within their scope ICT knowledge are limited in the training program available to them. A lot it needs are considered difficult to find assignments in class and prepare for exams online search, database search and web browsing subscription requirements and challenges; hence their low-level ICT makes it difficult to fully explore the information resources available for their access.

REFERENCES

Baskaran, C. (2020). Research Patterns on the Social Networks and Media: A Scientometric Portrait. Handbook of Research on Emerging Trends and Technologies in Library and Information Science, pp.189-207. IGI Global. doi:10.4018/978-1-5225-9825-1.ch014

Baskaran, V., Davis, K., Bali, R. K., Naguib, R. N., & Wickramasinghe, N. (2013). Managing information and knowledge within maternity services: Privacy and consent issues. *Informatics for Health & Social Care*, *38*(3), 196–210. doi:10.3109/17538157.2012.735732 PMID:23323681

Biradar, B. S., Rajashekar, G. R., & Kumar, B. T. (2006). A study of internet usage by students and faculties in Kuvempu University. *Library Herald*, *44*(4), 283–294.

Boekhorst, A. K. (2003). Becoming information literate in the Netherlands. *Library Review*, *52*(7), 298–309. doi:10.1108/00242530310487399

Das, K. (2019). The role and impact of ICT in improving the quality of education: An overview. *International Journal of Innovative Studies in Sociology and Humanities*, *4*(6), 97–103.

Jebamalar, A. A., Seethai, R. M., & Vanathi, B. (2013). Use of ICT based Resources and Services among the users of Arts and Science Colleges Affiliated to Manonmaniam Sundaranar University, Tirunelveli. *Journal of Advances in Library and Information Science*, *2*(2), 45–49.

Rohatgi, A., Scherer, R., & Hatlevik, O. E. (2016). The role of ICT self-efficacy for students' ICT use and their achievement in a computer and information literacy test. *Computers & Education*, *102*, 103–116. doi:10.1016/j.compedu.2016.08.001

Sutrisno, S., Andre, H. L., & Susilawati, S. (2021, February). The analysis of the facilities and ICT applications usage by the university's students. *Journal of Physics: Conference Series*, *1793*(1), 012050. doi:10.1088/1742-6596/1793/1/012050

Thanuskodi, S. (2012). Use of ICT for information creation among faculty members of Annamalai University, Tamil Nadu, India. *International Journal of Library Science*, *1*(4), 61–65.

Compilation of References

Abdallah, A. K. (2018). Parents Perception of E-Learning in Abu Dhabi Schools in United Arab Emirates. *IJASOS- International E-Journal of Advances in Social Sciences, 4*(10), 30–41. doi:10.18769/ijasos.415513

Abinew, A. A., & Vuda, S. (2013). A case study of acceptance and use of electronic library services in universities based on SO-UTAUT model. *International Journal of Innovative Research in Computer and Communication Engineering, 1*(4), 903–910. http://ijircce.com/upload/2013/june/17_A%20Case%20Study.pdf

Adeleke, D. S., & Emeahara, E. N. (2016). Relationship between information literacy and use of electronic information resources by postgraduate students of the University of Ibadan. *Library Philosophy & Practice*. http://eds.b.ebscohost.com/eds/detail/detail?vid=0&sid=9a7caa11-b75

Adeniran, P. (2013). Usage of Electronic Resources by Under Graduates at the Redeemer's University, Nigeria. *International Journal of Library and Information Science*, *5*(10), 319–324.

Ahmad, A. (2011). Rising of social network websites in India overview. *International Journal of Computer Science and Network Security*, *11*(2), 155–158.

Ajayi, S. A. Oludare, Shorunke, A &Modupe, Aboyade, A. (2014) The Influence of electronic resources use on students' reading culture in Nigerian Universities: A case study of AdelekeUniversite Ede, Osun State, *Library Philosophy and Practice (e-journal),* http://digitalcommons.unl.edu/ libphilprac /1182

Ajegbomogun, F. O., Okunlaya, R. O. A., & Alawiye, M. K. (2017). Analytical study of E-learning resources in national open University of Nigeria. *Education and Information Technologies*, *22*(5), 2403–2415. doi:10.100710639-016-9548-z

Ajuwon, G. A. (2003). Computer and internet use by first year clinical and nursing Students in a Nigerian teaching hospital. *BMC Medical Informatics and Decision Making*, *3*(1), 10. doi:10.1186/1472-6947-3-10 PMID:14498997

Akçayır, G. (2017). Why do faculty members use or not use social networking sites for education? *Computers in Human Behavior*, *71*, 378–385. doi:10.1016/j.chb.2017.02.028

Akuffo, M., & Budu, S. (2019, May). Use of Electronic Resources by Students in a Premier Postgraduate Theological University in Ghana. *South African Journal of Information Management, 21*(1). doi:10.4102ajim.v21i1.1026

Akullo, W. N., Akello, C., & Mulumba, O. (2016). *The Role of Social Media In Promoting Library Services in Secondary School in Uganda.* 517-525. SCECSAL. https://www.scecsal.org/publications/papers2016/051_akullo_2016.pdf

Akussah, M., Asante, E., & Adu-Sarkodee, R. (2015). Impact of Electronic Resources and Usage in Academic Libraries in Ghana: Evidence from Koforidua Polytechnic & All Nations University College, Ghana. *Journal of Education and Practice*, *6*(33), 33–38. https://eric.ed.gov/?id=EJ1083497

Al-Alwan, A. F. (2014). Modeling the relations among parental involvement, school engagement and academic performance of high school students. *International Education Studies*, *7*(4), 47–56. doi:10.5539/ies.v7n4p47

Al-Aufi, A. (2007). *Networked Research and Scholarly Communication in a Developing Arabic Country: An Investigation of Sultan Qaboos Autonomous institutions.* [Unpublished Doctoral Dissertation, Curtin Autonomous institutions of Technology, Australia].

Al-Aufi, A., & Fulton, C. (2014). Use of social networking tools for informal scholarly communication in humanities and social sciences disciplines. *Procedia: Social and Behavioral Sciences*, *147*, 436–445. doi:10.1016/j.sbspro.2014.07.135

Alfonzo, P. (2019). Chapter 4. Instagram in the Library. *Library Technology Reports*, *55*(2). https://journals.ala.org/index.php/ltr/article/view/6960/9415

Ali, M., Wolski, M., & Richardson, J. (2017). Strategies for using ResearchGate to improve institutional research outcomes. *Library Review*, *66*(8/9), 726–739. doi:10.1108/LR-07-2017-0060

Ali, N. (2005). The use of electronic resources at IIT Delhi Library: A study of search behaviors. *The Electronic Library*, *23*(6), 691–700. doi:10.1108/02640470510635773

Aljuboori, A. F., Fashakh, A. M., & Bayat, O. (2020). The impacts of social media on Autonomous institutions students in Iraq. *Egyptian Informatics Journal*, *21*(3), 139–144. doi:10.1016/j.eij.2019.12.003

Almaiah, M. A., Al-Khasawneh, A., & Althunibat, A. (2020). Exploring the critical challenges and factors influencing the E-learning system usage during COVID-19 pandemic. *Education and Information Technologies*, *25*(6), 5261–5280. doi:10.100710639-020-10219-y PMID:32837229

Al-Mukhaini, E. M., Al-Qayoudhi, W. S., & Al-Badi, A. H. (2014). Adoption of social networking in education: A study of the use of social networks by higher education students in Oman. *Journal of International Education Research*, *10*(2), 143–154. doi:10.19030/jier.v10i2.8516

Alqurashi, E. (2019). Predicting student satisfaction and perceived learning within online learning environments. *Distance Education*, *40*(1), 133–148. doi:10.1080/01587919.2018.1553562

Al-Rahmi, W. M., & Othman, M. S. (2013). Evaluating student's satisfaction of using social media through collaborative learning in higher education. *International Journal of Advances in Engineering and Technology*, *6*(4), 1541–1551.

Altınay, Z., Ossiannilsson, E., Kalaç, M. O., Başarı, G., Aktepebaşı, A., & Altınay, F. (2016). Establishing a framework on OER practices for ICT competence of disabled citizens. *The Turkish Online Journal of Educational Technology*, *15*(3), 68–72.

Alvi, M., & Gupta, M. (2020). Learning in times of lockdown: How Covid-19 is affecting education and food security in India. *Food Security*, *12*(4), 793–796. doi:10.100712571-020-01065-4 PMID:32837637

Amankwah, P. B. (2014). *Use of Electronic Resources by undergraduate students of the Ghana Institute of Management and Public Administration (GIMPA).* [Doctoral dissertation, University of Ghana, Legon]. http://197.255.68.203/handle/123456789/6989

American Library Association. (1989). *Presidential committee on information literacy: final report.* ALA. https://www.ala.org/ala/acrl/acrlpubs/whis/presidential.cfm

Ameyaw, S., Kwakye, J. O., & Asante, E. (2016). Electronic Resources Usage in PrivateUniversities: Evidence from Valley View University, Ghana. *ADRR Journal of Arts and Social Science, 14*(4), 1-15. https://journals.adrri.org/index.php/adrrijass/article/view/266

Amy, L. (2011). Mirror, Mirror on my Facebook Wall: Effects of Exposure to Facebook on Self-Esteem. *Cyberpsychology, Behavior, and Social Networking*, *14*(1-2), 79–83. Advance online publication. doi:10.1089/cyber.2009.0411 PMID:21329447

Ananda, S. K. (2017, January-March). Use and Awareness of Electronic Information Resources among UG & PG students of T John College, Bangalore: A Study. *International Journal of Library and Information Studies*, *7*(1), 20–25.

Andersen, N. (2002). New Media and New Media Literacy: The horizon has become the landscape - new media are here. Cable in the Classroom.

Anwar, M & Tang, Z. (2020). Social Media and Libraries: The Perfect Combination of the Next Generation. *International Journal of Library and Information Studies*, *10*(3).

Asante, R.K.B. (2018). Exploration of the forms of mobile phone attachment among traders in Ghana. *Mobile media & Communication*, *7* (1), 24-40. doi:10.1177/2050157918764015

Bahrini, R., & Qaffas, A. A. (2019). Impact of information and communication technology on economic growth: Evidence from developing countries. *Economies*, *7*(1), 21. doi:10.3390/economies7010021

Bakri, A., & Willet, P. (2008). The Malaysian Journal of Library and Information Science 2001- 2006: A Bibliometric study. *Malaysian Journal of Library and Information Science*, *13*(1), 1–13.

Bala, S., & Lal, P. (2016, July). Usage of Electronic Resources and their Impact on Reading Culture: Study of Punjab Agricultural University. *International Journal of Digital Library Services*, *6*(3), 59–66.

Bali, M., Cronin, C., & Jhangiani, R. S. (2020). Framing open educational practices from a social justice perspective. *Journal of Interactive Media in Education*, *2020*(1), 1–12. doi:10.5334/jime.565

Bandyopadhay, A. (1999). Accessing SCI-tech literature: Commercial document delivery services and online full-text databases. *Collection Building*, *18*(1), 10–15. doi:10.1108/01604959910256481

Barbosa, S. D. J., Silveira, M. S., & Gasparini, I. (2016). What publications metadata tell us about the evolution of a scientific community: The case of the Brazilian Human–Computer Interaction conference series. *Scientometrics*, *110*(1), 275–300. doi:10.100711192-016-2162-4

Baskaran, C. (2020). Research enrichment of the faculty members in Alagappa University, India: the metrics based on WoS (Web of Science) and SCOPUS. *Library Philosophy and Practice (e-journal)*. 3726. https://digitalcommons.unl.edu/libphilprac/3726

Baskaran, C. (2020). *Research Patterns on the Social Networks and Media: A Scientometric Portrait.* Handbook of Research on Emerging Trends and Technologies in Library and Information Science, 189-207. IGI Global. https://www.igi-global.com/chapter/research-patterns-on-the-social-networks-and-media/241564

Baskaran, C. (2020). Research Patterns on the Social Networks and Media: A Scientometric Portrait. Handbook of Research on Emerging Trends and Technologies in Library and Information Science, pp.189-207. IGI Global. doi:10.4018/978-1-5225-9825-1.ch014

Baskaran, C. (2021). An Exponential trend and Mapping of Research in Web 2.0, *Library Philosophy and Practice (e-journal)*, 5008. https://digitalcommons.unl.edu/libphilprac/5008

Baskaran, C. (2021). An Exponential trend and Mapping of Research in Web 2.0, *Library Philosophy and Practice (e-journal)*. 5008. https://digitalcommons.unl.edu/libphilprac/5008

Baskaran, C. (2021). An Exponential trend and Mapping of Research in Web 2.0. *Library Philosophy and Practice (e-journal),* 5008. https://digitalcommons.unl.edu/libphilprac/5008

Baskaran, C. (2021). An Exponential trend and Mapping of Research in Web 2.0. *Library Philosophy and Practice (e-journal).* 5008. https://digitalcommons.unl.edu/libphilprac/5008

Baskaran, C. (2019). Scholarly information share through Social Networks (SNs) and Medias among social science scholars in selected State Universities in Kerala. *International Journal of Library and Information Studies*, *9*(3), 83–92.

Baskaran, C. (2019). Scholarly information share through Social Networks (SNs) and Medias among social science scholars in selected State Universities in Tamil Nadu. *International Journal of Library and Information Studies*, *9*(3), 83–92.

Baskaran, C. (2019). Scholarly information share through Social Networks (SNs) and medias among social science scholars in selected State Universities in Tamilnadu. *International Journal of Library and Information Studies*, *9*(3), 83–92.

Baskaran, C. (2019). Scholarly information share through Social Networks (SNs) and Medias among social science scholars in selected State Universities in Tamilnadu. *International Journal of Library and Information Studies*, *9*(3), 83–92.

Baskaran, C. (2019). Scholarly information share through Social Networks (SNs) and Medias among social science Students and Scholars in selected State Autonomous Institutions in Tamilnadu. *International Journal of Library and Information Studies*, *9*(3), 83–92.

Baskaran, V., Davis, K., Bali, R. K., Naguib, R. N., & Wickramasinghe, N. (2013). Managing information and knowledge within maternity services: Privacy and consent issues. *Informatics for Health & Social Care*, *38*(3), 196–210. doi:10.3109/17538157.2012.735732 PMID:23323681

Bennett, S., Maton, K., & Kervin, L. (2008). The 'digital natives' debate: A critical review of the evidence. *British Journal of Educational Technology*, *39*(5), 775–786. doi:10.1111/j.1467-8535.2007.00793.x

Bhat, N. A., & Ganaie, S. A. (2016). E-resources: Use and search strategies adopted by users of Dr YS Parmar University of Horticulture and Forestry. *Collection Building*, *35*(1), 16–21. doi:10.1108/CB-08-2015-0015

Bhatt, R. K., & Kumar, A. (2014). Student opinion on the use of social networking tools by libraries. *The Electronic Library*, *32*(5), 594–602. https://www.emerald.com/insight/content/doi/10.1108/EL-09-2012-0110/full/html. doi:10.1108/EL-09-2012-0110

Bianchi, A., & Phillips, J.G. (2005). Psychological predictors of problem mobile phone use. *cyberpsychology & Behavior and Social Networking*, *8*(1), 39-51. doi:10.1089/cpb.2005.8.39

Bicen, H., & Cavus, N. (2010). The most preferred social network sites by students. *Procedia: Social and Behavioral Sciences*, *2*(2), 5864–5869. doi:10.1016/j.sbspro.2010.03.958

Bilawar, P. B., & Pujar, S. M. (2011). *Information literacy models: correlation and conceptual model for higher education.*

Bin Salamh, F. M. (2019). Protecting organization reputations during a crisis: Emerging social media in risk and crisis communication. *Indian Journal of Science and Technology*, *12*(18), 1–7. doi:10.17485/ijst/2019/v12i18/144603

Biradar, B. S., Rajashekar, G. R., & Kumar, B. T. (2006). A study of internet usage by students and faculties in Kuvempu University. *Library Herald*, *44*(4), 283–294.

Bittner, S., & Müller, A. (2011). Social networking tools and research information systems: Do they compete. *Webology*, *8*(1), 1-8. https://www.webology.org/2011/v8n1/a82.html

Blank, G., & Reisdorf, B. C. (2012). The participatory web: A user perspective on Web 2.0. *Information Communication and Society*, *15*(4), 537–554. doi:10.1080/1369118X.2012.665935

Boase, J. (2013). Implications of software-based mobile media for social research. *Mobile Media & Communication, 1*(1), 57–62. doi:10.1177/2050157912459500

Boase, J., & Ling, R. (2013). Measuring Mobile Phone Use: Self-Report Versus Log Data. *Journal of Computer-Mediated Communication, 18*(4), 508–519. doi:10.1111/jcc4.12021

Boekhorst, A. K. (2003). Becoming information literate in the Netherlands. *Library Review, 52*(7), 298–309. doi:10.1108/00242530310487399

Boud, D. (2001). Using journal writing to enhance reflective practice. In L. M. English & M. A. Gillen (Eds.), *Promoting Journal Writing in Adult Education. New Directions in Adult and Continuing Education. No. 90* (pp. 9–18). Jossey-Bass. doi:10.1002/ace.16

Boud, D., Keogh, R., & Walker, D. (1985). Writing and Reflection. In *Reflection: Turning Experience into Learning*. Nichols Publishing Company. https://craftingjustice.files.wordpress.com/2017/04/david-boud-rosemary-keogh-david-walker-reflection_-turning-experience-into-learning-routledge-1985-pp-1-165.pdf

Bouhnik, D., & Marcus, T. (2006). Interaction in distance-learning courses. *Journal of the American Society for Information Science and Technology, 57*(3), 299–305. doi:10.1002/asi.20277

Boyd, D. M., & Ellison, N. B. (2007). Social network sites: Definition, history, and scholarship. *Journal of Computer-Mediated Communication, 13*(1), 210–230. doi:10.1111/j.1083-6101.2007.00393.x

Brennan, N. (2017). Fifteen Years of Research Information Management in Ireland. *In Proceedings of the EuroCRIS Membership Meeting*, Dublin, Berlin.

Brenner, J., & Smith, A. (2013). *72% of online adults are social networking site users.* Pew Research. http://pewInternet.org/Reports/2013/social-networking-sites.aspx

Brown, M., & Croft, B. (2020). Social annotation and an inclusive praxis for open pedagogy in the college classroom. *Journal of Interactive Media in Education, 2020*(1), 1–8. doi:10.5334/jime.561

Buckingham, D. (2010). In B. Bachmair (Ed.), *Defining Digital Literacy Medienbildung in neuen Kulturräumen* (pp. 59–71). doi:10.1007/978-3-531-92133-4_4

Buhl, M., Andreasen, L. B., & Pushpanadham, K. (2018). Upscaling the number of learners, fragmenting the role of teachers: How do massive open online courses (MOOCs) form new conditions for learning design? *International Review of Education, 64*(2), 179–195. doi:10.100711159-018-9714-1

Bundy, A. (Ed.). (2004). *Australian and New Zealand information literacy framework. Principles, standards and practice* (2nd ed.). Australian and New Zealand Institute for Information Literacy.

Burnett, K., Bonnici, L. J., Miksa, S. D., & Kim, J. (2007). Frequency, Intensity and Topicality in Online Learning: An Exploration of the Interaction Dimensions that Contribute to Student Satisfaction in Online Learning. *Journal of Education for Library and Information Science, 48*(1), 21–35.

Cabero-Almenara, J., Gutiérrez-Castillo, J. J., Palacios-Rodríguez, A., & Barroso-Osuna, J. (2020). Development of the teacher digital competence validation of DigCompEdu check-in questionnaire in the university context of Andalusia (Spain). *Sustainability (Basel), 12*(15), 6094. doi:10.3390u12156094

Canty, N. (2012). *Social Media in Libraries: It's like, Complicated.* Research Gate. https://www.researchgate.net/publication/249371034

Cavus, N., & Ibrahim, D. (2008). *A mobile tool for learning English words*. Libes Proxy. http://libezproxy.open.ac.uk/login?url=http://search.ebscohost.com/login.aspx?direct=true&db=eric&AN=ED504283&site=ehost-live&scope=site

Central Michigan University. (2016). Research readiness self-assessment (RRSA). CMU. http://rrsa.cmich.edu/cgi-bin/rrsalib.cgi/

Çetinkaya Aydın, G. (2020). COVID-19 Salgını Sürecinde Öğretmenler. *Tedmem*. https://tedmem.org/covid-19/covid-19-salgini-surecinde-ogretmenler

Chakrabarti, A. (2016, April-June). Social Media & Libraries: A Symbolic Relationship for 21st Century Librarianship. *International Journal of Digital Library Services*, *6*(2).

Chan, L. (1999). Electronic journals and academic libraries. *Library Hi Tech*, *17*(1), 10–16. doi:10.1108/07378839910267145

Chan, T. T. W., Lam, A. H. C., & Chiu, D. K. W. (2020). From Facebook to Instagram: Exploring user engagement in an academic library. *Journal of Academic Librarianship*, *46*(6), 102229. doi:10.1016/j.acalib.2020.102229 PMID:34173399

Chen. (2005). Using wikis and weblogs to support reflective learning in an introductory engineering design course. *Proceeding of the 2005 American Society for Engineering Education Annual Conference & Exposition*. https://citeseerx.ist.psu.edu/viewdoc/summary?doi=10.1.1.127.5481

Chen, B., & Bryer, T. (2012). Investigating instructional strategies for using social media in context of social media. *The Electronic Library*, *29*(6), 762–776. doi:10.1108/02640471111187999

Choukhande Vaishali, G. (2009). Information Literacy curriculum in academic libraries. *Information Age*, *3*(3), 5.

Claro, M., Preiss, D. D., San Martín, E., Jara, I., Hinostroza, J. E., Valenzuela, S., Cortes, F., & Nussbaum, M. (2012). Assessment of 21st century ICT skills in Chile: Test design and results from high school level students. *Computers & Education*, *59*(3), 1042–1053. doi:10.1016/j.compedu.2012.04.004

Collins, M. P., & Berge, Z. L. (1996). *Mailing lists as a venue for adult learning. Continuing and Distance Education Research Conference*, Pennsylvania State University, University Park, PA.

Conole, G. (2012). Fostering social inclusion through open educational resources (OER). *Distance Education*, *33*(2), 131–134. doi:10.1080/01587919.2012.700563

Çubukçu, A & Bayzan, B. (2013). Türkiye'de Dijital Vatandaşlık Algısı ve Bu Algıyı İnternetin Bilinçli, Güvenli ve Etkin Kullanımı ile Artırma Yöntemleri. *Middle Eastern & African Journal of Educational Research*, 5-148.

Dalvi, M. (2010). In Information literacy for effective use of e-resources: A study of selected colleges in Mumbai Tiruchirappalli. *International Conference on e-resources in higher education: Issues, Developments, Opportunities and Challenges*, 82-120. Bharathidasan University.

Daniel, S. J. (2020). Education and the COVID-19 pandemic. *Prospects*, *49*(1-2), 91–96. doi:10.100711125-020-09464-3 PMID:32313309

Das, A. (2011). Emergence of open educational resources (OER) in India and its impact on lifelong learning. *Library Hi Tech News*, *28*(5), 10–15. doi:10.1108/07419051111163848

Das, K. (2019). The role and impact of ICT in improving the quality of education: An overview. *International Journal of Innovative Studies in Sociology and Humanities*, *4*(6), 97–103.

Dastforoush, M. T., & Venkatesha, Y. (2011). Dependency on electronic and print journals: A case study. *SRELS Journal of Information Management*, *48*(4), 441–448.

Data Reportal. (2021). *Global Social Media Stats October 2021*. Data Reportal. https://datareportal.com/reports/digital-2021-october-global-statshot

de Souza e Silva, A., & Frith, J. (2010). Locative Mobile Social Networks: Mapping Communication and Location in Urban Spaces. *Mobilities*, *5*(4), 485–505. doi:10.1080/17450101.2010.510332

de Souza Rodrigues, M. A., Chimenti, P., & Nogueira, A. R. R. (2020). An exploration of eLearning adoption in the educational ecosystem. *Education and Information Technologies*, *26*(1), 585–615. doi:10.100710639-020-10276-3

Devi, C. B., & Roy, N. R. (2012). Internet Use among University Students: A Case Study of Assam University Silchar. *Pratidhwani- A journal of Humanities and Social Science*. http://www.thecho.in/files/abstract,-nil-ratan.pdf

Dhanavandan, S. S. (2020). Open Access Repositories in the World: An Overview. *Library Philosophy and Practice, 3805*.

Dhanavandan, S., Esmail, S. M., & Nagarajan, M. (2012). Access and awareness of ICT resources and services in medical college libraries in Puducherry. *Library Philosophy and Practice (e-journal), 750*.

Dhanavandan, S., Saravanan, G., & Esmail, S. M. (2011). Internet-based resources and services among librarians in Tamil Nadu, India. *Library philosophy and Practice*. Digital Commons. https://digitalcommons.unl.edu/libphilprac/471/

Di Bitonto, P., Pesare, E., Rossano, V., & Roselli, T. (2015). Smart learning environments using social network, gamification and recommender system approaches in e-health contexts. *Smart Innovation. Systems and Technologies*, *41*, 491–500. doi:10.1007/978-3-319-19875-0_44

Digital Marketing Institute. (2021). *Social Media: What Countries Use It Most & What Are They Using?* Digital Marketing Institute. https://digitalmarketinginstitute.com/blog/social-media-what-countries-use-it-most-and-what-are-they-using

Dilek-Kayaoglu, H. (2008). Use of electronic journals by faculty at Istanbul University, Turkey: The results of a survey. *Journal of Academic Librarianship*, *34*(3), 239–247. doi:10.1016/j.acalib.2008.03.007

Divya, V., & Subaveerapandiyan, A. (2019). *Use of Social Media in Libraries and Higher Education*. Research Gate. https://www.researchgate.net/publication/349552956

Doğantan, D. E. (2020). An interactive instruction model design with role play technique in distance education: A case study in open education system. *Journal of Hospitality, Leisure, Sport and Tourism Education*, *27*, 100268. doi:10.1016/j.jhlste.2020.100268

Donelan, H. (2016). Social media for professional development and networking opportunities in academia. *Journal of Further and Higher Education*, *40*(5), 706–729. doi:10.1080/0309877X.2015.1014321

Doraswamy, M. (2005). *Familiarity and Use of The Available Electronic Information Resources by The Students in Siddhartha Engineering College Library, Vijay Wada: A Survey*. The National Convention On Library and Information Networking (NACLIN), Bangalore.

DuBois, J. M., & Antes, A. L. (2018). Five dimensions of research ethics: A stakeholder framework for creating a climate of research integrity. *Academic Medicine*, *93*(4), 550–555. doi:10.1097/ACM.0000000000001966 PMID:29068823

Durak, G., Çankaya, S., & İzmirli, S. (2020). COVID-19 pandemi döneminde Türkiye'deki üniversitelerin uzaktan eğitim sistemlerinin incelenmesi. *Necatibey Eğitim Fakültesi Elektronik Fen ve Matematik Eğitimi Dergisi*, *14*(1), 787–809. doi:10.17522/balikesirnef.743080

Eisenberg, M. B., & Johnson, D. (2002). *Learning And Teaching İnformation Technology – Computer Skills İn Context*. ERIC Digest, ED465377.

E-learning Task Force. (2002). *Get on with IT*. Department for Education and Skills.

Elkaseh, A. M., Wong, K. W., & Fung, C. C. (2016). Perceived ease of use and perceived usefulness of social media for e-learning in Libyan higher education: A structural equation modelling analysis. *International Journal of Information and Education Technology (IJIET), 6*(3), 192–199. doi:10.7763/IJIET.2016.V6.683

Elsayed, A. (2015). The use of academic social networks among Arab researchers: A survey. *Social Science Computer Review, 34*(3), 1–14. doi:10.1177/0894439315589146

Eshet, Y. (2004). Digital Literacy: A Conceptual Framework for Survival Skills in the Digital era. [Norfolk, VA: Association for the Advancement of Computing in Education] [AACE]. *Journal of Educational Multimedia and Hypermedia, 13*(1), 93–106.

ETS. (2003). *Succeeding in the 21st century. What higher education must do to address the gap in information and communication technology proficiencies. Assessing literacy for today and tomorrow*. Educational Testing Service.

Evans, C. (2014). Twitter for teaching: Can social media be used to enhance the process of learning? *British Journal of Educational Technology, 45*(5), 902–915. doi:10.1111/bjet.12099

Faisal, S. (2015). *Use of Social Media in Libraries. Capacity Building of SCERT, DIET and School Librarians to Modernise the Library.* Regional Institute of Education (NCERT). https://www.academia.edu/11980034/Use_of_Social_Media_in_Libraries

Farjon, D., Smits, A., & Voogt, J. (2019). Technology integration of pre-service teachers explained by attitudes and beliefs, competency, access, and experience. *Computers & Education, 130*, 81–93. doi:10.1016/j.compedu.2018.11.010

Fasae, J., & Aladeniyi, A. (2012). Internet Use by Students of Faculty of Science In Two Nigerian Universities. *Library Philosophy and Practice*. https://unllib.unl.edu/LPP

Ferdig, R. E., & Trammell, K. D. (2004). Content delivery in the Blogosphere. The academic staff at Egerton University, Kenya. *Library Management, 384*(5), 276–284. doi:10.1108/LM-02-2017-0018

Ferro, E., Helbig, N. C., & Gil-Garcia, J. R. (2011). The role of IT literacy in defining digital divide policy needs. *Government Information Quarterly, 28*(1), 3e10.

Fisher, S. (1995). Access to information. *Library Management, 16*(5).

Folorunso, O., Vincent, R. O., Adekoya, A. F., & Ogunde, A. O. (2010). Diffusion of innovation in social networking sites among university students. *International journal of computer science and security, 4*(3), 361-372.

Foo, S., Majid, S., Mokhtar, I. A., Zhang, X., Chang, Y. K., Luyt, B., & Theng, Y. L. (2014). Information literacy skills of secondary school students in Singapore. *Aslib Journal of Information Management, 66*(1), 54–76. doi:10.1108/AJIM-08-2012-0066

Gadd, E., & Weedon, R. (2017). Copyright ownership of e-learning and teaching materials: Policy approaches taken by UK universities. *Education and Information Technologies, 22*(6), 3231–3250. doi:10.100710639-017-9583-4

Gikas, J., & Grant, M. M. (2013). Mobile computing devices in higher education: Student perspectives on learning with cell phones, smartphones & social media. *The Internet and Higher Education, 19*, 18–26. doi:10.1016/j.iheduc.2013.06.002

Glister, P. (1997). *Digital Literacy*. John Wiley&Sons Inc.

Gu, F., & Widén-Wulff, G. (2011). Scholarly communication and possible changes in the Jackson, C. (2011). Your students love social media… and so can you. *Teaching Tolerance, 39*, 38–41.

Gui, M. (2007). *Formal and Substantial Internet Information Skills: The Role of Socio-Demographic Differences on the Possession of Different Components of Digital Literacy,* 12, 9–3. First Monday. https://www.firstmonday.org/htbin/cgiwrap/bin/ojs/index.php/fm/article/view/2009/1884

Gulati, S. (2008). Technology-enhanced learning in developing nations: A review. *International Review of Research in Open and Distance Learning*, *9*(1). doi:10.19173/irrodl.v9i1.477

Gupta, A. (2015). Five years of India HCI: A scientometric analysis. In *Proceedings of the 7th international conference on HCI (India HCI),* (pp. 56–61). ACM.

Hammond, N. V., & Trapp, A. L. (2001). *How can the web support the learning of Psychology. Learning and Teaching on the World Wide Web*. Academic Press.

Hargittai, E. (2010). Digital na(t)ives? Variation in internet skills and uses among members of the "net generation". *Sociological Inquiry, 80*(1), 92e113.

Hariharan, A., Chitra, H. M., Mymoon, M., Abdunnasar, A., & Suryakala, A. (2006). Enhancing the usage of library resources at SERC library with particular reference to e- resource. In Harish Chandra (Ed). Initiatives in Libraries and Information Centers in the Digital Era, (13-17), Chennai: Society for the Advancement of Library and Information science.

Hazar, E. (2019). A Comparison between European digital competence framework and the Turkish ICT curriculum. *Online Submission*, *7*(4), 954–962. doi:10.13189/ujer.2019.070406

He, J. (2020). Research and practice of flipped classroom teaching mode based on guidance case. *Education and Information Technologies*, *25*(4), 2337–2352. doi:10.100710639-020-10137-z

Hockings, C., Brett, P., & Terentjevs, M. (2012). Making a difference-inclusive learning and teaching in higher education through open educational resources. *Distance Education*, *33*(2), 237–252. doi:10.1080/01587919.2012.692066

Hodgkinson-Williams, C., & Paskevicius, M. (2012). The role of postgraduate students in co-authoring open educational resources to promote social inclusion: A case study at the University of Cape Town. *Distance Education*, *33*(2), 253–269. doi:10.1080/01587919.2012.692052

Hultgren, A. K. (2014). Internationalization of Higher Education in East Asia. *Journal of Interactive Media in Education*, *2014*(2), 1–2. doi:10.5334/jime.aa

Humphreys, L. (2007). Mobile Social Networks and Social Practice: A Case Study of Dodgeball. *Journal of Computer-Mediated Communication*, *13*(1), 341–360. doi:10.1111/j.1083-6101.2007.00399.x

Humphreys, L. (2010). Mobile social media: Future challenges and opportunities. *Mobile Media & Communication*, *1*(1), 20–25. doi:10.1177/2050157912459499

Inamdar, S. A. N. (2004). Information Handling in Libraries. *Keralan Library and Society*, *17*(2), 64–70.

International ICT Literacy Panel. (2002). *Digital transformation: A framework for ICT literacy (A report of the International ICT Literacy Panel).* Princeton, NJ: Educational Testing Service. http://www.ibertic.org/evaluacion/sites/default/files/biblioteca/32_digitaltransformation.pdf

Iroaganachi, M. A., & Izuagbe, R. (2018). Access to Online Databases: Predicate for Faculty Research Output, *Library Philosophy and Practice (e-journal*). https://www.researchgate.net/publication/330243888_Access_to_Online_Databases_Predicate_for_Faculty_Research_Output

ISTE. (1998). *The national educational technology standards for students*. International Society for Technology in Education.

Jacovi, M., Soroka, V., Gilboa-Freedman, G., Ur, S., Shahar, E., & Marmasse, N. (2006). The chasms of CSCW: A citation graph analysis of the CSCW conference. *In Proceedings of the 20th anniversary ACM conference on computer supported cooperative work,* (pp. 289–298). ACM.

Jaiswal, S. (2011). E-resources, their types. *National Seminar on digital resources*. Hyderabad.

James Madison University. (2016). *Information literacy test, JMU Center for Assessment and Research Studies and JMU Libraries*. James Madison University. www.madisonassessment.com/assessment-testing/information-literacy-test/

Jebamalar, A. A., Seethai, R. M., & Vanathi, B. (2013). Use of ICT based Resources and Services among the users of Arts and Science Colleges Affiliated to Manonmaniam Sundaranar University, Tirunelveli. *Journal of Advances in Library and Information Science*, *2*(2), 45–49.

Jeng, W., He, D., & Jiang, J. (2015). User participation in an academic social networking service: A survey of open group users on Mendeley. *Journal of the Association for Information Science and Technology*, *66*(5), 890–904. doi:10.1002/asi.23225

Jingyu, W., Yuping, B., Yihzong, W., Zhihui, L., Xiangzheng, D., Islam, M., & Managi, S. (2020). Measuring inclusive wealth of China: Advances in sustainable use of resources. *Journal of Environmental Management*, *264*, 110328. doi:10.1016/j.jenvman.2020.110328 PMID:32224293

Johnson, S., Evensen, O. G., Gelfand, J., Lammers, G., Sipe, L., & Zilper, N. (2012). *Key issues for e-resource collection development; A guide for libraries*. International Federation of Library Associations and Institutions.

Jones, B., & Flannigan, B. L. (2006). *Connecting the digital dots: Literacy of the 21st century.* Educause Quarterly. http://www.nmc.org/pdf/Connecting%20the%20Digital%20Dots.pdf

Jones, P. H. (2008). Turning the page: Learning about the future of electronic books from students today. Toronto: University of Toronto Libraries, 38.

Josserand, E., Schmitt, A., & Borzillo, S. (2017). Balancing present needs and future options: How employees leverage social networks with clients. *The Journal of Business Strategy*, *38*(1), 14–21. doi:10.1108/JBS-01-2016-0003

Junco, R., Heiberger, G., & Loken, E. (2010). The effect of Twitter on college student engagement and grades. *Journal of Computer Assisted Learning*, *27*(2), 119–132. doi:10.1111/j.1365-2729.2010.00387.x

Kagan, S. (1998). *Normative Ethics.* Westview Press. https://www.jstor.org/stable/2660152

Kalichman, M. (2010). *Introduction: What is Research Ethics? Resources for Research Ethics Education.* Online Ethics. https://onlineethics.org/cases/resources-research-ethics-education/introduction-what-research-ethics

Kaplan, A. M., & Haenlein, M. (2010). Users of the world, unite! The challenges and opportunities of social media. *Business Horizons*, *53*(1), 59–68. doi:10.1016/j.bushor.2009.09.003

Karip, E. (2020). COVID-19: Okulların Kapatılması ve Sonrası. *Tedem.* https://tedmem.org/vurus/covid-19-okullarin-kapatilmasi-ve-sonrasi,

Karpagam, R. (2014). Global research output of Nano-biotechnology research: A scientometrics study. *Current Science*, *106*(11), 1490–1499.

Kasimani, C., & Kasilingam, K. (2018). The Impact of Social Media in the Library Service. *International Journal of Information Studies*, *11*(3), 80–86.

Katabalwa, A. S. (2016). Use of electronic journal resources by postgraduate students at the University of Dar es Salaam. *Library Review*, *65*(6/7), 445–460. doi:10.1108/LR-11-2015-0108

Kaur, B., & Verma, R. (2006.) Use of electronic resources at Tiete library, Patiala: A case study. *ILA Bulletin*, *12*(3), 18-20.

Keegan, D. (1996). *Foundations of distance education*. Psychology Press.

Kerns, S. C. (2002). Information literacy instruction: Theory and practice. *Journal of the Medical Library Association: JMLA*, *90*(3), 351. https://www.ncbi.nlm.nih.gov/pmc/articles/PMC116414/

Khot, N., & Bansode, N. (2017). Use of social networking sites by research scholars of Shivaji University, Kolhapur. *International Journal of Research in Library Science*, *3*(1), 81–90.

Kim, D., Lee, I. H., & Park, J. H. (2019). Latent class analysis of non-formal learners' self-directed learning patterns in open educational resource repositories. *British Journal of Educational Technology*, *50*(6), 3420–3436. doi:10.1111/bjet.12746

Kim, S. (2020). How a company's gamification strategy influences corporate learning: A study based on gamified MSLP (Mobile social learning platform). *Telematics and Informatics*, *57*, 101505. doi:10.1016/j.tele.2020.101505

King, T. W. (2017). Postgraduate students as OER capacitators. *Open Praxis*, *9*(2), 223. doi:10.5944/openpraxis.9.2.566

Klausen, M. H. (2016). *Even minor integrations can deliver great value—A case study*. In *Proceedings of the CRIS2016 Conference*, St Andrews, UK.

Kuhlthau, C. C., Maniotes, L. K., & Caspari, A. K. (2007). *Guided Inquiry: Learning in the 21stcentury*. Libraries Unlimited. https://iasl-online.org/resources/Documents/IASL2009_2_KN_KUHLTHAU.pdf

Kuliya, M., & Usman, S. (2020). Perceptions of E-learning among undergraduates and academic staff of higher educational institutions in north-eastern Nigeria. *Education and Information Technologies*, *26*(2), 1787–1811. doi:10.100710639-020-10325-x

Kumar, K. S., Prabu, M. M., Kalaiyarasan, G., Ramnath, R., Kumar, N. S., Mookkiah, M., Kumar, M. A., & B, R. (2021). *Open Educational Practices For Research Effectiveness To Research Scholar*, *3*(4),113-121. Zenodo. doi:10.5281/ZENODO.4681771

Kumar, K. S., & Mahendraprabu, M. (2021).. . *Perception of Oer and Oep*, *1*(1), 1–26. doi:10.5281/zenodo.4705868#.YH8vr1I2Qz4.mendeley

Kumar, N., & Lalita, L. (2017). Usage of electronic resources at University of Delhi: A case study. *Collection Building*, *36*(3), 96–107. doi:10.1108/CB-11-2016-0031

Kuss, D. J., Griffiths, M. D., Karila, K., & Billieux, J. (2014). Internet addiction: A systematic review of epidemiological research for the last decade. *Current Pharmaceutical Design*, *20*(25), 4026–4052. doi:10.2174/13816128113199990617 PMID:24001297

Lawal, V. (2012). A contextual study of the information literacy of aspirant barristers in Nigeria. *South African Journal of Library and Information Science*, *78*(2), 101–111.

Lawrence, J. E., & Tar, U. A. (2018). Factors that influence teachers' adoption and integration of ICT in teaching/learning process. *Educational Media International*, *55*(1), 79–105. doi:10.1080/09523987.2018.1439712

Lee, H. E., Park, J. H., & Song, Y. (2014). *Research collaboration networks of prolific institutions in the HCI field in Korea: An analysis of the HCI Korea conference proceedings*. In Proceedings of HCI Korea, (pp. 434–441). ACM.

Lee, D. H. (2013). Smartphones, mobile social space, and new sociality in Korea. *Mobile Media & Communication*, *1*(3), 269–284. doi:10.1177/2050157913486790

Lenhart, A., Arafeh, S., & Smith, A. (2008). *Writing, technology and teens*. Pew Internet & American Life Project.

Lenhart, A., Madden, M., & Hitlin, P. (2005). *Teens and technology*. Pew Charitable Trusts.

Levula, A., & Harré, M. (2016). Social Networks and Mental Health: An egocentric perspective. *Mental Health Review (Brighton)*, *21*(3), 161–173. doi:10.1108/MHRJ-10-2015-0029

Lin, Y. (2022). Social Media for collaborative planning: A typology of support functions an challenges. *Cities (London, England)*, *125*, 103641. doi:10.1016/j.cities.2022.103641

Littman, J., Chudnov, D., Kerchner, D., Peterson, C., Tan, Y., Trent, R., Vij, R., & Wrubel, L. (2016). API-based social media collecting as a form of web archiving. *International Journal on Digital Libraries*, *19*(1), 21–38. doi:10.100700799-016-0201-7

Liu, L., Zhang, L., & Ye, P. (2018). Research on the application of SNS in University Libraries. *The Electronic Library*, *36*(2), 369–386. doi:10.1108/EL-05-2016-0120

Lladós, J. (2017). Explaining teaching uses of Wikipedia through faculty personal and contextual features. *Online Information Review*, *41*(5), 728–743. doi:10.1108/OIR-10-2016-0298

Lo, C.-K., Tlili, A., & Huang, X. (2022). education sciences The Use of Open Educational Resources during the COVID-19 Pandemic : A Qualitative Study of Primary School Mathematics Teachers in Hong Kong. *Education Sciences*, *12*(744), 1–14. doi:10.3390/educsci12110744

Madhusudhan, M. (2010). Use of Electronic Resources by Research Scholars of Kurukshetra University. *The Electronic Library*, *28*(4), 492–506. doi:10.1108/02640471011033684

Madhusudhan, M. (2012). Use of social networking sites by research scholars in the University of Delhi: A study. *The International Information & Library Review*, *44*(2), 100–113. doi:10.1080/10572317.2012.10762919

Mahendraprabhu, K., Sathish Kumar, M., & Mani, P. S. K. (2021). Open Educational Resources and their Educational Practices in Higher Education Open Educational Resources and their Educational Practices in Higher Education. *Mukt Shabd Journal*, *10*(2), 527-540-. doi:10.5281/zenodo.4682079

Major, L., Warwick, P., Rasmussen, I., Ludvigsen, S., & Cook, V. (2018). Classroom dialogue and digital technologies: A scoping review. *Education and Information Technologies*, *23*(5), 1995–2028. doi:10.100710639-018-9701-y

Manca, S., & Ranieri, M. (2017). Exploring digital scholarship. a study on use of social media for scholarly communication among Italian academics. In A. Esposito (Ed.), *Research 2.0 and the Impact of Digital Technologies on Scholarly Inquiry* (pp. 116–141). IGI Global. doi:10.4018/978-1-5225-0830-4.ch007

Manca, S., & Ranieri, M. (2017). Networked Students and Scholarship and motivations for social media use in scholarly communication. *International Review of Research in Open and Distance Learning*, *18*(2). Advance online publication. doi:10.19173/irrodl.v18i2.2859

Mander, J. (2017). *Daily time spent on social networks rises to over 2 hours*. Global Web Index. https://blog.globalwebindex.com/chart-of-the-day/daily-time-spent-on-social-networks

Mansou, E. (2015). The use of social networking sites (SNSs) by the faculty members of the school of library and information science, PAAET, Kuwait. *The Electronic Library*, *33*(3), 524–546. doi:10.1108/EL-06-2013-0110

Markauskaite, L. (2005). *From a static to dynamic concept: A model of ICT literacy and an instrument for self-assessment.* In P. Goodyear, D. G. Sampson, D. J.-T. Yang, Kinshunk, T. Okamoto, R. Hartley, & N.-S. Chen (Eds.), ICALT 2005. The *5th IEEE international conference on advanced learning technologies Proceedings* (pp. 464–466). IEEE Computer Society. 10.1109/ICALT.2005.161

Markauskaite, L., Reimann, P., Reid, D., & Goodwin, N. (2006). *Exploring the fit of an information technology course to the ICT literacy of trainee teachers.* In C. M. Crawford, R. Carlsen, K. McFerrin, J. Price, R. Weber, & D. A. Willis (Eds.), *SITE 2006 – society for information technology & teacher education. 17th International conference. Proceedings* (pp. 2960–2967). Chesapeake, VA: AACE.

Markauskaite, L. (2007). Exploring the structure of trainee teachers' ICT literacy: The main components of, and relationships between, general cognitive and technical capabilities. *Educational Technology Research and Development, 55*(6), 547–572. doi:10.100711423-007-9043-8

Martin, B. (2018). Faculty technology beliefs and practices in teacher preparation through a TPaCK lens. *Education and Information Technologies, 23*(5), 1775–1788. doi:10.100710639-017-9680-4

Martinson, B. C., Anderson, M. S., & De Vries, R. (2005). Scientists behaving badly. *Nature, 435*(7043), 737–738. doi:10.1038/435737a PMID:15944677

Meenakshi, S. P., Sivakumar, P., & Ilango, P. (2017). Effective health promotion interventions for targeted audience using social networking sites: A systematic review. *Indian Journal of Science and Technology, 10*(36), 1–09. doi:10.17485/ijst/2017/v10i36/112402

Mentegomery, C. H. (2002). Comparing Library and User related cost of print and electronic journal collections. *D-Lib Magazine : the Magazine of the Digital Library Forum, 8*(10), 1–15. doi:10.1045/october2002-montgomery

Miller, R. C., & Tsai, C. J. (2020). Scholarly Publishing in the wake of COVID-19. *International Journal of Radiation Oncology, Biology, Physics, 108*(2), 491–495. doi:10.1016/j.ijrobp.2020.06.048 PMID:32890542

Mingers, J., & Leydesdorff, L. (2015). A review of theory and practice in scientometrics. *European Journal of Operational Research, 246*(1), 1–19. doi:10.1016/j.ejor.2015.04.002

Miyazoe, T., & Anderson, T. (2013). Interaction Equivalency in an OER, MOOCS and Informal Learning Era. *Journal of Interactive Media in Education, 2013*(2), 1–15. doi:10.5334/2013-09

Mohamed, H. K., & Sumitha, E. (2011). Perception and Use of Social Networking Sites by the Students of Calicut University. *DESIDOC Journal of Library and Information Technology, 31*(4).

Mohammadyari, S., & Singh, H. (2015). Understanding the effect of e-learning on individual performance: The role of digital literacy. *Computers & Education, 82*, 11–25. doi:10.1016/j.compedu.2014.10.025

Monirul Islam, Md. (2022). *Social Media Adaptation and Its Impact in Bangladesh.* Research Gate. https://www.researchgate.net/publication/358470840_Social_Media_Adaptation_and_Its_Impact_in_Bangladesh

Moore, M. (1989). Three types of interaction. *American Journal of Distance Education, 3*(2), 1–6. doi:10.1080/08923648909526659

Moore, M. G., & Kearsley, G. (1996). *Distance education: A systems view.* Wadsworth Publishing.

Mountifield, H. M., & van Brakel, P. A. (1994). Network-based electronic journals: A new source of information. *South African Journal of Library and Information Science, 62*(1), 28–33.

Mubin, O., Al Mahmud, A., & Ahmad, M. (2017). HCI down under: Reflecting on a decade of the OzCHI conference. *Scientometrics*, *112*(1), 367–382. doi:10.100711192-017-2293-2

Nadez, G. (2012). Use of social networks for academic purposes: A case study. *The Electronic Library*, *31*(6), 781–791. doi:10.1108/EL-03-2012-0031

Nalimov, V. V., & Mulchenko, Z. M. (1969b). *Naukometriya. Izuchenie Razvitiya Nauki kak Informatsionnogo Protsessa. [Scientometrics. Study of the Development of Science as an Information Process].* Nauka.

Nalimov, V., & Kordon, V. a. Ya., & Korneeva. (1971). Geograficheskoe Raspredelenie Nauchnoi Informatsii. [Geographic Distribution of Scientific Information.] [V. V. Nalimov trans.] Informatsionnye Materialy. Moscow:an SSSR Nauchnyi Sovet po Kompleksnoi Probleme Kibernetiki. [Moscow: Soviet Academy of Science, Scientific Council on Cybernetics.] 2: 3–37.

Nalimov, V. V., & Mulchenko, B. M. (1969). *Scientometrics. Studies of science as a process of information*. Science.

Nalimov, V. V., & Mulchenko, Z. M. (1969a). Eshche raz k voprosu o kontseptsii eksponentsial'nogo rosta.[A word to add on the exponential growth concept.] Nauchno-Tekhnicheskaya Informatsiya. *Seriya*, *2*(8), 12–14.

Naqvi, S. H. (2014). *The Use of E-Resources by Postgraduate Engineering Students at Jamia Millia Islamia University*. PARIPEX Indian Journal of Research., doi:10.15373/22501991/July2014/33

Natrajan, K, K. (2010). Use and user perception of electronic resources. *Annals of Lib. & Inf. Studies*, *57*(3), 59-64.

Nichols, D. M., & Cunningham, S. J. (2015). A scientometric analysis of 15 years of CHINZ conferences. *Proceedings of the 15th New Zealand conference on human–computer interaction*, pp. 73–80. 10.1145/2808047.2808060

Nishi, F. & Ali, N. (2013). Witness and use of e-journals by IIT Delhi and Delhi University Library System. *International Journal of Digital Library Services, 32*(2), 57-64. http://www.ijodls.in/uploads/3/6/0/3/3603729/23-42.pdf

Oakleaf, M., & Kaske, N. (2009). Guiding questions for assessing information literacy in higher education. *Portal (Baltimore, Md.)*, *9*(2), 273–28.

Ogdanov, E., Limpens, F., Li, N., El Helou, S., Salzmann, C., & Gillet, D. (2012). A social media Platform in higher education. In *Proceedings of the 2012 IEEE Global Engineering Education Conference (EDUCON).* IEEE.

Ololube, N. P., Ubogu, A. E., & Egbezor, D. E. (2007). ICT and distance education programs in a sub-Saharan African country: A theoretical perspective. *Journal of Information Technology Impact*, *7*(3), 181–194.

Ozdamar-Keskin, N., Ozata, F. Z., Banar, K., & Royle, K. (2015). Examining digital literacy competences and learning habits of open and distance learners. *Contemporary Educational Technology*, *6*(1), 74–90. doi:10.30935/cedtech/6140

P.C. B., & Baskaran, C. (2020). The Faculty Members And Research Scholars Attain Academic And Research Thoughts By Accessing Electronic Information In The Universities Of Kerala, India. *Library Philosophy and Practice (e-Journal), 3739*, 1–27. https://digitalcommons.unl.edu/libphilprac/3739

Padilla, S., Methven, T. S., & Chantler, M. J. (2014). Is British HCI important? A topic-based comparison with CHI. In *Proceedings of the 28th international BCS human computer interaction conference HCI*, pp. 365–370. 10.14236/ewic/HCI2014.2

Panneerdas, C. (2022). Utilization of E- Resources by Faculty Members and Students in Management Institutions, Coimbatore: A Study, *Library Philosophy and Practice (e-journal)*, 6802. https://digitalcommons.unl.edu/libphilprac/6802

Parabhoi, L., & Pathy, S. (2017). India, Social Media and its Application of Library Services in India. *KIIT Journal of Library and Information Management, 4*(1).

Parameshwar, S., & Patil, D. B. (2009). Use of electronic resources in university libraries of Karnataka: A case study of Gulbarga University Library. *Keralan Journal of Information Science and Services*, *3*(1), 13-21. 5.

Parveen, N. (2011). Use of social networking site (Facebook) in making awareness among the library and information science professionals of university libraries of UP: A case study. *International Journal of Digital Library Services*, *1*(1), 9–17.

Patra, S. K., Bhattacharya, P., & Verma, N. (2006). Bibliometric study of literature on bibliometrics. *DESIDOC Bulletin of Information Technology*, *26*(1), 27–32. doi:10.14429/dbit.26.1.3672

Patsula, P. J. (2002). *Practical guidelines for selecting media: An international perspective*. The Usableword Monitor.

Pearson, V., Lister, K., McPherson, E., Gallen, A. M., Davies, G., Colwell, C., Bradshaw, K., Braithwaite, N., & Collins, T. (2019). Embedding and sustaining inclusive practice to support disabled students in online and blended learning. *Journal of Interactive Media in Education*, *2019*(1), 1–10. doi:10.5334/jime.500

Peeraer, J., & Van Petegem, P. (2011). ICT in Teacher Education in an Emerging Country: Vietnam's Baseline Situation at The Start Of "The year of ICT". *Computers & Education*, *56*(4), 974–982. doi:10.1016/j.compedu.2010.11.015

Perryman, B. E. (2011). Use of social networking as an instructional aide in higher education [Doctoral dissertation, Oklahoma State University]. https://hdl.handle.net/11244/7193

Perryman, L.-A., Hemmings-Buckler, A., & Seal, T. (2014). Learning from TESS-India's Approach to OER Localisation Across Multiple Indian States. *Journal of Interactive Media in Education*, *2014*(2), 1–11. doi:10.5334/jime.af

Pew Research Center. (2021). *Social media users more likely to interact with people who are different from them.* Pew Research. https://www.pewresearch.org/internet/2019/08/22/social-media-users-more-likely-to-interact-with-people-who-are-different-from-them/

Pitchaipandi, P. (2020). Impact and Usage of Social Media among the Post Graduate Students of Arts in Alagappa University, Karaikudi, India. Measuring and Implementing Altmetrics in Library and Information Science Research. IGI Global. doi:10.4018/978-1-7998-1309-5.ch010

Prabakaran, T. (2013). Use of E-Resources Among Faculty Members of Engineering Colleges in Cuddalore District. *Journal of Advances in Library and Information Science, 2 (2), pp-71-75.*

Prabhakar, S.V.R., & Rani, M.S.V. (2017). Influence of Social Networking Sites on Library and Information centers. [IJLIS]. *International Journal of Library and Information Science*, *6*(1), 83–87.

Price, D. S. (1963). *Little science, big science*. Columbia University Press. doi:10.7312/pric91844

Procter, R., Williams, R., Stewart, J., Poschen, M., Snee, H., Voss, A., & Asgari-Targhi, M. (2010). Adoption and use of web 2.0 in scholarly communications. *Philosophical Transactions of the Royal Society A: Mathematical, Physical and Engineering Sciences*, *368*(1926), 4039–4056. https://doi.org/ doi:10.1098/rsta.2010.0155

Pugoy, D.L., R. A., D.L. Habito, C., & B. Figueroa, R. (2016). Hybrid online/offline mobile solutions for accessing open educational resources in areas with poor internet connectivity. *Asian Association of Open Universities Journal*, *11*(2), 182–196. doi:10.1108/AAOUJ-09-2016-0030

Rabeth, K. (2021). *A peek into the digital emergence of Bangladesh.* WARC. https://www.warc.com/newsandopinion/opinion/a-peek-into-the-digital-emergence-of-bangladesh/en-gb/4397

Rachman, Y. B. (2020). Examining academic libraries' use of instagram. *College & Undergraduate Libraries*, *27*(2–4), 462–473. doi:10.1080/10691316.2021.1915221

Radhakrishnan, S., & Baskaran, C. (2019). Authorship Productivity and Applicability of Lotka's Law in Phytochemistry Literature. *Library Philosophy and Practice (e-journal).* 3618. https://digitalcommons.unl.edu/libphilprac/3618

Radhakrishnan, S., & Baskaran, C. (2020). Phytochemistry Literature: An Altmetrics Analysis (2020). *Library Philosophy and Practice (e-journal).* 4048. https://digitalcommons.unl.edu/libphilprac/4048

Rao, S. S. (1998). Trends in publishing and delivery of electronic information. *New Library World, 99*(1145).

Repman, J., & Logan, S. (1996). Interactions At a Distance: Possible Barriers and Collaborative Solutions. *TechTrends, 41*(6), 35–38. doi:10.1007/BF02818861

Richardson, J., & Ting, E. E. (1999) *Making the most of interaction: what instructors do that most affect students' perceptions of their learning.* ALN. https://www.aln.org/conference/proceedings/1999/ppt/99_richardson.ppt

RielJ.ChristianS.HinsonB. (2012). Charting digital literacy: A framework for information technology and digital skills education in the community college. *Available at* SSRN 2781161. doi:10.2139/ssrn.2781161

Roeder, I., Severengiz, M., Stark, R., & Seliger, G. (2017). Open Educational Resources as a Driver for Manufacturing-related Education for Learning of Sustainable Development. *Procedia Manufacturing, 8,* 81–88. doi:10.1016/j.promfg.2017.02.010

Rohatgi, A., Scherer, R., & Hatlevik, O. E. (2016). The role of ICT self-efficacy for students' ICT use and their achievement in a computer and information literacy test. *Computers & Education, 102,* 103–116. doi:10.1016/j.compedu.2016.08.001

Rosie, A. (2000). Online pedagogies and the promotion of "deep learning". *Information Services & Use, 20*(2/3), 109–116. doi:10.3233/ISU-2000-202-306

Roslina Wan Othman, W., Fariha Mohd Apandi, Z., & Haslinda Ngah, N. (2016). The uses of social media on student's communication and self concepts among TATIUC students. *Indian Journal of Science and Technology, 9*(17). doi:10.17485/ijst/2016/v9i17/88730

Ruleman, A. B. (2012). Comparison of student and faculty technology use. *Library Hi Tech News, 29*(3), 3. doi:10.1108/07419051211241877

Ruths, D., & Jurgen, P. (2014). Social media for large studies of behavior. *Science, 346*(6213), 1063–1064. doi:10.1126cience.346.6213.1063 PMID:25430759

Ryan, T., Chester, A., Reece, J., & Xenos, S. (2014). The uses and abuses of Facebook: A review of Facebook addiction. *Journal of Behavioral Addictions, 3*(3), 133–148. doi:10.1556/JBA.3.2014.016 PMID:25317337

s., A., M., A., & B., J. (2019). Use of Social Media in Marketing of Library Information Services in Polytechic University of the Philippines. *IJARIIE, 5*(6), 534-545. http://ijariie.com/AdminUploadPdf/Use_of_Social_Media_in_Marketing_of_Library_Information_Services_in_Polytechnic_University_of_the_Philippines_ijariie11028.pdf

Sahu, R. (2018). *Social Media for Effective Library Services.* JMIR Publications. doi:10.2196/jmir.2912

Salomon, D. (2013). Moving on from Facebook: Using Instagram to connect with undergraduates and engage in teaching and learning. *College & Research Libraries News, 74*(8), 408–412. doi:10.5860/crln.74.8.8991

Saravanakumar, M., & SuganthaLakshmi, T. (2012). Social media marketing. *Life Science Journal, 9,* 4444–4451. doi:10.7537/marslsj090412.670

Sathish Kumar, K. M. M. (2020). Awareness of swayam programme among the Postgraduate students. In LAP LAMBERT Academic Publishing, 1 (1). doi:10.1108/AAOUJ-07-2016-0022

Sathish Kumar, K., Mahendraprabu, M., Kalaiyarasan, G., Ramnath, R., Sasi Kumar, N., & Mani, M. (2021). Social Media as an Open Educational Practice Tools and Challenges. *Library Philosophy and Practice, 2021*(1), 1–24. doi:10.5281/zenodo.4705256

Sathish Kumar, K., Mahendraprabu, M., Kalaiyarasan, G., Ramnath, R., Sasi Kumar, N., & Mookkiaha, M. (2021). A vision of teaching learning practices in mathematics education through open educational resources. *International Journal of Education and Teaching, 1*(2), 1–30. doi:10.51483/IJEDT.1.2.2021.30-36

Shafique, F., Anwar, M., & Bushra, M. (2010). Exploitation of Social media among University students: A case study. *Webology, 7*(2), 34–47.

Sharma, C. (2009). Use and Impact of E-Resources at Guru Gobind Singh Indraprastha University (India): A Case Study. *E-JASL 1999-2009,* 1-10, 123. https://digitalcommons.unl.edu/ejasljournal/123

Sharma, H.P. (2008). Moving beyond library automation: Role of e-resources in academic libraries. *University news, 46*(34), 6-10.

Sharma, J.N. & Gupta, S. (2013). Management of libraries in modern era: A view. *Lib. & Inf.Tech, 9*(1), 7–20.

Shashikiran, M., & Maadu, M. S. (2021). *Library Services Through Social Media.* Research Gate. https://www.researchgate.net/publication/353417222_library_services_through_social_media

Sher, A. (2009). Assessing the relationship of student-instructor and student-student interaction to student learning and satisfaction in Web-based Online Learning Environment. *Journal of Interactive Online Learning, 8*(2).

Shinde, M. D. B. (2016). Social Media Applications And Its Impact on Libraries. *Prime International Research Journal, 3*(2), 9–16.

Singh, K. P., & Gill, M. S. (2011). *Using social media for social research: An introduction.* Use of social networking sites by the research scholars: Social Media Research Group. https://assets.publishing.service.gov.uk/government/uploads/system/uploads

Singh, G., Mittal, R., & Ahmad, M. A. (2007). bibliometric study of literature on digital libraries. *The Electronic Library, 25*(3), 342–348. doi:10.1108/02640470710754841

Smith, K. (2019). 123 Amazing social media statistics and facts. *brandwatch,* https://www.brandwatch.com/blog/amazing-social-media-statistics-and-facts/

Sobreira, D., Santos de Oliveira, D., & García-Peñalvo, F. J. (2020). The Use of Instagram as a Digital Marketing Tool by the Brazilian Library Councils in times of Covid-19. *Eighth International Conference on Technological Ecosystems for Enhancing Multiculturality,* (pp. 582–587). ACM. 10.1145/3434780.3436599

Somi, N. G., & De Jager, K. (2005). The role of academic libraries in the enhancement of information literacy: A study of Fort Hare Library. *South African Journal of Library and Information Science*, 259–266.

Srinivasa Rao Y. (2018). Scholarly Communication Cycle: SWOT Analysis. *AIJR Abstracts*, 9.

Stern, D. M., & Willits, M. D. (2011). *Social media killed the LMS: Re-imagining the traditional learning management system in the age of blogs and online social networks.* In Educating Educators with Social Media Emerald Group Publishing Limited. doi:10.1108/S2044-9968(2011)0000001020/full/pdf

Sutko, D. M., & Silva, A. D. S. (2011). Location-aware mobile media and urban sociability. *New Media & Society, 13*(15), 807–823. doi:10.1177/1461444810385202

Sutrisno, S., Andre, H. L., & Susilawati, S. (2021, February). The analysis of the facilities and ICT applications usage by the university's students. *Journal of Physics: Conference Series*, *1793*(1), 012050. doi:10.1088/1742-6596/1793/1/012050

Sweeney, L. (1997). The future of academic journals: Considering the current situation in academic libraries. *New Library World*, *98*(1132), 4–15. doi:10.1108/03074809710155569

Tahir, M., Mahmood, K., & Shafique, F. (2010). Use of electronic information resources and facilities by humanities scholars. *The Electronic Library*, *28*(1), 122–136. doi:10.1108/02640471011023423

Tella, A., & Adu, E. O. (2009). Information communication technology (ICT) and curriculum development: The challenges for education for sustainable development. *Indian Journal of Science and Technology*, *2*(3), 55–59. doi:10.17485/ijst/2009/v2i3.12

Thanuskodi, S. (2012). Use of ICT for information creation among faculty members of Annamalai University, Tamil Nadu, India. *International Journal of Library Science*, *1*(4), 61–65.

Thelwall, M., & Kousha, K. (2014). Academia.edu: Social network or academic network? *Journal of the Association for Information Science and Technology*, *65*(4), 721–723. doi:10.1002/asi.23038

Thoms, B., & Eryilmaz, E. (2014). How media choice affects learner interactions in distance learning classes. *Computers & Education*, *75*, 112–126. doi:10.1016/j.compedu.2014.02.002

Tiew, W. S. (2000). Characteristics of self-citations in Journal of Natural Rubber Research 1988-1997: A ten-year bibliometric study. *Malaysian Journal of Library and Information Science*, *5*(1), 95–104.

Tyler, L. (2005). ICT literacy: *Equipping students to succeed in an information-rich, technology-based society.* USA: Educational Testing Service. Retrieved from https://www.ets.org/Media/Tests/ICT_Literacy/pdf/ICT_Equipping_Students_to_Succeed.pdf

Uşun, S. (2003). A Review of Communication Elements and Learner Support Services in Turkish Distance Education System. *The Turkish Online Journal of Educational Technology*, *2*(3).

Van Raan, A. (1997). Scientometrics: State-of-the-art. *Scientometrics*, *38*(1), 205–218. doi:10.1007/BF02461131

Vancauwenbergh, S. (2017). Research data management at the Flemish universities: State of affairs. In *Proceedings of the CRIS2018 Conference*, Umeå, Sweden.

Vancauwenbergh, S. (2017). The transition of FRIS 1.0 towards FRIS 2.0: Focus on semantic interoperability. In *Proceedings of the EuroCRIS Membership Meeting*, Dublin, Berlin.

Verkuyl, M., Lapum, J. L., St-Amant, O., Bregstein, J., & Hughes, M. (2020). Healthcare students' use of an e-textbook open educational resource on vital sign measurement: A qualitative study. *Open Learning*, *38*(1), 22–37. doi:10.1080/02680513.2020.1835623

Walsh, A. J. (2020). Removing barriers: Textbook affordability and OER at Sinclair community college. *RSR. Reference Services Review*, *48*(3), 385–396. doi:10.1108/RSR-03-2020-0011

Wang, M., Ran, W., Liao, J., & Yang, S. J. (2010). A Performance-Oriented Approach to E-Learning in the Workplace. *Journal of Educational Technology & Society*, *13*(4), 167–179.

Wang, X. (2014). Using attitude functions, self-efficacy, and norms to predict attitudes and intentions to use mobile devices to access social media during sporting event attendance. *Mobile Media & Communication*, *3*(1), 75–90. doi:10.1177/2050157914548932

Whyte, S., Schmid, E. C., van Hazebrouck Thompson, S., & Oberhofer, M. (2014). Open educational resources for CALL teacher education: The iTILT interactive whiteboard project. *Computer Assisted Language Learning*, *27*(2), 122–148. doi:10.1080/09588221.2013.818558

Williams, A., & Woodacre, M. (2016). The possibilities and perils of academic social networking sites. *Online Information Review*, *40*(2), 282–294. doi:10.1108/OIR-10-2015-0327

Work, S., Haustein, S., Bowman, T. D., & Larivière, V. (2015). Social Media in Scholarly Communication. A Review of the Literature and Empirical Analysis of Twitter Use by SSHRC Doctoral Award Recipients, Study commissioned by the Social Sciences and Humanities Research Council, Montreal, Canada.

Xia, J. (2013). Let us take a Yale open course: A Chinese view of open educational resources provided by institutions in the West. *Journal of Computer Assisted Learning*, *29*(2), 122–137. doi:10.1111/j.1365-2729.2012.00477.x

Yebowaah., F., Dedzoe., F., & Plockey, D. (2017). Awareness and Use of Electronic Resources in University Libraries: A Case Study of University for Development Studies Library. *Library Philosophy and Practice (e-journal)*. 1562, 1-32.

Yusuf, M. O. (2006). Problems and Prospects of Open and Distance Education in Nigeria. *Turkish Online Journal of Distance Education*, *7*(1), 22–29.

Zhang, X., Tlili, A., Nascimbeni, F., Burgos, D., Huang, R., Chang, T. W., Jemni, M., & Khribi, M. K. (2020). Accessibility within open educational resources and practices for disabled learners: a systematic literature review. In Smart Learning Environments 7(1),1-19. Springer. doi:10.118640561-019-0113-2

Zhang, L., Ye, P., & Liu, Q. (2011). A survey of the use of electronic resources at seven universities in Wuhan, China. *Program*, *45*(1), 67–77. doi:10.1108/00330331111107402

Zhang, Z., & Kenny, R. (2010). Learning in an online distance education course: Experiences of three international students. *International Review of Research in Open and Distance Learning*, *11*(1), 17–36. doi:10.19173/irrodl.v11i1.775

Zhiwei, T., & Muhammad, A. (2021). Why libraries are giving too much importance to social media? A review note. *Library Philosophy and Practice (e-Journal)*. https://digitalcommons.unl.edu/libphilprac/5175

Zhu, C. (2012). Student satisfaction, performance and knowledge construction in online collaborative learning. *Journal of Educational Technology & Society*, *15*(1), 127–136.

About the Contributors

C. Baskaran is presently serving as the Librarian/Professor in Alagappa University, Karaikudi, India. He successfully completed a Major Research Project has funded by ICSSR, New Delhi and currently also taking up a Major Research Project from same funding agency (ICSSR). He has authored three books in Library and Information Science. He is a research guide in the Dept. of Library and Information Science, Alagappa University and he guided and awarded eight Ph.D scholars and eight Scholars pursuing Ph.D under his guidance. He has published more than 100 articles to his credit. He was an adjudicator and examiner for Viva-voce exams of Many Ph.D in several Universities. Moreover, he is an editor of a Journal and Editorial advisory board member for journals of Library and Information Science. He visited and Presented papers at International Conferences in Malaysia and Greece. He has been holding charge additionally SWAYAM /MOOCs Co-ordinator of the University. He is a recipient of EET-Award of Research Excellence-2019.

Dhanavandan S. is working as Deputy Librarian at Central University of Tamil Nadu, Thiruvarur-610005 from 2017 onwards, Before joining in CUTN, he served as Assistant Librarian at 'The Gandhigram Rural Institute – Deemed to be University', Gandhigram. He is having more than twenty three years in the LIS profession. Six Ph.D degrees were awarded under his guidance. He has published more than 140 articles in National and International Journals and more than 175 papers presented and published in National and International Conferences. He authored/edited more than 35 books in Library and Information Science and contributed 65 Chapters in edited books. He has attended more than 50 Seminars/Workshops, Webniars and training programmes. Three times he has received Cash Award (more than One Lakh) for the best articles from Konkuk University, South Korea. He served as Chairperson and Resource Person in many workshops, conferences and also organized fifteen workshops and conferences.

***

Santhiya Gurusamy is currently pursuing a Doctor of Philosophy (Ph. D) at the Department of Library and Information Science, Alagappa University, Karaikudi. My areas of interests Include Metric studies, especially Organic farming and Scientometrics. She has Published and Presented Research Papers, in the various field of Library and Information science in few National and International Conference and Seminars.

Sanmati Jinendran Jain works as Semi Professional Assistant (Library) at Central Library, Central University of Tamil Nadu, Thiruvarur. Prior to joining CUTN, he worked as a Jr. Technician (Library) at the Indian Institute of Information Technology, Design, and Manufacturing (IIITDM), Kancheepuram and as an Officer (Information Resource Centre) at Tata Consultancy Services (TCS), Chennai. He has qualified UGC-NET and is currently pursuing his Ph.D. from IGNOU, New Delhi. He holds B.Sc. (Biochemistry) from Vels University, Chennai, and M.Lib.I.Sc from the University of Madras, Chennai. He is the recipient of Prof. Parvataneni Gangadhara Rao Memorial Medal for securing the first rank in M.Lib.I.Sc (Library & Information Science) Programme. His areas of interest are the Application of ICT in Libraries, Open Access and Open Educational Resources.

R. Jeyshankar is presently working as Associate Professor & Head in the Department of Library and Information Science, Pondicherry University. He was selected as an UGC Research Awardee (2012-2014) by UGC and recently received the UGC - DAAD Fellowship 2022 (Indo-German Research Collaborative Project). He has guided 17 Ph.Ds, 49 M.Phils and 60 M.L.I.Sc students and he is guiding 6 Ph.D. research scholars at present. He has successfully completed a minor project with "Alagappa University Research Fund (AURF) and Two major projects from "Indian Council Social Science Research (ICSSR) and University Grants Commission (UGC)". He has edited 5 conference proceedings till date. He has written 20 chapters in the textbooks of Library and Information Science and published more than 72 articles in the National & International SCOPUS, Web of Science, CARE list, refereed journals and also his h-index-14, i10 index-20 in Google scholar and Scopus in h-index 6. He visited Srilanka, South Korea & Taiwan for attended and presented papers in international conferences and international project.

R. Jeyshankar is presently working as Associate Professor & Head in the Department of Library and Information Science, Pondicherry University. He was selected as an UGC Research Awardee (2012-2014) by UGC and recently received the UGC - DAAD Fellowship 2022 (Indo-German Research Collaborative Project). He has guided 17 Ph.Ds, 49 M.Phils and 60 M.L.I.Sc students and he is guiding 6 Ph.D. research scholars at present. He has successfully completed a minor project with "Alagappa University Research Fund (AURF) and Two major projects from "Indian Council Social Science Research (ICSSR) and University Grants Commission (UGC)". He has edited 5 conference proceedings till date. He has written 20 chapters in the textbooks of Library and Information Science and published more than 72 articles in the National & International SCOPUS, Web of Science, CARE list, refereed journals and also his h-index-14, i10 index-20 in Google scholar and Scopus in h-index 6. He visited Srilanka, South Korea & Taiwan for attended and presented papers in international conferences and international project.

Palanivel K. has qualified M.SC in Library and Information Science from University of Madras and Ph. D in Library and Information Science from Alagappa University. Presently he is working as Librarian and Information Assistant, Anna Centenary Library, Chennai in India. Previously, he has worked as Graduate Trainee in Anna University, MIT Campus, Chennai, and Asst. Librarian in SRM Dental College, Ramapuram, Chennai, and Library Reader in St. Johns International School, Chennai. He has many articles in peer reviewed Journals from National and International Journals. H has attended many Conferences, Seminars and workshops. He has Qualified UGC- NET(National Eligibility Test) in Library and Information Science.

Sathish Kumar K. is a Research Scholar in the Department of Education at Alagappa University, Karaikudi. He has more than eight years of research experience in educational technology and six years in Open and Distance Mode. He has presented more than ten papers in National and International Seminars/Conferences. He has authored three books, seven Self-Learning Materials, fifty open access PowerPoint presentations, and published more than 20 research articles in various Scopus, Web of Science, and peer-reviewed journals. He has served as an editorial board member and reviewer in several journals. His areas of interest include Open Access, Open Educational Resources, Open Educational Practices, and Online Learning.

Burcu Karabulut Coşkun was born in 1983 in Ankara/Turkey. Graduated from Baskent University Faculty of Education, Department of Computer Education and Instructional Technology. Got MS degree from Usak University, Department of Primary Education and a Ph.D. degree from Gazi University Department of Computer Education and Instructional Technology. Still working as a Dr instructor at Kastamonu University Faculty of Education, Department of Computer Education and Instructional Technology and the coordinator of Distance Education Application and Research Center of the University.

Anil Kumar T. works as Assistant Librarian at the Knowledge Resource Centre, Indian Institute of Corporate Affairs, Gurgaon, and Haryana. Before joining IICA, he worked as Professional Assistant (Library) at Cochin University of Science and Technology, Kochi; Indian School of Business, Hyderabad; and IISER Mohali. He hols Masters' in Library and Information Science from the University of Madras and B.A. (English) from the University of Calicut. His areas of areas of interest are Library Automation, Content Management System and Digital Libraries.

Chandrasekaran M. has qualified with MA., M.Phil., and Ph.D in Philosophy from Bharathidasan University, Tiruchirappalli, India. He has also qualified Post Graduate in Library and Information Science and he has been working as a Library Professional in the Alagappa University, Karaikudi, India. Presently, he is the Ph.D research scholar in the Dept. of Library and Information Science, Alagappa University, Karaikudi. He has participated and Presented papers in many National and International Conferences and Journals.

Prasad M. was qualified MBA, MA, MLIS, M.Phil, B.Ed and Ph.D in Library and Information Science from Alagappa University, Karaikudi, India. He has qualified UGC-NET (Library and Information Science) in 2014. He was started career as Library Trainee at PSG College of Technology, after he was joined as Assistant Librarian, PSG Institute of Management, Librarian, PET Engineering College, Chief Librarian, Narayanaguru College of Engineering, Head, UGC scale Assistant Librarian, Amrita University, University Librarian & Head, Vel Tech University, Chennai, and presently working as the College Librarian, Kumaraguru College of Technology, Coimbatore. He has published many articles in National and International Journals, He was attended many National and international conferences attended and organized many workshops.

Prasad M. was qualified MBA, MA, MLIS, M.Phil, B.Ed and Ph.D in Library and Information Science from Alagappa University, Karaikudi, India. He has qualified UGC-NET (Library and Information Science) in 2014. He was started career as Library Trainee at PSG College of Technology, after he was joined as Assistant Librarian, PSG Institute of Management, Librarian, PET Engineering College,

Chief Librarian, Narayanaguru College of Engineering, Head, UGC scale Assistant Librarian, Amrita University, University Librarian & Head, Vel Tech University, Chennai, and presently working as the College Librarian, Kumaraguru College of Technology, Coimbatore. He has published many articles in National and International Journals, He was attended many National and international conferences attended and organized many workshops.

Muhammad Mezbah-ul-Islam is a currently Professor and former Chairman Dept. of Information Science and Library Management, University of Dhaka. He has received graduation & post-graduation from the University of Dhaka and was awarded Ph.D. from North-Eastern Hill University (NEHU) as a SAARC Scholar. He produced more than 70 publications and has 24 years of education and research experience. He is an active researcher and has worked for the National and International levels. He has worked hard to enhance the quality of the academic and research program in all possible ways so Dept. enjoys a leading position. His on-going research includes OSILS, IL, E-Learning, LAM, Digital Curation, Etc. he has acted as Member Secretary Foundation of Research on Education, Research Board, Planning and Development (FREPD, Dhaka University, Dhaka, Chairman, Dept. of Information Science and Library Management, University of Dhaka and So on.

M. Muniyasamy has completed Master of Library and Information Science and M.Phil from Alagappa University, Karaikudi, India. He is currently Ph.D research scholar at Department of Library and Information Science, Alagappa University, Karaikudi. His areas of interests include open access, open educational resources, and webometrics. He has published and Presented many papers in various peer reviewed Journals in LIS and Conference proceedings. He was published a research paper in Scopus Indexed Journal.

Sivagami Murugappan presently pursuing Doctor of Philosophy in Library and Information Science at Alagappa University, Karaikudi. Researcher interested and well-known in Scientometrics, Bibliometrics, Data Analyzing scientometric tools, Visualization tools, Content Analysis, Social Media Handling, Exploring AI tools for research and promoting reading habits. Received Best Paper Award in Two Days National Seminar on Innovative Librarianship: A Foresight on Aatmanirbhar Baharat' on 29th & 30th July 2022 organized by Central Library, Central University of Tamil Nadu. Thiruvaur Presented and participated in two international conference and five national seminars papers.

P. Natarajan has been working as Assistant Professor and Head of the Department of Visual Communication in SRM Arts and Science College Chennai for the last 25 years. He has a total academic experience of 28 years. He was a gold medalist from M.S University Tirunelveli and 3 times UGC/NET cleared with rich experience. He is a member of various academic boards and Chairman of the University of Madras Exam board. Has been a pioneer in the syllabus framing of major universities and autonomous colleges.

Seher Özcan graduated from Balıkesir University, Faculty of Education, Department of Computer and Instructional Technology Education in 2005. She worked as a teacher of information technologies in a public school between 2005-2010. In 2009, she received my MS degree in Department of Computer and Instructional Technology Education at Gazi University and in 2019, her PhD degree in the same department and university. She started to work as an instructor in Department of Informatics at Gazi

University in 2010. Seher Ozcan currently work an Assistant Professor in the department of Computer Education and Instructional Technologies (CEIT) of Gazi Faculty of Education of Gazi University. At the same time, She is working as an e-learning specialist at Gazi University Distance Education Application and Research Center. Her research area includes ICT use in education, teaching, theories, applications of distance education, designing and developing story board and content for e-learning, instructional design, design principles for e-learning and interaction applications for e-learning, She national and international articles, book chapters and projects on these issues.

Muthumari P. has received PhD degree in Library and Information Science from Alagappa University. He has completed his Post Graduation in Library and Information Science in Madurai Kamaraj University, Madurai. He was worked as a teaching Assistant for several years in the Dept. of Library and Information Science in Alagappa University, Karaikudi. He has worked as College Librarian at JCT College of Engineering and Technology, Coimbatore. Presently, he has been working as a Librarian in the Government Arts college, Paramakudi. He has attended and Presented papers in many National and International Conferences and Journals.

Pitchaipandi P. is currently pursuing a Doctor of Philosophy (Ph.D.) at the Department of Library and Information Science, Alagappa University, Karaikudi, Tamil Nadu. He did his MLISC and M.Phil from the Madurai Kamaraj University, Madurai. He started his career as a Library Trainee at the S.R.M. Institute of Science and Technology, Chennai. He was acted as Project Associate (ICSSR) for period of two years in the Alagappa Univerity, Karaikudi. His areas of specialization include Social Networks/Media, User Studies, Digital Libraries, and ICT. He has published and presented Research papers in various fields of Library and Information Science in many national and International conferences and Seminars.

Ramesh P. has completed his Master in Library and Information Science, in Bharathidasan University, Tiruchirappalli and Ph.D (LIS) in Alagappa University. He has more than 22 Year of experience as Librarian Sri Ramakrishna Polytechnic College, Coimbatore, India. He has Qualified State Level Eligibility Test in Library and Information Science. He has organized many book exhibition programmes and workshops. He has published many articles and book chapters. He has attended many conferences and seminars.

Radhakrishnan S. is currently working as Librarian and Information Assistant Grade-I, in Anna Centenary Library, Chennai in India. He has completed MA, MLIS, and Ph.D in Library and Information Science. He has 18 years of experience in the field of Library services. He has Qualified UGC- NET (National Eligibility Test) in Library and Information Science. He has attended tem National and International Conferences and published seven Research Articles in the Journals.

Saravanan S. is presently working as College Librarian in Government Arts College. Ariyalur, India. He has more than 15 years of professional experiences across various academic and corporate institutions. Currently, he is a Part-time Research Scholar at Alagappa University, pursuing his Doctoral degree in Library and Information Science. His area of specialization includes Digital Library, Library services through social networking, Scientometric etc. He has attended and presented papers in various national and international conferences, seminars and have the credit of publishing in e-book chapter and acted as reviewer of the Journals/Books.

Ebru Solmaz graduated from Gazi University Faculty of Education, Department of Computer and Instructional Technology Education in 2007. She worked as an information technologies teacher in a public school between 2008-2010. In 2009, she received her MS degree in Department of Computer and Instructional Technology Education at Gazi University and in 2014 her doctoral degree in the same department and university. She started to work as an instructor in Department of Informatics at Gazi University in 2010. She currently works as instructor in Distance Education Application and Research Center at Gazi University. Her research area includes ICT use in education, technology integration, programming learning and teaching, distance education. She has national and international articles and book chapters on these issues.

Index

P

R

S

T

W

Y

www.ingramcontent.com/pod-product-compliance
Lightning Source LLC
LaVergne TN
LVHW080529300926
849162LV00007B/1567
* 9 7 8 1 6 6 8 4 8 8 0 5 8 *